T0142845

Lecture Notes in Information Systems and Organisation

Volume 32

Lecture Notes in Information Systems and Organization—LNISO—is a series of scientific books that explore the current scenario of information systems, in particular IS and organization. The focus on the relationship between IT, IS and organization is the common thread of this collection, which aspires to provide scholars across the world with a point of reference and comparison in the study and research of information systems and organization. LNISO is the publication forum for the community of scholars investigating behavioral and design aspects of IS and organization. The series offers an integrated publication platform for high-quality conferences, symposia and workshops in this field. Materials are published upon a strictly controlled double blind peer review evaluation made by selected reviewers.

LNISO is abstracted/indexed in Scopus

More information about this series at http://www.springer.com/series/11237

Fred D. Davis · René Riedl ·
Jan vom Brocke · Pierre-Majorique Léger ·
Adriane Randolph · Thomas Fischer
Editors

Information Systems and Neuroscience

NeuroIS Retreat 2019

Editors
Fred D. Davis
Information Systems and Quantitative Sciences (ISQS)
Texas Tech University
Lubbock, TX, USA

Jan vom Brocke
Department of Information Systems
University of Liechtenstein
Vaduz, Liechtenstein

Adriane Randolph
Department of Information Systems
Kennesaw State University
Kennesaw, GA, USA

René Riedl
University of Applied Sciences Upper Austria
Steyr, Oberösterreich, Austria

Johannes Kepler University Linz
Linz, Oberösterreich, Austria

Pierre-Majorique Léger
Department of Information Technology
HEC Montréal
Montreal, QC, Canada

Thomas Fischer
Department for Digital Business
University of Applied Sciences Upper Austria
Steyr, Oberösterreich, Austria

ISSN 2195-4968 ISSN 2195-4976 (electronic)
Lecture Notes in Information Systems and Organisation
ISBN 978-3-030-28143-4 ISBN 978-3-030-28144-1 (eBook)
https://doi.org/10.1007/978-3-030-28144-1

This Springer imprint is published by the registered company Springer Nature Switzerland AG
The registered company address is: Gewerbestrasse 11, 6330 Cham, Switzerland

Preface

NeuroIS is a field in Information Systems (IS) that makes use of neuroscience and neurophysiological tools and knowledge to better understand the development, adoption, and impact of information and communication technologies. The NeuroIS Retreat is a leading academic conference for presenting research and development projects at the nexus of IS and neurobiology (see http://www.neurois.org/). This annual conference has the objective to promote the successful development of the NeuroIS field. The conference activities are primarily delivered by and for academics, though works often have a professional orientation.

Since 2018, the conference is taking place in Vienna, Austria, one of the world's most beautiful cities. In 2009, the inaugural conference was organized in Gmunden, Austria. Established on an annual basis, further conferences took place in Gmunden from 2010 to 2017. The genesis of NeuroIS took place in 2007. Since then, the NeuroIS community has grown steadily. Scholars are looking for academic platforms to exchange their ideas and discuss their studies. The NeuroIS Retreat seeks to stimulate these discussions. The conference is best characterized by its workshop atmosphere. Specifically, the organizing committee welcomes not only completed research but also work in progress. A major goal is to provide feedback for scholars to advance research papers, which then, ultimately, have the potential to result in high-quality journal publications.

This year is the fifth time that we publish the proceedings in the form of an edited volume. A total of 40 research papers are published in this volume, and we observe diversity in topics, theories, methods, and tools of the contributions in this book. The 2019 keynote presentation entitled "How to Tell Your NeuroIS Story to an MIS Audience" was given by David Gefen, Prof. of MIS and Provost Distinguished Research Professor at the LeBow College of Business, Drexel University, USA. Moreover, Karin VanMeter, biologist and guest lecturer at the Austrian Biotech University of Applied Sciences, gave a hot topic talk entitled "The Importance of the Autonomic Nervous System for Information Systems Research."

Altogether, we are happy to see the ongoing progress in the NeuroIS field. More and more IS researchers and practitioners have been recognizing the enormous potential of neuroscience tools and knowledge.

Lubbock, USA	Fred D. Davis
Steyr, Austria	René Riedl
Vaduz, Liechtenstein	Jan vom Brocke
Montreal, Canada	Pierre-Majorique Léger
Kennesaw, USA	Adriane Randolph
Steyr, Austria	Thomas Fischer
June 2019	

David Gefen—Keynote: How to Tell Your NeuroIS Story to an MIS Audience

The philosophy of science and methodology of neuroscience, NeuroIS included, is different from that of the more "traditional" philosophies of science and methodologies in MIS such as surveys, design science, archival data analysis, and various types of ethnographic research. Telling a neuroscience research story and making the claim for its contribution to such an audience can be challenging. This talk will present the case, make suggestions, and open the floor to an honest conversation of those issues.

Karin VanMeter—Hot Topic Talk: The Importance of the Autonomic Nervous System for Information Systems Research

The autonomic nervous system (ANS), also referred to as the "involuntary nervous system," is the part of the peripheral nervous system supplying internal organ systems and glands. It consists of three portions, the sympathetic, parasympathetic, and enteric divisions, all of which largely regulate bodily functions unconsciously. The ANS plays a major role in homeostasis and adaptive functions and thus a response to internal and external stimuli. Examples of external stimuli are changes in light, temperature, and general environment. The sympathetic branch regulates metabolic resources and coordinates the emergency response—"fight or flight." The parasympathetic division is responsible for "rest and digest," while the enteric branch is considered separately because of its location.

While sympathetic activity is increased during the day, the parasympathetic activity becomes more active during the night when regeneration occurs at the cellular and organ level, as well as the mental level. From an Information Systems (IS) perspective, the ANS is critical, for example, due to its role in stress processes. This talk describes the fundamentals of the functioning of the ANS. Because reviews of the literature revealed that measures of ANS activity (e.g., pupil dilation, heart rate, blood pressure, skin conductance) play a significant role in NeuroIS research, this talk deals with a fundamental NeuroIS research domain.

Contents

Circadian Rhythms and Social Media Information-Sharing

Rob Gleasure

Abstract Large amounts of information are shared through social media. Such communication assumes users are sufficiently aligned, not only in terms of their interests but also in terms of their emotional and cognitive states. It is not clear how this emotional and cognitive alignment is achieved for social media, given one-to-one interactions are infrequent and discussion often spans loosely connected individuals. This study argues that circadian rhythms play an important physiological role in aligning users for information-sharing, as information shared at different times of the day is likely to encounter users with common physiological states. Data are gathered from Twitter to examine patterns of sentiment and text complexity in social media, as well as how these patterns affect information-sharing. Results suggest the timing of a social media post, relative to collective patterns of sentiment and text complexity, is a better predictor of information-sharing than the sentiment and text complexity of the post itself. Put differently, information is more likely to be shared when it is posted at times of the day when other users are primed for emotion and concentration, independent of whether that posted information is itself emotional or demanding in concentration.

Keyword Circadian · Social media · Sentiment · Text complexity · Twitter

1 Introduction

Social media provides an important means of gathering and distributing information. Yet the sheer volume of information limits what individuals can consume and share, i.e. the amount of information users may 'convey' significantly exceeds the amount of information upon which they may 'converge' [c.f. 1]. Key determinants of convergence and information-sharing have been identified as *sentiment* [2, 3] and *text complexity* [4, 5]. These qualities influence a recipient's motivation and capability to engage with particular pieces of information. The influence of *sentiment* and *text*

R. Gleasure (✉)
Department of Digitalization, Copenhagen Business School, Copenhagen, Denmark
e-mail: rg.digi@cbs.dk

F. D. Davis et al. (eds.), *Information Systems and Neuroscience*,
Lecture Notes in Information Systems and Organisation 32,
https://doi.org/10.1007/978-3-030-28144-1_1

complexity on information-sharing is not absolute; rather, their impact depends on their alignment with the needs of recipients at some particular time. Failure to match the *sentiment* of recipients may result in posts appearing out of sync or 'tone deaf' [6, 7]. Similarly, more complex information is often less welcome when discussion is adversarial [4, 5] and more welcome when discussion is collaborative [8–10].

This need for alignment between communicators and recipients is typically developed over the course of one-to-one symbolic interactions [11] and physiological mirroring [12]. Yet social media-based information-sharing is rarely one-to-one and often occurs between individuals who do not frequently interact [13]. Hence it is not obvious how users achieve the alignment to interact effectively.

This study proposes the alignment of social media users relies partly on common circadian rhythms, i.e. daily light-entrained physiological oscillations that help to ensure individuals are most active during the day and most restful at night [14, 15]. Studies have shown circadian rhythms produce predictable patterns in the sentiment of social media posts. Notably, an extensive study by Golder and Macy [16] found consistent circadian patterns in social media sentiment across countries, seasons, and days of the week. Previous research has also shown that information-sharing on social media is disproportionally between individuals in geographical proximity [17], hence in similar time zones. Thus, there is an intuitive role for circadian rhythms as a mechanism for creating alignment between social media users.

2 Social Media and Circadian Rhythms

Circadian rhythms encourage us to be active at the times best suited for our environment, e.g. to crave food and increase in activity when food sources are typically plentiful [18]. Circadian rhythms regulate a range of biological processes, from hormonal changes, to body temperature, to mood [14, 18–21]. These roughly 24-h cycles are coded into the cells of most living things, creating a natural clock that oscillates between wakefulness and restfulness—even when environments are artificially manipulated to make days seem longer or shorter [14, 22, 23].

For mammals such as humans, daily circadian cycles are entrained by light through the suprachiasmatic nucleus (SCN), which fires to dorsomedial areas of the hypothalamus and links to neural pathways involved in the release of mood and effort-related hormones such dopamine [24], serotonin [18], and cortisol [25]. The SCN simultaneously inhibits the pineal gland from secreting melatonin, the hormone that accumulates to promote sleep states [22]. This results in dual-process cycle (see [18]) where (i) the ascending arousal system triggers hormones to promote activity/inhibit the release of sleep-inducing melatonin via the pineal gland, while (ii) the competing homeostatic sleep system gradually builds up pressure until it can overwhelm sleep-inhibitors and produce enough melatonin to inhibit the SCN, resulting in a 'flip flop' switch between wake-sleep transitions. A summary of documented daily circadian hormonal patterns is illustrated in Fig. 1.

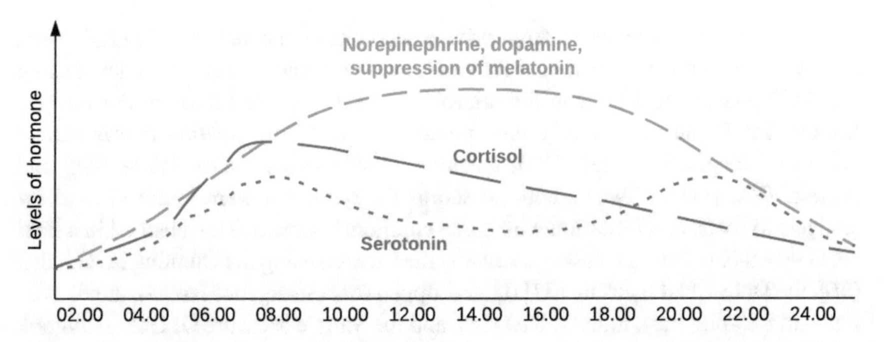

Fig. 1 Typical circadian levels of dopamine, serotonin, cortisol, and melatonin suppression

The role of these hormones in regulating engagement and energy means these patterns are relevant for social media information-sharing in two ways.

First, increased engagement and energy are linked to higher levels of emotional affect [26]. Hence circadian rhythms tend to influence the mood of individuals at different times of the day in a way that harmonizes that mood with other social actors [19], even in where no interaction has occurred.

Second, increased engagement and energy are associated with an individual's willingness to engage in challenging behaviors [27]. Communication via social media changes the nature of communication, wherein individuals must decide which communications to ignore, which to prioritize, and which to share with others [28, 29]. More complex communications increase mental load for the recipient [30], increasing the pressure on specific intrinsic and extrinsic rewards [31].

Circadian hormone patterns have been used to predict collective shifts in mood and information-processing in social media use. This includes daily contribution patterns to Wikipedia [32], seasonal changes in depression-related information search [33], and changes in word volume variation [34]. Most comprehensively, Golder and Macy [16] found strikingly consistent daily *sentiment* patterns on Twitter across countries, seasons, and days of the week.

Thus, circadian rhythms may conceivably have a direct impact on the *sentiment* and *text complexity* of social media posts, as well as subsequent information-sharing behaviors of users (as users will be in different, common physiological states at different times of the day). It may further moderate the relationship between *sentiment/text complexity* and information-sharing by extending alignment between the communicator and the recipients.

3 Method

Data were gathered from Twitter Data on 8th August and 6th December 2018. For both dates, 1000 English-language tweets were gathered from US social media users in each of the 50 states at 1-h intervals (total N = 2,400,000). Duplicates and retweets were removed, as were tweets from private accounts or accounts with no followers, and tweets with no text. *Sentiment* for each tweet was analyzed at a word level using the AFINN sentiment lexicon for microblogs [35], accessed through the tidytext library[1] for R (an open source data processing platform). *Sentiment* was scored according to positive affect (*PA*), negative affect (*NA*), *valence* (PA − NA) and *arousal* (PA + NA). Tweets with no scores for *sentiment* were removed to allow analysis to focus on discussion with some emotional content. This resulted in a final set of 404,946 tweets. *Text complexity* was then scored using the Gunning FOG index [36], the Dale-Chall measure [37] (later dropped for convergence issues), the Flesch-Kincaid Reading Ease Index (FRE) [38], and the Simple Measure Of Gobbledygook (SMOG) [39] (accessed via the quanteda library[2]).

4 Findings

Data show reliable circadian patterns of *sentiment* and *text complexity*, consistent with existing research (see Figs. 2 and 3) [c.f. 16]. The predicted *sentiment* and *text complexity* at different times were estimated using separate locally weighted regression (LOESS) curves for each measure of *sentiment* and *text complexity*. These curves were tested against the patterns and effect size of comparative polynomials to ensure reliability. A series of negative binomial regressions (see Tables 1 and 2) also compared the impact of a tweet's *sentiment* and *text complexity* with the predicted *sentiment* and *text complexity* based on the time of day it was posted, i.e. the qualities of the tweet versus the daily aggregate qualities of Twitter discussion at the time of posting. Hierarchical models were introduced that predicted information-sharing by adding the *sentiment/text complexity* of a tweet (model 1), then the circadian predicted

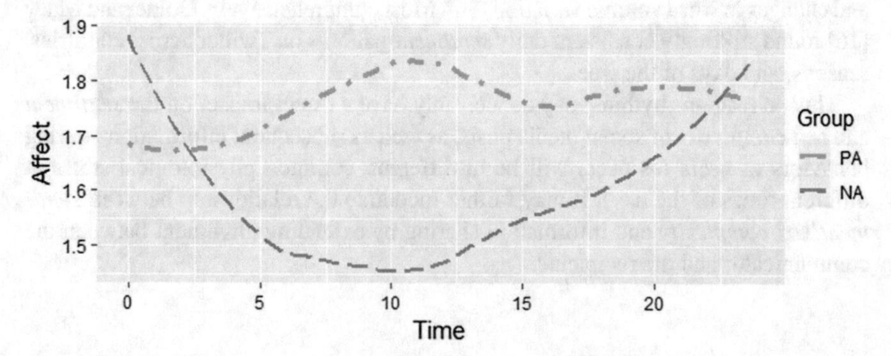

Fig. 2 LOESS curves for *positive affect* (*PA*) and *negative affect* (*NA*) based on avg. *sentiment* for time

[1]Tidytext version 0.1.8, available at https://cran.r-project.org/web/packages/tidytext/index.html.
[2]quanteda ver. 1.3.4, available at https://cran.r-project.org/web/packages/quanteda/index.html.

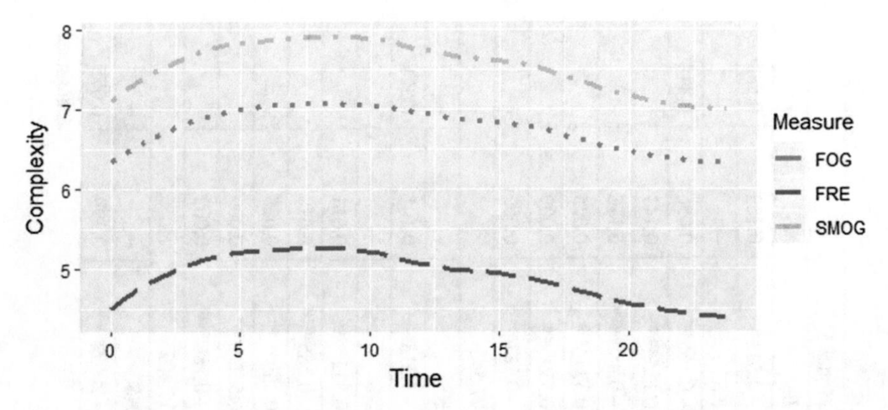

Fig. 3 LOESS curves for FOG, FRE, and SMOG, based on avg. text complexity for time

sentiment/text complexity at the time that tweet was posted (model 2), then finally the interaction term (model 3).

5 Discussion

Findings from this study support previous observations of circadian patterns in the *sentiment* of social media discussion. They also extend these patterns to *text complexity*, the first study to do so, to the author's knowledge.

More importantly, findings from this study suggest collective circadian patterns of *sentiment* and *text complexity* provide stronger predictions of information-sharing than the *sentiment* and *text complexity* of individual posts. Put differently, information is more likely to be shared when it is posted at times of the day when other users are primed for emotion and concentration, independent of whether that posted information is itself emotional or demanding in concentration.

More broadly, this study provides an explanatory physiological mechanism for how loosely connected individuals can achieve the emotional and cognitive alignment required for information-sharing. This has obvious practical implications for social media, e.g. perhaps posted information should be delayed for users in other time zones. However, this finding also has implications beyond social media discussion. For example, the circadian model proposed in this study may help to explain communication and relationship-building difficulties in distributed organizational teams.

Table 1 Results of negative binomial regression for circadian predicted sentiment on retweets

	Model 1			Model 2			Model 3		
	B	*SE*	*exp*	*b*	*SE*	*exp*	*B*	*SE*	*Exp*
Arousal	0.026**	0.006	1.026	0.024***	0.006	1.024	−0.531**	0.189	0.588
Predicted arousal				0.817***	0.134	2.257	NS	–	–
Ar * Predicted Ar							−0.165**	0.056	0.848
Hashtags	0.131***	0.014		0.135***	0.014		0.135***	0.141	
Mentions	−0.291***	0.018		−0.289***	0.018		−0.289***	0.178	
Urls	0.319***	0.025		0.336***	0.025		0.335***	0.025	
Log (followers)	0.557***	0.009		0.559***	0.009		0.559***	0.009	
Log (activity)	−0.179***	0.009		−0.181***	0.009		−0.181***	0.009	
AIC	77,604			77,563			77,563		
Valence	−0.011**	0.004	0.989	−0.010	0.004	0.990	NS	–	–
Predicted valence				−1.122***	0.101	0.320	−1.118***	0.101	0.321
Val * Predicted Val							NS	–	–
Hashtags	0.131***	0.014		0.133***	0.014		0.133***	0.014	
Mentions	−0.291***	0.018		−0.285***	0.018		−0.285***	0.018	
Urls	0.312***	0.025		0.339***	0.025		0.339***	0.025	
Log (followers)	0.557***	0.009		0.562***	0.009		0.563***	0.009	
Log (activity)	−0.181***	0.009		−0.187***	0.009		−0.188***	0.009	
AIC	77,615			77,492			77,492		
PA	NS	–	–	NS	–		NS	–	–
Predicted PA				−2.785***	0.296	0.053	−2.469***	0.399	0.075

(continued)

Table 1 (continued)

	Model 1			Model 2			Model 3		
	B	*SE*	*exp*	*b*	*SE*	*exp*	*B*	*SE*	*Exp*
PA * Predicted PA							NS	–	–
Hashtags	0.128***	0.014		0.123***	0.014		0.123***	−0.014	–
Mentions	_0.294***	0.018		−0.288***	0.018		−0.288***	0.018	
Urls	0.308***	0.025		0.309***	0.025		0.309***	0.025	
Log (followers)	0.556***	0.009		0.559***	0.009		0.559***	0.009	
Log (activity)	−0.179***	0.009		−0.185***	0.009		−0.185***	0.009	
AIC	77,623			77,535			77,535		
NA	0.029***	0.069	1.029	0.027***	0.006	1.027	NS	–	–
Predicted NA				1.182***	0.122	3.277	1.159***	0.153	3.218
NA * Predicted NA							NS	–	–
Hashtags	0.134***	0.014		0.138***	0.014		0.138***	0.014	–
Mentions	−0.287***	0.018		−0.285***	0.018		−0.285***	0.018	
Urls	0.319***	0.025		0.346***	0.025		0.346***	0.025	
Log (followers)	0.558***	0.009		0.563***	0.009		0.563***	0.009	
Log (activity)	−0.182***	0.009		−0.186***	0.009		−0.186***	0.009	
AIC	77,602			77,512			77,514		

*$p < 0.05$, **$p < 0.01$, ***$p < 0.001$, †$p < 0.1$, NS = not significant

Table 2 Results of negative binomial regression for circadian predicted sentiment on retweets

	Model 1			Model 2			Model 3		
	B	*SE*	*exp*	*b*	*SE*	*Exp*	*B*	*SE*	*Exp*
FOG	0.017***	0.003	1.017	−0.012***	0.003	1.012	NS	–	
LOESS FOG				−0.259***	0.046	1.308	−0.238**	0.084	1.287
FOG * LOESS							NS	–	
Hashtags	0.129***	0.014		0.133***	0.014		0.133***	0.141	
Mentions	−0.288***	0.018		−0.287***	0.018		−0.287***	0.178	
Urls	0.286***	0.025		0.303***	0.025		0.303***	0.025	
Log (followers)	0.553***	0.009		0.555***	0.009		0.555***	0.009	
Log (activity)	−0.178***	0.009		−0.179***	0.009		−0.179***	0.009	
AIC	77,603			77,574			77,576		
FRE	0.018***	0.003	1.018	0.019***	0.003	1.019	NS	–	
LOESS FRE				−0.251***	0.049	1.299	−0.229**	0.079	1.267
FRE * LOESS							NS	–	
Hashtags	0.132***	0.014		0.136***	0.014		0.136***	0.014	
Mentions	−0.285***	0.018		−0.284***	0.018		−0.285***	0.018	
Urls	0.279***	0.025		0.294***	0.025		0.294***	0.025	
Log (followers)	0.552***	0.009		0.554***	0.009		0.554***	0.009	
Log (activity)	−0.177***	0.009		−0.178***	0.009		−0.178***	0.009	
AIC	77,594			77,570			77,572		
SMOG	0.015***	0.004	1.015	0.016***	0.004	1.016	NS	–	
LOESS SMOG				−0.372***	0.055	1.464	−0.338**	0.125	1.416

(continued)

Table 2 (continued)

	Model 1			Model 2			Model 3		
	B	*SE*	*exp*	*b*	*SE*	*Exp*	*B*	*SE*	*Exp*
SMOG * LOESS							NS	–	
Hashtags	0.131***	0.014		0.135***	0.014		0.135***	0.014	
Mentions	−0.289***	0.018		−0.288***	0.018		−0.288***	0.018	
Urls	0.286***	0.025		0.305***	0.025		0.305***	0.025	
Log (followers)	0.552***	0.009		0.555***	0.009		0.555***	0.009	
Log (activity)	−0.177***	0.009		−0.179***	0.009		−0.179***	0.009	
AIC	77,609			77,567			77,569		

*$p < 0.05$, **$p < 0.01$, ***$p < 0.001$, †$p < 0.1$, NS = not significant

References

1. Dennis, A. R., Fuller, R. M., & Valacich, J. S. (2008). Media, tasks, and communication processes: A theory of media synchronicity. *MIS Quarterly, 32*(3), 575–600.
2. Ferrara, E., & Yang, Z. (2015). Quantifying the effect of sentiment on information diffusion in social media. *PeerJ Computer Science, 1,* e26. https://doi.org/10.7717/peerj-cs.26.
3. Stieglitz, S., & Dang-Xuan, L. (2013). Emotions and information diffusion in social media—Sentiment of microblogs and sharing behavior. *Journal of Management Information Systems, 29*(4), 217–248.
4. Murthy, D. (2011). Twitter: Microphone for the masses? *Media, Culture and Society, 33*(5), 779–789.
5. Speed, E., & Mannion, R. (2017). The rise of post-truth populism in pluralist liberal democracies: Challenges for health policy. *International Journal of Health Policy and Management, 6*(5), 249–251.
6. Bollen, J., Mao, H., & Zeng, X. (2011). Twitter mood predicts the stock market. *Journal of Computational Science, 2*(1), 1–8.
7. Taylor, Z. W. (2017). Speaking in tongues: Can international graduate students read international graduate admissions materials? *International Journal of Higher Education, 6*(3), 99–108.
8. Dabbagh, N., & Kitsantas, A. (2012). Personal learning environments, social media, and self-regulated learning: A natural formula for connecting formal and informal learning. *The Internet and Higher Education, 15*(1), 3–8.
9. Guille, A., Hacid, H., Favre, C., & Zighed, D. A. (2013). Information diffusion in online social networks: A survey. *ACM Sigmod Record, 42*(2), 17–28.
10. Korda, H., & Itani, Z. (2013). Harnessing social media for health promotion and behavior change. *Health Promotion Practice, 14*(1), 15–23.
11. Blumer, H. (1986). *Symbolic interactionism: Perspective and method.* University of California Press.
12. Neumann, R., & Strack, F. (2000). "Mood contagion": The automatic transfer of mood between persons. *Journal of Personality and Social Psychology, 79*(2), 211–223.
13. Enjolras, B., Steen-Johnsen, K., & Wollebæk, D. (2013). Social media and mobilization to offline demonstrations: Transcending participatory divides? *New Media & Society, 15*(6), 890–908.
14. Aschoff, J. (1965). Circadian rhythms in man. *Science, 148*(3676), 1427–1432.
15. Crowley, S. J., Acebo, C., & Carskadon, M. A. (2007). Sleep, circadian rhythms, and delayed phase in adolescence. *Sleep Medicine, 8*(6), 602–612.
16. Golder, S. A., & Macy, M. W. (2011). Diurnal and seasonal mood vary with work, sleep, and daylength across diverse cultures. *Science, 333*(6051), 1878–1881.
17. Yardi, S., & Boyd, D. (2010). Tweeting from the town square: Measuring geographic local networks. In *International AAAI Conference on Weblogs and Social Media*, Washington DC, USA.
18. Saper, C. B., Scammell, T. E., & Lu, J. (2005). Hypothalamic regulation of sleep and circadian rhythms. *Nature, 437*(7063), 1257–1263.
19. McClung, C. A. (2007). Circadian genes, rhythms and the biology of mood disorders. *Pharmacology & Therapeutics, 114*(2), 222–232.
20. Murray, G., Nicholas, C. L., Kleiman, J., Dwyer, R., Carrington, M. J., Allen, N. B., & Trinder, J. (2009). Nature's clocks and human mood: The circadian system modulates reward motivation. *Emotion, 9*(5), 705–716.
21. Pontes, A. L. B. D., Engelberth, R. C. G. J., Nascimento, E. D. S., Jr., Cavalcante, J. C., Costa, M. S. M. D. O., Pinato, L. … Cavalcante, J. D. S. (2010). Serotonin and circadian rhythms. *Psychology & Neuroscience, 3*(2), 217–228.
22. Bell-Pedersen, D., Cassone, V. M., Earnest, D. J., Golden, S. S., Hardin, P. E., Thomas, T. L., & Zoran, M. J. (2005). Circadian rhythms from multiple oscillators: Lessons from diverse organisms. *Nature Reviews Genetics, 6*(7), 544–556.

23. Czeisler, C. A., Shanahan, T. L., Klerman, E. B., Martens, H., Brotman, D. J., Emens, J. S. … Rizzo, J. F. (1995). Suppression of melatonin secretion in some blind patients by exposure to bright light. *New England Journal of Medicine, 332*(1), 6–11.
24. Korshunov, K. S., Blakemore, L. J., & Trombley, P. Q. (2017). Dopamine: A modulator of circadian rhythms in the central nervous system. *Frontiers in Cellular Neuroscience, 11*(91), 1–17.
25. Dimitrov, S., Benedict, C., Heutling, D., Westermann, J., Born, J., & Lange, T. (2009). Cortisol and epinephrine control opposing circadian rhythms in T cell subsets. *Blood, 113*(21), 5134–5143.
26. Watson, D., & Tellegen, A. (1985). Toward a consensual structure of mood. *Psychological Bulletin, 98*(2), 219–235.
27. Kahn, W. A. (1990). Psychological conditions of personal engagement and disengagement at work. *Academy of Management Journal, 33*(4), 692–724.
28. Lee, S. K., Lindsey, N. J., & Kim, K. S. (2017). The effects of news consumption via social media and news information overload on perceptions of journalistic norms and practices. *Computers in Human Behavior, 75,* 254–263.
29. Oeldorf-Hirsch, A., & Sundar, S. S. (2015). Posting, commenting, and tagging: Effects of sharing news stories on Facebook. *Computers in Human Behavior, 44*, 240–249.
30. Petty, R. E., & Cacioppo, J. T. (1986). The elaboration likelihood model of persuasion. In L. Berkowitz (Ed.), *Advances in experimental social psychology* (pp. 123–205), New York: Academic Press.
31. Lee, W., Reeve, J., Xue, Y., & Xiong, J. (2012). Neural differences between intrinsic reasons for doing versus extrinsic reasons for doing: An fMRI study. *Neuroscience Research, 73*(1), 68–72.
32. Yasseri, T., Sumi, R., & Kertész, J. (2012). Circadian patterns of Wikipedia editorial activity: A demographic analysis. *PLoS ONE, 7*(1), e30091.
33. Dzogang, F., Lansdall-Welfare, T., & Cristianini, N. (2016). Seasonal fluctuations in collective mood revealed by Wikipedia searches and Twitter posts. In *IEEE International Conference on Data Mining Workshop (SENTIRE)*, Barcelona.
34. Dzogang, F., Lightman, S., & Cristianini, N. (2017). Circadian mood variations in Twitter content. *Brain and Neuroscience Advances, 1,* 1–14.
35. Nielsen, F. Å. (2011). A new ANEW: Evaluation of a word list for sentiment analysis in microblogs. ESWC2011 Workshop on 'Making Sense of Microposts': Big things come in small packages, Heraklion, Crete.
36. Gunning, R. (1952). *The technique of clear writing*. UK: McGraw-Hill.
37. Chall, J. S., & Dale, E. (1995). *Readability revisited: The new Dale-Chall readability formula.* Massachusetts: Brookline Books.
38. Kincaid, J. P., Fishburn, R. P., Rogers, R. L., & Chissom, B. S. (1975). *Derivation of new readability formulas for navy enlisted personnel.* Technical Report Research Branch Report 8-75, Millington, Tennesse, Naval Air Station.
39. McLaughlin, G. H. (1969). SMOG grading-a new readability formula. *Journal of Reading, 12*(8), 639–646.

Does a Social Media Abstinence Really Reduce Stress? A Research-in-Progress Study Using Salivary Biomarkers

Eoin Whelan

Abstract There is much scientific evidence in recent years indicating that our 'always on' culture powered by platforms such as Facebook, LinkedIn, Instagram, Twitter, and WhatsApp, is leading to negative health outcomes, particularly stress. To mitigate social media induced stress, people are being advised to abstain from using social media for a period of time. However, the effectiveness of such breaks is open to question. As many people are heavily dependent on social media, the inability to access these platforms for a period of time could actually create stress and anxiety. To determine if and how social media abstinence relates to stress, this project will investigate the role of passion as a mediating variable. Stress will be measured using a combination of the salivary biomarkers cortisol and alpha amylase, with psychological scales. Ultimately, this study aims to determine the boundary conditions under which an abstinence from social media use will either increase or decrease stress levels in working professionals.

Keywords Social media · Stress · Abstinence · Cortisol · Alpha amylase

1 Introduction

Combining biomarkers of stress (cortisol and alpha amylase) with psychological measures, the objective of this study is to determine how an abstinence from social media use affects the wellbeing of working professionals. This proposed study will focus on working professionals as they are a population who are heavily dependent on social media for work, family, and leisure activities, yet are understudied in terms of the resulting health implications [1].

Social media use is increasing across society, and its association with mental wellbeing remains unclear. Some recent social media studies report on its harmful association with stress [2], anxiety [3], and depression [4]. Other studies conclude the health implications of social media use to be minimally detrimental [5, 6], and

E. Whelan (✉)
Business Information Systems, National University of Ireland, Galway, Ireland
e-mail: eoin.whelan@nuigalway.ie

F. D. Davis et al. (eds.), *Information Systems and Neuroscience*,
Lecture Notes in Information Systems and Organisation 32,
https://doi.org/10.1007/978-3-030-28144-1_2

even beneficial in some cases [7, 8]. Notwithstanding the lack of scientific clarity, many national and organisational health policies have emerged, as well as a sizable digital detox industry, advising social media users to abstain from use for a period of time. For example, the Royal Society for Public Health now advocate a "Scroll Free September". While such policies may be well intentioned, they are in essence untested medical interventions.

Given the contemporary nature of the phenomena, there is a limited body of knowledge pertaining to the health implications of social media abstinence. Many people are heavily dependent on social media [9, 10]. Thus, removing access to a person's social media accounts for a period of time may actually increase stress and anxiety. Indeed, recent studies have validated the link between the inability to access digital technology and stress [11], anxiety [3, 12], and sleep difficulties [13]. In the context of working professionals, one study found some employees worked for longer stretches when online distractions were blocked, a consequence of which was increased stress [14]. Likewise, the withdrawal symptoms of craving and boredom have been reported by participants while abstaining from social media [15]. Involuntary abstinence has also been studied, with research showing that participants who lost their smartphones reported negative feelings, such as boredom, anxiety, and loneliness [16]. On the bright side, an experiment with 1095 participants in Denmark, demonstrated that taking a one week break from Facebook had positive effects on life satisfaction and emotions, and such benefits were significantly greater for heavy Facebook users [17].

To determine why people respond differently to a social media abstinence, this study will examine the passion a worker has for social media, and if that explains increases of decreases in stress levels. Vallerand and colleagues [18–20] have conceptualised the passion a person feels for an activity, such as social media, as a duality. The dual model of passion (DMP) posits that an individual can have a strong inclination toward a self-defining activity that is loved, but that activity is comprised of both harmonious and obsessive dimensions [21, 22]. Both forms of passion describe a "*strong inclination toward an activity that people like, that they find important, and in which they invest time and energy*" [21, p. 756]. However, the opposing dimensions of passion differ in how they become internalised in the identity of an individual. A harmonious passion is adaptive and reflects a level of control to engage in the activity. The internalisation of the activity into the person's identity is autonomous [20]. A person demonstrating harmonious passion is not compelled to do the activity and can stop at any time. Harmonious individuals observe the activity as a supplement to a well-balanced lifestyle and are not consumed by a sense of "I must, I need to" engage with the activity. They are able to bound the activity (e.g., set limits), set personal goals which are consistent with their own strengths and weaknesses, and can align and/or prioritise the activity, thus, reducing conflict with other life domains (e.g., work, family). In other words, the respective activity is in "harmony" with other aspects of person's life [23].

In cases of obsessive passion, the internalisation is driven by intrapersonal or interpersonal pressures, such as heightened self-esteem or social acceptance within a specific group [24]. People demonstrating obsessive passion experience an internal compulsion to engage in the activity even when not appropriate to do so, as it goes beyond the person's self-control [23]. Obsessive passion is maladaptive and is related to negative emotions such as shame [21]. The activity dominates the person's identity to the extent it conflicts with other aspects of the person's life [20]. IS scholars have drawn from the DMP to shed light on the effects of online gaming [24, 25], social media use [26], and internet activities [27].

To extend state-of-the-art knowledge of the health effects of social media abstinence, this proposed study will combine physiological data with phycological data to measure stress. Both approaches are susceptible to validity issue, such as subjectivity, social desirability, and common method bias for psychological measures [28, 29], and construct reliability [30]. Physiological approaches are particularly well equipped for measuring constructs people are unable to accurately self-report, such as stress [28]. Previous technostress studies have used cortisol measures to determine the stress effects of systems breakdown [31], extensive media use [1], and interruptions [32]. That is not to say that such physiological approaches are better than traditional self-reported methods. As physiological data cannot be manipulated by the subject, or susceptible to social desirability bias, triangulation with additional data sources can result in a more holistic representation of research constructs [28, 30]. As advocated by Tams et al. [30], to improve validity and reliability, this study will combine physiological data with self-reported psychological measures of well-being.

2 Proposed Methods

Sixty volunteers who are fulltime working professionals will be recruited for this study. It is envisioned participants will be recent graduates of NUI Galway's MBA program. If required, snowball sampling techniques will also be employed to recruit more participants (e.g., postings on Facebook and newspapers). Working professionals are the focus of this study as some organisations have or are considering implementing social media blocking apps to reduce distractions and stress [14]. Applicants will be screened for suitability against the following criteria;

- Regularly using social media at least 1 h per day.
- Not required by employer to use social media for work purposes.
- No recent infections, or suffering from a chronic illness, or a heavy smoker, or receiving hormonal replacement treatment (all affect hormonal stress measures).

Selected volunteers will be randomly split into a control group of 30, who do not abstain from using social media, and an experimental group of 30, who will abstain from social media use. The ecological momentary intervention (EMI) method will be adopted to achieve the project objectives i.e. experimental intervention in the natural setting during participants' everyday lives. This involves gathering data over; (a) a 2 day baseline phase where all participants use their social media as normal, (b) for the experimental group, a 2 day intervention phase where access to social media is blocked, while experimental group continue use as normal (c) a 2 day post intervention where social media can be used as normal again by all participants. To ensure consistency, data gathering for each phase will take place on the same workdays, Tuesday and Wednesday over 3 weeks.

Prior to commencing the EMI, participants will complete psychometric tests to measure variables which previous studies suggest may influence reaction to social media abstinence e.g. personality, emotional characteristics, fear of missing out, preoccupation with social media, work-family segmentation preferences, boredom proneness. Harmonious and obsessive passion for social media will be measured using the 14 item DMP scale [21]. For the physiological measures of stress, each day participants will passively droll into a small vial they will be provided with. Following best practice in saliva collection and analysis [1, 31], samples will be collected immediately upon waking in the morning before feet touch the floor, 30 min after waking, at noon (before lunch), and right before bedtime. Participants will be instructed how to freeze their saliva samples. This will result in 1,440 samples. Samples will be sent securely to the Biomarker Lab in Anglia Ruskin University for analysis. Stress is measured by calculating the change in levels of salivary cortisol and alpha amylase from morning to evening (i.e. the diurnal slope). The project is focusing specifically on cortisol and alpha amylase as previous studies found links between these stress biomarkers and digital technology use [1, 11].

Across all three phases of the EMI, the smartphone app ‘Moment’ will be used to objectively track social media usage. For the experimental group in the abstinence phase, the ‘Freedom’ app will used to block access to social media, which participants can override if necessary. From these apps, the research team will have documented evidence if the experimental group successfully completed the social media abstinence. Participants will also complete a daily questionnaire, to be completed at a specific time during each work day, designed to measure psychological perceptions of wellness including stress, anxiety, mood, and life satisfaction using well established scales [33–35]. The questionnaire will also require participants to reflect on their behaviours during the abstinence. This will allow the researchers to determine if any compensatory behaviours emerged. In keeping with the objective of the project, the questionnaires will be paper based as opposed to computer mediated.

3 Expected Outcomes

This project will determine;

- The efficacy of a social media abstinence as a workplace health intervention.
- If rebound effects are prevalent when users end a social media abstinence.
- If possessing a harmonious or obsessive passion for social media explains why different groups of people respond differently to a social media abstinence.

- If a more nuanced intervention is needed to reduce the harmful effects of social media use on well-being, whilst also developing the framework for such interventions.

References

1. Afifi, T. D., Zamanzadeh, N., Harrison, K., & Callejas, M. A. (2018). WIRED: The impact of media and technology use on stress (cortisol) and inflammation (interleukin IL-6) in fast paced families. *Computers in Human Behavior, 81,* 265–273.
2. Reinecke, L., Aufenanger, S., Beutel, M. E., Dreier, M., Quiring, O., Stark, B., et al. (2017). Digital stress over the life span: The effects of communication load and internet multitasking on perceived stress and psychological health impairments in a german probability sample. *Media Psychology, 20*(1), 90–115.
3. Hartanto, A., & Yang, H. (2016). Is the smartphone a smart choice? The effect of smartphone separation on executive functions. *Computers in Human Behavior, 64,* 329–336.
4. Lin, L. Y., Sidani, J. E., Shensa, A., Radovic, A., Miller, E., Colditz, J. B., et al. (2016). Association between social media use and depression among US young adults. *Depression and Anxiety, 33,* 323–331.
5. Twenge, J.M., Joiner, T. E., Rogers, M. L., & Martin, G. N. (2018). Increases in depressive symptoms, suicide-related outcomes, and suicide rates among U.S. adolescents after 2010 and links to increased new media screen time. *Clinical Psychological Science, 6*(1), 3–17.
6. Orben, A., & Przybylski, A. K. (2019). The association between adolescent well-being and digital technology use. *Nature Human Behaviour, 3,* 173–182.
7. Chen, H. T., & Li, X. (2017). The contribution of mobile social media to social capital and psychological well-being: Examining the role of communicative use, friending and self-disclosure. *Computers in Human Behavior, 75,* 958–965.
8. Park, N., & Lee, H. (2012). Social implications of smartphone use: Korean College students' smartphone use and psychological well-being. *Cyberpsychology, Behavior, and Social Networking, 15*(9).
9. Sha, P., Sariyska, R., Riedl, R., Lachmann, B., & Montag, C. (2019). Linking internet communication and smartphone use disorder by taking a closer look at the Facebook and WhatsApp applications. *Addictive Behaviors Reports* (Forthcoming).
10. Turel, O. (2015). An empirical examination of the "vicious cycle" of Facebook addiction. *Journal of Computer Information Systems, 55*(3), 83–91.
11. Tams, S., Legoux, R., & Leger, P.-M. (2018). Smartphone withdrawal creates stress: A moderated mediation model of nomophobia, social threat, and phone withdrawal context. *Computers in Human Behavior, 81,* 1–8.
12. Cheever, N. A., Rosen, L. D., Carrier, L. M., & Chavez, A. (2014). Out of sight is not out of mind: The impact of restricting wireless mobile device use on anxiety levels among low, moderate and high users. *Computers in Human Behavior, 37,* 290–297.
13. Russo, M., Bergami, M., & Morandin, G. (2018). Surviving a day without smartphones. *MIT Sloan Management Review, 59*(2), 6–9.
14. Mark, G., Czerwinski, M., & Iqbal, S. T. (2018). Effects of Individual Differences in Blocking Workplace Distractions. In *Proceedings of the 2018 CHI Conference on Human Factors in Computing Systems*, Montreal.
15. Stieger, S., & Lewetz, D. (2018). A week without using social media: Results from an ecological momentary intervention study using smartphones. *Cyberpsychology, Behavior, and Social Networking, 21*(10), 618–624.
16. Hoffner, C. A., Lee, S., & Park, S. J. (2016). "I miss my mobile phone!": Self-expansion via mobile phone and responses to phone loss. *New Media & Society, 18,* 2452–2468.

17. Tromholt, M. (2016). The Facebook experiment: Quitting Facebook leads to higher levels of well-being. *Cyberpsychology, Behavior, and Social Networking, 19*(11).
18. Rousseau, F. L., & Vallerand, R. J. (2003). Le rôle de la passion dans le bien-être subjectif des aînés. *Revue Québécoise de Psychologie, 24,* 197–211.
19. Vallerand, R. J., Salvy, S. J., Mageau, G. A., Elliot, A. J., Denis, P. L., Grouzet, F. M. E., et al. (2007). On the role of passion in performance. *Journal of Personality, 75*(3), 505–534.
20. Vallerand, R. J. *The psychology of passion: A dualistic model. Series in Positive Psychology* (403 p). Oxford: Oxford University Press.
21. Vallerand, R. J., Mageau, G. A., Ratelle, C., Léonard, M., Blanchard, C., Koestner, R., et al. (2003). Les passions de l'âme: On obsessive and harmonious passion. *Journal of Personality and Social Psychology, 85*(4), 756–767.
22. Vallerand, R. J. (2008). On the psychology of passion: In search of what makes people's lives most worth living. *Canadian Psychology [Psychologie Canadienne], 49*(1), 1–13.
23. Paradis, K. F., Cooke, L. M., Martin, L. J., & Hall, C. R. (2013). Too much of a good thing? Examining the relationship between passion for exercise and exercise dependence. *Psychology of Sport and Exercise, 14*(4), 493–500.
24. Utz, S., Jonas, K. J., & Tonkens, E. (2012). Effects of passion for massively multiplayer online role-playing games on interpersonal relationships. *Journal of Media Psychology, 24*(2), 77–86.
25. Przybylski, A. K., Weinstein, N., Ryan, R. M., & Rigby, C. S. (2009). Having to versus wanting to play: Background and consequences of harmonious versus obsessive engagement in video games. *CyberPsychology & Behavior, 12*(5), 485–492.
26. Orosz, G., Vallerand, R. J., Bothe, B., Tóth-Király, I., & Paskuj, B. (2016). On the correlates of passion for screen-based behaviors: The case of impulsivity and the problematic and non-problematic Facebook use and TV series watching. *Personality and Individual Differences, 101,* 167–176.
27. Tosun, L. P., & Lajunen, T. (2009). Why do young adults develop a passion for internet activities? The associations among personality, revealing "True Self" on the internet, and passion for the internet. *CyberPsychology & Behavior, 12*(4), 401–406.
28. Dimoka, A., Banker, R., Benbasat, I., Davis, F., Dennis, A., Gefen, D., et al. (2012). On the use of neurophysiological tools in IS research: Developing a research agenda for neurois. *MIS Quarterly, 36*(3), 679–702.
29. Podsakoff, P. M., MacKenzie, S. B., Lee, J. Y., & Podsakoff, N. P. (2003). Common method biases in behavioral research: A critical review of the literature and recommended remedies. *Journal of Applied Psychology, 88*(5), 879–903.
30. Tams, S., Hill, K., Ortiz de Guinea, A., Thatcher, J., & Grover, V. (2014). NeuroIS—Alternative or complement to existing methods? Illustrating the holistic effects of neuroscience and self-reported data in the context of technostress research. *Journal of the Association for Information Systems, 15*(10), 723–753.
31. Riedl, R., Kindermann, H., Auinger, A., & Javor, A. (2012). Technostress from a neurobiological perspective: System breakdown increases the stress hormone cortisol in computer users. *Business and Information Systems Engineering*, 61–69.
32. Tams, S., Thatcher, J., & Ahuja, M. (2015). The impact of interruptions on technology usage: Exploring interdependencies between demands from interruptions, worker control, and role-based stress. In *Lecture Notes in Information Systems and Organisation—Information Systems and Neuroscience. Gmunden Retreat on NeurolS* (pp. 19–25).
33. Cohen, S., Kamarck, T., & Mermelstein, R. (1983). A global measure of perceived stress. *Journal of Health and Social Behavior, 24*(4), 385–396.
34. Diener, E., Emmons, R., Larsen, R. J., & Griffin, S. (1985). The satisfaction with life scale. *Journal of Personality Assessment, 49*(1), 71–75 [Internet]. Available from: http://www.ncbi.nlm.nih.gov/pubmed/16367493.
35. Rosenberg, M., Schooler, C., Schoenbach, C., & Rosenberg, F. (1995). Global self-esteem and specific self-esteem: Different concepts, different outcomes. *American Sociological Review, 60*(1), 141–156.

Multicommunicating During Team Meetings and Its Effects on Team Functioning

Ann-Frances Cameron, Shamel Addas and Matthias Spitzmuller

Abstract This research-in-progress examines the phenomenon of multicommunicating during team meetings (Meeting MC). Drawing upon social interdependence theory, multilevel theorizing, and research on multitasking, we examine the positive and negative effects of Meeting MC on individual team members' reactions, as well as on team processes and team outcomes. We propose a two-phase experimental approach to investigate the individual-level affective, cognitive, and behavioral responses in other team members, as well as the how these individual-level effects of Meeting MC spill over and affect team-level functioning and performance. This research advances our understanding of Meeting MC and how it affects individuals and groups. It also provides guidelines to managers and decision makers to leverage the beneficial aspects of Meeting MC while limiting and mitigating its detrimental effects.

Keywords Multicommunicating · Team meetings · Meeting MC · Team processes · Team performance · NeuroIS · Physiological measures

1 Introduction

Workplace teams are increasingly popular and have become one of the main structures used to perform organizational work [1, 2]. Team research has shown that effective team performance is largely determined by the processes team members use to interact with one another in order to achieve their goals [3, 4]. Our research-in-progress focuses on team processes performed within the context of meetings,

A.-F. Cameron
HEC Montreal, Montreal, Canada
e-mail: ann-frances.cameron@hec.ca

S. Addas (✉) · M. Spitzmuller
Smith School of Business, Queen's University, Kingston, Canada
e-mail: shamel.addas@queensu.ca

M. Spitzmuller
e-mail: matthias.spitzmuller@queensu.ca

F. D. Davis et al. (eds.), *Information Systems and Neuroscience*,
Lecture Notes in Information Systems and Organisation 32,
https://doi.org/10.1007/978-3-030-28144-1_3

defined as "communicative event[s] involving three or more people who agree to assemble for a purpose ostensibly related to the functioning of an organization or group" [5]. Meetings are a ubiquitous team tool that can benefit team members [6, 7], but they can also be detrimental to individual wellbeing and team effectiveness [8, 9].

Meeting Multicommunicating (Meeting MC) is one key behavior that can influence meeting effectiveness. It is defined as "being simultaneously engaged both in an organizational meeting and in one or more technology-mediated secondary conversation(s)" [10]. Meeting MC can involve various forms of secondary conversations such as texting, checking email, or mobile phone use during face-to-face or technology-mediated meetings [6, 11–16]. While evidence from neuro- and cognitive psychology research shows the task performance detriments associated with multitasking [17–19], the consequences of Meeting MC are expected to be more complex because individuals engage not only in secondary tasks, but must also balance "different media, conversations, and communication partners" [20].

Our research examines secondary conversations that occur with others who are outside of the meeting. This type of Meeting MC is quite common and is often used for conversation leveraging (gathering information in the secondary conversations to serve the meeting) [21]. Extant research has shed light on the effects of MC on individual outcomes [20, 22]. Our study complements this research by focusing on how the actions of a person engaged in multiple conversations during a meeting (herein termed the MCer) affect other team members and team processes and performance. We address the following research questions: (i) how do Meeting MC trigger individual-level affective, cognitive, and behavioral responses in the other team members in the meeting who are not engaging in Meeting MC?, and (ii) how do these individual-level effects spill over and affect team-level functioning and performance? Given the complexity of Meeting MC outlined above, we leverage multiple theoretical frameworks to address this phenomenon, namely social interdependence theory, multilevel theorizing, and research on multitasking.

2 Theoretical Development

The basic premise of social interdependence theory is that the goal structure of a team determines how team members will interact, which in turn influences the outcomes of the situation [23–25]. Teams with congruent goals tend to exhibit "effective" actions that promote perceptions of joint goal achievement. Alternatively, teams with incongruent goals tend to display "bungling" actions and self-interested behaviors that decrease perceptions of joint goal accomplishment [24, 25].

A team's goal structure and its effective and bungling actions influence team functioning through three processes, namely cathexis, inducibility, and substitutability. Cathexis refers to the willingness to invest psychological energy in others. Inducibility refers to one's willingness to be influenced by others [24]. Substitutability is the degree to which one's actions can be performed by other members [24].

In our research, social interdependence theory is applied to the Meeting MC context, in which an MCer is simultaneously working toward multiple goals (e.g., being involved in a team meeting while also engaging in a secondary conversation). Congruent Meeting MC refers to situations where the MCer is engaging in a secondary conversation that is pertinent to the meeting goals [10, 26]. Incongruent Meeting MC refers to situations where the secondary conversation is unrelated to the meeting goals (e.g., pertaining to another work project, a personal issue, etc.). A third option also exists, with the goal congruence of the Meeting MC being unknown to the other team members. Unknown goal congruence—while not covered by social interdependence theory—is practically important to examine, as other team members do not always know the content of the MCer's secondary conversations [22].

2.1 Individual-Level Effects of Using Meeting MC

We propose that Meeting MC can induce affective, cognitive, and behavioral responses in the other team members and that many of these responses will differ based on the goal congruence of the Meeting MC. Additionally, we propose that Meeting MC can have negative effects on the other team members through distraction, and this effect will exist regardless of goal congruence.

Specifically, we predict that congruent Meeting MC will lead to positive affective responses in the other team members, as the MCer is bringing new relevant information to the meeting. Congruent Meeting MC also induces cognitive responses in the other team members, such as increasing the other team members' perceptions of the MCer's capabilities and motivation. Thus, we expect that they will invest more psychological energy in their relationships with the MCer (cathexis), particularly in terms of willingness to work with the MCer on subsequent tasks and to help the MCer as needed (e.g., directing prosocial behaviors at the MCer). Further, through inducibility, other team members are expected to develop higher levels of trust towards the MCer. Finally, through substitutability, goal congruent Meeting MC will be perceived by other team members as evidence that the MCer is working toward the common good. Consequently, other team members will be more likely to feel ownership of the MCer's teamwork tasks, and thus more willing to adapt and shift roles with the MCer as needed.

Whereas incongruent Meeting MC may be considered an effective action by the MCer, it would be perceived as a bungling action by the other team members. This is because the MCer is focusing on their own productivity rather than contributing to the joint goals of the team. Thus, incongruent Meeting MC will lead to other team members experiencing negative affective responses (e.g., feelings of frustration or anger). Cognitively, the other team members may perceive the MCer as rude [21] and unprofessional. Hence, related to cathexis, we would predict lower willingness to work together and to help the MCer, as well as less prosocial behaviors and more counterproductive behaviors targeted at the MCer (e.g., incivility or aggressiveness). Through inducibility, other team members are expected to develop lower level of trust towards the MCer. They are less likely to be influenced by the MCer, which reduces the MCer's influence on team discussions and meeting outcomes. Finally, goal incongruent Meeting MC would reduce substitutability, with other team members being less willing to adapt their work roles to emerging needs of the MCer.

With unknown goal congruence, social interdependence theory does not help us to understand the effects of Meeting MC on other team members' responses. However, the fundamental attribution error [27] would suggest that when the content of the MCer's secondary conversations are unknown, other team members might make internal attributions and therefore judge the MCer more harshly than when the goals are known to be congruent. Preliminary results of one of the authors' pretest video vignette studies support this proposition. Thus, Meeting MC with unknown goal congruence may engender generally negative responses in the other team members (decreased interest, increased perceptions of rudeness, decreased trust, and decreased prosocial behaviors to help the 'important' MCer), similar to those associated with goal-incongruent Meeting MC.

Drawing upon the multitasking literature, we argue that Meeting MC will also have a negative distraction effect that will materialize irrespective of goal congruence. Meeting MC, much like any form of multitasking, reduces task processing efficiency and effectiveness [2, 11, 28, 29]. Whereas these negative outcomes occur due to attention switching and directly affect the MCer, we argue that the other team members will also be influenced negatively via a distraction effect. Regardless of goal congruence, meeting participants can become distracted by the activities of the MCer (e.g., wondering what the MCer is doing and whether it is meeting-related). This effect is consistent with evidence from the literature on multitasking in the classroom, which shows that laptop usage by a student distracts others around them [30]. We predict that these distractions negatively influence the quantity and quality of information contributed to the meeting by the other team members.

2.2 *Team-Level Effects of Using Meeting MC*

Our research will examine how the individual-level effects of Meeting MC spill over to influence team-level outcomes. We posit that Meeting MC will influence team outcomes via two types of emergence processes: dynamic interactions between team members during the meeting [31, 32] that affect intra-team trust (an inducibility-related construct) and team adaptation (a substitutability-related construct) and emotional contagion processes [33] that affect team cohesion (a cathexis-related construct).

Meeting MC may trigger dynamic interactions that shape a team-level response to the behavior [cf. 31, 32]. For example, a team member who notices incongruent Meeting MC may aggressively challenge the MCer by questioning why they are engaging in secondary conversations or asking them to stop the behavior. Such conflicts can affect both the degree and emergence of intra-team trust [34]. Intra-team trust represents the shared generalized perceptions of trust among team members

[35]. The nature of emergence of this construct follows a direct consensus compositional model [36]. Also, the referent in our case is a specific team member, namely the MCer, rather than the team as a whole [34].

Furthermore, Meeting MC may influence team adaptation, a substitutability-related construct defined as adjustments to relevant role configurations in the team in response to unforeseen changes [28]. We posit that team adaptation will increase by both congruent and incongruent Meeting MC. For congruent Meeting MC, the increased individual willingness to adapt will emerge to the group level through a compositional process [37]. Team members will develop a shared responsibility to help with the MCer's teamwork tasks and create adaptive mechanisms to recalibrate who performs what task. For incongruent Meeting MC, we expect a cross-level effect on team adaptation. Team members are likely to react to the MCer's actions by redesigning their roles and withholding responsibility from the MCer as a punitive act [38]. Hence, the team will reconfigure their roles and structures to take over responsibility from the MCer.

Meeting MC may also influence team outcomes via less overt social processes. More specifically, emotional contagion research indicates that individuals can transmit their affective experiences [e.g., 33] and stress perceptions to others, along with their accompanying subjective feelings [39]. Thus, we predict that individual affective reactions of specific team members to Meeting MC will spill over to influence the affective experiences of other team members [40]. For example, one team member might notice and become annoyed by the MCer's incongruent Meeting MC. This feeling of annoyance (although not necessarily its cause) may be expressed and transferred implicitly (e.g., through facial or vocal gestures that get mimicked). Positive feelings (e.g., excitement) elicited by goal congruent Meeting MC can similarly be transmitted via contagion. These emotional contagion processes are likely to influence team cohesion, which is defined as "the extent to which group members are socially integrated, possess shared feelings of unity, and are attracted to the group and each other" [6]. We propose that Meeting MC will influence all three facets of team cohesion, namely task cohesiveness, interpersonal cohesiveness, and team pride [7]. We expect that the emergence of these affective and cognitive responses has an isomorphic nature, meaning that individuals contribute a similar type and amount of elemental content to the group [41].

The final team-level outcome we examine is meeting effectiveness. Existing multitasking literature would suggest that engaging in multiple tasks during a meeting would negatively influence meeting effectiveness by increasing the quantity of information processed, causing dual task interference, and reducing the quality of the team's decision [e.g., 29]. Multicommunicating research further suggests that goal congruence plays a role. Through the relevant new information that the MCer brings to the meeting, goal congruent Meeting MC should increase meeting effectiveness [10]. We argue that the effects are more complex due to both the dynamic interactions and team outcomes (team cohesion, intra-team trust, and adaptation) outlined above. For example, whereas Meeting MC may negatively impact meeting effectiveness due to the distraction effect, goal-congruent Meeting MC may increase team cohesion and ultimately reshape the dynamic interactions and team performance.

In sum, our research suggests that dynamic interactions and emotional contagion are important processes that will translate the individual-level effects of Meeting MC into team level outcomes such as team cohesion, intra-team trust, and team adaptation. Further, these will have implications for overall team meeting effectiveness.

3 Proposed Methodology

A two-phase experimental approach will be used to investigate the effects of Meeting MC on team functioning and performance. Phase I will focus on the individual-level affective, cognitive, and behavioral responses in other team members. Phase II will explore how these individual-level effects of Meeting MC spill over and affect team-level functioning and performance.

In Phase I, three-person experiments using a hidden profile paradigm [42] will be employed in which each team member receives unique information, all of which will be needed to produce an optimal team decision during the meeting. In the first experimental condition, all team members in the control condition will be asked to focus on the meeting exclusively (condition 1: control group). In other groups, one participant in each team will be given a series of secondary tasks to complete during the meeting. These secondary conversations will occur via text message with a research assistant who is outside of the meeting. Some of these secondary conversations will be goal incongruent (condition 2: unrelated content), while others will be goal congruent (condition 3: information that is needed to make the optimal team decision). In conditions 2 and 3, other team members will be explicitly made aware of the goal congruence of the secondary conversations. To increase ecological validity, participants will bring their own text-enabled smartphone to the experiment. Meeting effectiveness will be measured by comparing the team's decision to the optimal decision. Post-meeting questionnaires using existing scales will be used to examine individual-level outcomes such as each individual's willingness to work with the MCer in the future.

Phase II will use the same three-person hidden profile experiments to examine how the individual-effects of Meeting MC influence the team's dynamic interactions and team-level outcomes. Phase II will have the same three conditions as Phase I; however, the experimental sessions will be longer, allowing time for the dynamic interactions to unfold during the team meeting. In addition, other team members will not be explicitly made aware of the goal congruence of the secondary conversations. Using one camera per participant, the meetings will be recorded and manually coded after the experiment to capture the dynamic interactions that occur during the meeting. Coding of the verbal statements during the meeting will occur using the INTERACT software and Advanced Interaction Analysis [act4team®, e.g., 43], which includes four main categories of interaction (problem-focused, procedural, socioemotional, and action-oriented statements). These interaction categories are then further subdivided in multiple sub-categories. Phase II will enable us to identify the dynamic processes that follow Meeting MC and their subsequent impact on team functioning and performance.

Table 1 summarizes the individual- and team-level outcomes of the study. As illustrated in the Table, we plan to use neuro-physiological measures to complement the traditional psychometric measurement for several of the key variables [44]. The reasons for this are to provide converging evidence, alleviate subjective biases, and provide complementary insights into the findings. To capture nuanced affective responses and stress of other team members in reaction to Meeting MC, each participant will have one mobile wrist unit to measure electrodermal activity. Physiological tools will be used (e.g., face reader to capture emotion, skin conductance to capture stress, and wearable eye trackers to capture participant gaze) during the meeting. In terms of prosocial attitudes and behaviors (a measure of cathexis), Volk and Becker [27] suggested that how people react to and behave toward others (especially prosocial behavior) is determined by their perceptions of the fairness of others' behaviors. These fairness perceptions produce prepotent response tendencies (automatized response patterns that support pro-social attitudes) that can be measured through skin conductance [30] or by tracking activity in the brain's limbic system [e.g., 45, 46]. Similarly, for the inducibility mechanism, neuro-physiological measures of trust will be developed. IS studies used fMRI to map the different individual trust dimensions to different areas of activation in the brain [47, 48]. Others have called for applying EEG to study trust mechanisms [49].

At the team-level of analysis, we seek to assess the neural correlates of team cohesion. Research in social neuroscience—while still in the early stages—has suggested that team processes such as team cohesion can be measured by mapping brain activity configurations: "it may be possible to compare configurations of brain activity patterns across teams to see which configuration might be associated with more cohesion, including the excessive cohesion that accompanies groupthink, as well as conflict in teams." [50, p. 287]. Others have echoed the need to use neuro-physiological

Table 1 Measurement of the key outcomes

Level of analysis	Outcomes	Measures	
		Psychometric	Neuro-physiological
Individual level	Stress and affective reactions of other team members	X	X
	Willingness to work with MCer (cathexis)	X	X
	Trust (inducibility)	X	X
	Willingness to adapt (substitutability)	X	
Team level	Team cohesion (cathexis)	X	X
	Intra-team trust (inducibility)	X	X
	Team adaptation (substitutability)	X	
	Meeting effectiveness	X	X

measures of team cohesion—including skin responses and EEG [51]—and aggregating the individual belief measures to form a statistical score that represents a shared belief [52]. For intra-team trust, there is preliminary evidence indicating that measuring the synchrony of the heart rate profiles of team members could be used to assess the building of team trust [53].

Finally, our research operationalizes meeting effectiveness as the quantity of information processing (number of pieces of new information assimilated by the team) and the team's decision quality [29]. Neuro-physiological tools will shed more light on these objective outcomes. Specifically, it will be used to determine whether shortcomings in the quantity of information used is due to lack of attention by the team or deliberate discounting of the information provided by the MCer.

4 Contributions and Conclusion

The present research-in-progress is expected to provide important theoretical and practical contributions. It advances our understanding of the positive and negative effects of Meeting MC on individual team members' outcomes and on team processes and performance. Our research will also provide practical contributions that enable managers and decision makers to leverage the beneficial aspects of Meeting MC while limiting and mitigating its detrimental effects.

Our hope is to benefit from the Neuro IS workshop by engaging with the Neuro IS community and getting feedback on how to execute our research in a way that provides accurate physiological measures of team members' reactions to Meeting MC.

References

1. Kozlowski, S. W. J., & Ilgen, D. R. (2006). Enhancing the effectiveness of work groups and teams. *Psychological Science in the Public Interest, 7,* 77–124.
2. Mathieu, J. E., Hollenbeck, J. R., van Knippenberg, D., & Ilgen, D. R. (2017). A century of work teams. *Journal of Applied Psychology, 102,* 452–467. https://doi.org/10.1037/apl0000128.
3. Cronin, M. A. (2015). Advancing the science of dynamics in groups and teams. *Organizational Psychology Review, 5,* 267–269. https://doi.org/10.1177/2041386615606826.
4. Ilgen, D. R., Hollenbeck, J. R., Johnson, M., & Jundt, D. (2005). Teams in organizations: From input-process-output models to IMOI models. *Annual Review of Psychology, 56,* 517–543.
5. Schwartzman, H. B. (1989). *The meeting*. New York, NY: Plenum Press.
6. Dennis, A. R., Rennecker, J. A., & Hansen, S. (2010). Invisible whispering: Restructuring collaborative decision making with instant messaging. *Decision Sciences, 41,* 845–886. https://doi.org/10.1111/j.1540-5915.2010.00290.x.
7. Sonnentag, S., & Volmer, J. (2009). Individual-level predictors of task-related teamwork processes: The role of expertise and self-efficacy in team meetings. *Group & Organization Management, 34,* 37–66. https://doi.org/10.1177/1059601108329377.
8. Nixon, C. T., & Littlepage, G. E. (1992). Impact of meeting procedures on meeting effectiveness. *Journal of Business and Psychology, 6,* 361–369. https://doi.org/10.1007/BF01126771.

9. Rogelberg, S. G., Leach, D. J., Warr, P. B., & Burnfield, J. L. (2006). "Not another meeting!" Are meeting time demands related to employee well-being? *Journal of Applied Psychology, 91,* 83–96. https://doi.org/10.1037/0021-9010.91.1.83.
10. Cameron, A.-F., Barki, H., Ortiz de Guinea, A., Coulon, T., & Moshki, H. (2018). Multicommunicating in meetings: Effects of locus, topic relatedness, and meeting medium. *Management Communication Quarterly, 32,* 303–336. https://doi.org/10.1177/0893318918759437.
11. Camacho, S., Hassanein, K., & Head, M. (2015). Understanding the effect of techno-interruptions in the workplace. In A. Rocha, A. M. Correia, S. Costanzo, & L. P. Reis (Eds.), *New contributions in information systems and technologies* (pp. 1065–1071). Springer International Publishing.
12. Chudoba, K. M., Watson-Manheim, M. B., Crowston, K., & Lee, C. S. (2011). Participation in ICT-enabled meetings. *Journal of Organizational and End User Computing (JOEUC), 23,* 15–36. https://doi.org/10.4018/joeuc.2011040102.
13. Kleinman, L. (2007). Physically present, mentally absent: Technology use in face-to-face meetings. In *CHI'07 Extended Abstracts on Human Factors in Computing Systems* (pp. 2501–2506). New York, NY, USA: ACM.
14. Stephens, K. K. (2012). Multiple conversations during organizational meetings: Development of the multicommunicating scale. *Management Communication Quarterly, 26,* 195–223. https://doi.org/10.1177/0893318911431802.
15. Stephens, K. K., & Davis, J. (2009). The social influences on electronic multitasking in organizational meetings. *Management Communication Quarterly, 23,* 63–83. https://doi.org/10.1177/0893318909335417.
16. Washington, M. C., Okoro, E. A., & Cardon, P. W. (2014). Perceptions of civility for mobile phone use in formal and informal meetings. *Business and Professional Communication Quarterly, 77,* 52–64. https://doi.org/10.1177/1080569913501862.
17. Monsell, S. (2015). Task-set control and task switching. In J. Fawcett, A. Kingstone, & E. Risko (Eds.), *The handbook of attention* (pp. 139–172). MIT Press.
18. Ophir, E., Nass, C., & Wagner, A. D. (2009). Cognitive control in media multitaskers. *PNAS Proceedings of the National Academy of Sciences, 106,* 15583–15587. https://doi.org/10.1073/pnas.0903620106.
19. Watanabe, K., & Funahashi, S. (2014). Neural mechanisms of dual-task interference and cognitive capacity limitation in the prefrontal cortex. *Nature Neuroscience, 17,* 601–611. https://doi.org/10.1038/nn.3667.
20. Cameron, A.-F., & Webster, J. (2013). Multicommunicating: Juggling multiple conversations in the workplace. *Information Systems Research, 24,* 352–371.
21. Cameron, A.-F., & Webster, J. (2011). Relational outcomes of multicommunicating: Integrating incivility and social exchange perspectives. *Organization Science, 22,* 754–771. https://doi.org/10.1287/orsc.1100.0540.
22. Reinsch, N. L. J. R., Turner, J. W., & Tinsley, C. H. (2008). Multicommunicating: A practice whose time has come? *Academy of Management Review, 33,* 391–403.
23. Deutsch, M. (1949). A theory of cooperation and competition. *Human Relations, 2,* 129–152. https://doi.org/10.1177/001872674900200204.
24. Deutsch, M. (2012). A theory of cooperation—competition and beyond. *Handbook of theories of social psychology* (Vol. 2, pp. 275–294). London: SAGE Publications Ltd.
25. Johnson, D. W., & Johnson, R. T. (2005). New developments in social interdependence theory. *Genetic, Social, and General Psychology Monographs, 131,* 285–358.
26. Addas, S., & Pinsonneault, A. (2018). Email interruptions and individual performance: Is there a silver lining? *MIS Quarterly, 42,* 381–405.
27. Volk, S., & Becker, W. J. (2014). How insights from neuroeconomics can inform organizational research: The case of prosocial organizational behavior. *Schmalenbach Business Review, 66,* 65–86. https://doi.org/10.1007/BF03396919.
28. LePine, J. (2005). Adaptation of teams in response to unforeseen change: Effects of goal difficulty and team composition in terms of cognitive ability and goal orientation. *Journal of Applied Psychology, 90,* 1153–1167. https://doi.org/10.1037/0021-9010.90.6.1153.

29. Heninger, W. G., Dennis, A. R., & Hilmer, K. M. (2006). Individual cognition and dual-task interference in group support systems. *Information Systems Research, 17,* 415–424.
30. van't Wout, M., Kahn, R. S., Sanfey, A. G., & Aleman, A. (2006). Affective state and decision-making in the Ultimatum Game. *Experimental Brain Research, 169*, 564–568. https://doi.org/10.1007/s00221-006-0346-5.
31. Chen, G., & Kanfer, R. (2006). Toward a systems theory of motivated behavior in work teams. *Research in Organizational Behavior, 27,* 223–267. https://doi.org/10.1016/S0191-3085(06)27006-0.
32. Morgeson, F. P., & Hofmann, D. A. (1999). The structure and function of collective constructs: Implications for multilevel research and theory development. *Academy of Management Review, 24,* 249–265.
33. Barsade, S. G. (2002). The ripple effect: Emotional contagion and its influence on group behavior. *Administrative Science Quarterly, 47,* 644–675. https://doi.org/10.2307/3094912.
34. Fulmer, C. A., & Gelfand, M. J. (2012). At what level (and in whom) we trust: Trust across multiple organizational levels. *Journal of Management, 38,* 1167–1230. https://doi.org/10.1177/0149206312439327.
35. De Jong, B. A., & Elfring, T. (2010). How does trust affect the performance of ongoing teams? The mediating role of reflexivity, monitoring, and effort. *Academy of Management Journal, 53,* 535–549. https://doi.org/10.5465/amj.2010.51468649.
36. Chan, D. (1998). Functional relations among constructs in the same content domain at different levels of analysis: A typology of composition models. *Journal of Applied Psychology, 83,* 234–246.
37. Han, T. Y., & Williams, K. J. (2008). Multilevel investigation of adaptive performance: Individual- and team-level relationships. *Group & Organization Management, 33,* 657–684. https://doi.org/10.1177/1059601108326799.
38. Langfred, C. W. (2007). The downside of self-management: A longitudinal study of the effects tf conflict on trust, autonomy, and task interdependence in self-managing teams. *Academy of Management Journal, 50,* 885–900. https://doi.org/10.5465/amj.2007.26279196.
39. Westman, M. (2001). Stress and strain crossover. *Human Relations, 54,* 717–751. https://doi.org/10.1177/0018726701546002.
40. Ilies, R., Wagner, D. T., & Morgeson, F. P. (2007). Explaining affective linkages in teams: Individual differences in susceptibility to contagion and individualism-collectivism. *Journal of Applied Psychology, 92,* 1140–1148. https://doi.org/10.1037/0021-9010.92.4.1140.
41. Kozlowski, S. W. J., & Klein, K. J. (2000). A multilevel approach to theory and research in organizations: Contextual, temporal, and emergent processes. In K. J. Klein & S. W. J. Kozlowski (Eds.), *Multilevel theory, research, and methods in organizations* (pp. 3–90). San Francisco, CA: Jossey-Bass Inc.
42. Lu, L., Yuan, Y. C., & McLeod, P. L. (2012). Twenty-five years of hidden profiles in group decision making: A meta-analysis. *Personality and Social Psychology Review, 16,* 54–75. https://doi.org/10.1177/1088868311417243.
43. Lehmann-Willenbrock, N., & Kauffeld, S. (2010). The downside of communication: Complaining cycles in group discussions. In S. Schuman (Ed.), *The handbook for working with difficult groups: How they are difficult, why they are difficult and what you can do about it* (pp. 33–54). San Francisco, CA: Wiley.
44. Tams, S., Hill, K., de Guinea, A. O., Thatcher, J., & Grover, V. (2014). NeuroIS—Alternative or complement to existing methods? Illustrating the holistic effects of neuroscience and self-reported data in the context of technostress research. *Journal of the Association for Information Systems, 15,* 723–753.
45. Sanfey, A. G., Rilling, J. K., Aronson, J. A., Nystrom, L. E., & Cohen, J. D. (2003). The neural basis of economic decision-making in the Ultimatum Game. *Science, 300,* 1755–1758. https://doi.org/10.1126/science.1082976.
46. Tabibnia, G., Satpute, A. B., & Lieberman, M. D. (2008). The sunny side of fairness: Preference for fairness activates reward circuitry (and disregarding unfairness activates self-control circuitry). *Psychological Science, 19,* 339–347. https://doi.org/10.1111/j.1467-9280.2008.02091.x.

47. Dimoka, A. (2010). What does the brain tell us about trust and distrust? Evidence from a functional neuroimaging study. *MIS Quarterly, 34,* 373–396. https://doi.org/10.2307/20721433.
48. Riedl, R., Hubert, M., & Kenning, P. (2010). Are there neural gender differences in online trust? An fMRI study on the perceived trustworthiness of Ebay offers. *MIS Quarterly, 34,* 397–428.
49. Müller-Putz, G. R., Riedl, R., & Wriessnegger, S. C. (2015). Electroencephalography (EEG) as a research tool in the information systems discipline: Foundations, measurement, and applications. *Communications of the Association for Information Systems, 37,* 911–948. https://doi.org/10.17705/1CAIS.03746.
50. Waldman, D. A., Stikic, M., Wang, D., Korszen, S., & Berka, C. (2015). Neuroscience and team processes. In *Organizational neuroscience* (pp. 277–294). Emerald Group Publishing Limited.
51. Volk, S., Ward, M. K., & Becker, W. J. (2015). An overview of organizational neuroscience. In *Organizational neuroscience* (pp. 17–50). Emerald Group Publishing Limited.
52. Shearer, D. A., Holmes, P., & Mellalieu, S. D. (2009). Collective efficacy in sport: The future from a social neuroscience perspective. *International Review of Sport and Exercise Psychology, 2,* 38–53. https://doi.org/10.1080/17509840802695816.
53. Mitkidis, P., McGraw, J. J., Roepstorff, A., & Wallot, S. (2015). Building trust: Heart rate synchrony and arousal during joint action increased by public goods game. *Physiology & Behavior, 149,* 101–106. https://doi.org/10.1016/j.physbeh.2015.05.033.

A Neuroimaging Study of How ICT-Enabled Interruptions Induce Mental Stress

Zhensheng Zhang and Hock-Hai Teo

Abstract In modern society, information and communication technologies (ICTs) provide individuals with social connectivity and facilitate their task execution either in daily life or work. While ICTs bring about numerous benefits, the technologies can expose individuals to frequent interruptions which disrupt thinking processes and potentially cause mental stress. Even though Information Systems research has investigated the effect of interruptions on stress, the neural mechanism underlying how ICT-enabled interruptions induce individuals' mental stress remains to be further revealed. Accordingly, this neuroimaging study aims to examine the neural activation associated with mental stress in response to ICT-enabled interruptions by means of functional magnetic resonance imaging (fMRI) and electroencephalography (EEG). Furthermore, this research distinguishes the neural activation patterns with regard to quantity and task relevancy of ICT-enabled interruptions.

Keywords Interruptions · Mental stress · ICT · fMRI · EEG · NeuroIS

1 Introduction

Information and communication technologies (ICTs) have been developed for individuals to share information, to enhance collaboration and to advance task performance, such as instant messaging and e-mail. By leveraging the ICT applications individuals tend to reach multiple interacting partners more easily [1]. Meanwhile, interruptions enabled by ICTs can cause individuals' distraction or low level of attention when engaging in tasks. For instance, approximate 4 min are necessary for workers to re-engage to their original tasks after an e-mail interruption [2]. In aver-

Z. Zhang · H.-H. Teo (✉)
National University of Singapore, Singapore, Singapore
e-mail: teohh@comp.nus.edu.sg

Z. Zhang
e-mail: zhang@u.nus.edu

Z. Zhang
Hebei University of Economics and Business, Shijiazhuang, China

F. D. Davis et al. (eds.), *Information Systems and Neuroscience*,
Lecture Notes in Information Systems and Organisation 32,
https://doi.org/10.1007/978-3-030-28144-1_4

age, workers in offices are interrupted by e-mails every 5 min [3]. Even worse, other research estimates that 40% of workers fail to reorient to original tasks [4]. Undergraduates are interrupted by e-mails, instant messages and other disruptions every 2 min when using computers [5]. In such cases, ICTs increasingly expose individuals to continuous interruptions that disrupt their cognitive processes and task procedures [6, 7]. More importantly, distraction caused by ICT-enabled interruptions during task execution may further induce individuals' mental stress.

Information Systems research has focused on stress induced by ICT-enabled interruptions by survey and experiment methods [8, 9]. Nowadays, neuroscience has promoted the research on individuals' response in the brain to interruptions including dual-task interference [10], mental load [11] and interruptibility [12, 13]. Despite the existing research evidence, it remains unclear that how individuals' mental stress in the brain is induced by ICT-enabled interruptions. To this end, we plan to employ functional magnetic resonance imaging (fMRI) and electroencephalography (EEG) to explore the neural mechanism underlying mental stress induced by ICT-enabled interruptions. In particular, the research captures the neural activation of mental stress and compares the neural activation patterns when individuals are exposed to distinct ICT-enabled interruptions concerning quantity and task relevancy.

2 Theoretical Background and Hypotheses Development

2.1 *ICT-Enabled Interruptions*

Interruption is an event which is unrelated to focal task, and it disrupts individuals' cognitive focus on the task [14]. Both individuals' internal mental processes and external environments can result in interruptions [15, 16]. In modern society, individuals face increasing external interruptions which occur in certain forms of notifications from external environments [3]. External interruptions have been studied as intrusions, discrepancies and distractions in individuals' attention [9, 17–19]. In this sense, ICTs may drive external interruptions to a higher level with additional unintended cues from various external sources. Especially, the advent of ICTs has increased the chance for individuals to be interrupted when concentrating on tasks.

In comparison to traditional interruptions, ICT-enabled interruptions have relatively less social presence which is the communicator's awareness of interacting partners [1, 20]. Correspondingly, ICTs-enabled interruptions can manifest into negative outcomes in terms of increased conflict with individuals' current tasks [21]. This research conceptualizes ICT-enabled interruptions as intrusive interruptions caused by receipt of instant messages or e-mails which distract individuals' concentration on current tasks and degrade their task performance. The ICT-enabled interruptions with limited social presence fragment individuals' cognitive processes, which leads to more severe distraction issues in terms of mental stress.

This research emphasizes quantity and task relevancy of ICT-enabled interruptions which manifest instantaneous responses in mental stress by creating ambiguity, overload or conflict. The quantity of ICT-enabled interruptions highlights the amount of interruptions occurred during task execution. Massive interruptions limit individuals' ability to maintain a continuous relationship with their tasks, which hinders priori expectations towards goals and produces feelings of stress subsequently [22–24]. The task relevancy of ICT-enabled interruptions represents the relationship between the interruptions and individuals' primary tasks. On-task ICT-enabled interruptions do not conflict with primary tasks, but support the completion of the primary task by complementing information [25, 26]. Thus, individuals' cognitive load to complete the primary tasks is alleviated [27]. On the contrary, off-task ICT-enabled interruptions conflict with primary tasks and impose higher demands on individuals' cognitive load for task execution.

2.2 Mental Stress

The transactional stress perspective suggests that mental stress is an embedded ongoing process that involves individuals' transaction with environment, judgments making and coping with issues [28]. The perspective considers frequency, severity and duration of the stressful conditions [29]. Especially, mental stress results from mismatch between individuals' abilities and the demands placed on them, or from mismatch between individuals' values and insufficient supplies to meet their needs [30–32]. This research highlights mental stress as overall transactional process which is in response to the amount and type of stressors [23, 28]. In particular, stress occurs when individuals perceive negative consequences from receiving interruptions [9]. To be specific, ICT-enabled interruptions can be recognized as the objective stressors caused by ICTs. Accordingly, this research justifies the relationships between the ICT-enabled interruptions and mental stress.

Neuroimaging studies have investigated the neural activities in response to mental stress. Prior fMRI studies identify the specific brain regions underlying stressful situations [33–37]. Particularly, negative affect like mental stress typically evokes neural activation in hippocampus, amygdala and right prefrontal cortex in humans [38–40]. With regard to the prior neuroscience studies, this research attempts to explain how ICT-enabled interruptions influence the brain regions involved in mental stress. Due to more attention distracted and more cognitive load required during task execution, ICT-enabled interruptions induce higher mental stress than without the interruptions. Accordingly, the neural activation in the relevant brain regions is higher with presence of ICT-enabled interruptions than without the interruptions. Meanwhile, EEG studies show alpha-band power suppression and theta-band power enhancement under the conditions of mental stress [41, 42]. The right hemispheric activation in the frontopolar region is shown to be associated with mental stress-related emotions in anxious subjects [43]. In terms of EEG alpha wave and theta wave, we suggest that alpha-band power suppresses and theta-band power enhances in response to ICT-enabled interruptions. The hypotheses are as followings:

H1: Hippocampus, amygdala and right prefrontal cortex are more activated in response to ICT-enabled interruptions than without the interruptions.

H2: Alpha-band power is lower in response to ICT-enabled interruptions than without the interruptions, while theta-band power is higher in response to ICT-enabled interruptions than without the interruptions.

To be more specific, this research distinguishes neural activities of mental stress induced by ICT-enabled interruptions with different quantity and task relevancy. When the quantity of ICT-enabled interruptions during task execution increases individuals distract more attention from ongoing tasks, which obstructs their goals achievement and induces more feelings of mental stress [22–24]. As far as task relevancy is concerned, on-task ICT-enabled interruptions facilitate individuals' task achievement by complementing information [25, 26]. In this case, individuals are less stressed when they work through less cognitive load. In comparison, off-task ICT-enabled interruptions distract individuals more from their tasks and impose higher demands on cognitive load for task execution, which results in higher mental stress. Taken together with prior neuroimaging research findings, we propose the following hypotheses:

H3: Quantity of ICT-enabled interruptions is positively associated with neural activation in hippocampus, amygdala and right prefrontal cortex.

H4: Quantity of ICT-enabled interruptions is negatively associated with alpha-band power and positively associated with theta-band power.

H5: Hippocampus, amygdala and right prefrontal cortex are less activated in response to on-task ICT-enabled interruptions than in response to off-task ICT-enabled interruptions.

H6: Alpha-band power is higher in response to on-task ICT-enabled interruptions than in response to off-task ICT-enabled interruptions, while theta-band power is lower in response to on-task ICT-enabled interruptions than in response to off-task ICT-enabled interruptions.

3 Research Methodology

3.1 Experimental Design

The main experiment is composed of fMRI session and EEG session. Each session consists of 4 task conditions: quantity of ICT-enabled interruptions (high vs. low) × task relevancy of ICT-enabled interruptions (on-task vs. off-task). The task for subjects is manipulated to read and comprehend an English text about 5000 words displayed on computer screen, and their understanding will be tested by 5 questions after scanning. Under the task conditions of ICT-enabled interruptions, the text-reading process is interrupted by messaging conversation dialogs which randomly appear on the screen (see Fig. 1). Each messaging conversation dialog displays for 10 s on the screen. Especially, quantity of ICT-enabled interruptions is represented by the number of messaging conversation dialogs displayed on the screen, and task relevancy of ICT-enabled interruptions is represented by whether the contents of messaging conversation dialogs facilitate subjects' text understanding or not. A pilot study finalizes the number and content of messaging conversation dialogs applied in the main experiment. In comparison, the baseline condition is designed as stimuli-free condition under which subjects carry out tasks without messaging conversation dialogs shown as ICT-enabled interruptions. Besides, psychometric data is collected as supplementary evidence to complement neuroimaging data. Following each task subjects complete a self-report questionnaire with 7-point Likert-type scale regarding the level of perceived stress during the task (1 = Least Stressful; 7 = Most Stressful).

Within-subject experiment is conducted with blocked trials design. The stimuli are presented sequentially within each condition, and the conditions are alternated as distinct blocks. 18 subjects will be recruited to ensure 80% statistical power of analysis with a threshold of $p < 0.05$ at voxel level [44]. The subjects will be healthy right-handed volunteers (9 males, 9 females) aged between 20 and 30 from National University of Singapore in return for SGD 30 compensation per person. The exclusion criteria include mental illness or psychological symptoms. The subjects will be pre-screened for fMRI and EEG safety. In the experiment to capture on-going neural activities in response to ICT-enabled interruptions, the subjects will go through all

Fig. 1 Manipulation of ICT-enabled interruptions

the task conditions and baseline conditions during fMRI session and EEG session. The protocol will be designed and performed in accordance with the guidelines for ethical human research as set forth by National University of Singapore.

3.2 Data Acquisition and Analysis

In the fMRI session, 3-Tesla blood oxygen level dependent (BOLD)-sensitive functional images will be acquired by GE Discovery MR750 3.0T scanner with Single-shot 2D EPI (GRE-EPI) functional imaging sequence. In the EEG session, EEG signals will be recorded by 64-channel BioSemi ActiveTwo Base system at a sampling frequency of 256 Hz. 14 passive electrodes will be used for collecting EEG signals from scalp at Fp1, Fp2, F3, F4, F7, F8, Fz, C3, C4, Cz, T3, T4, O1 and O2 positions based on the International 10–20 System, and FCz and AFz will be used as reference and ground electrodes respectively.

The scanned fMRI brain images will be pre-processed by using Matlab2010 and SPM8 toolbox. Then, whole brain analysis and regions of interest (ROI) analysis will be conducted. The pre-processed brain images will be analyzed by general linear model (GLM). In terms of EEG data analysis, Matlab2010 and EEGLAB toolbox will be used to process electrophysiological brain signals including artifacts removal, feature extraction and classification. In particular, signal power in alpha band (8–13 Hz) and theta band (4–8 Hz) for the channels will be analyzed to recognize EEG-based mental stress induced by ICT-enabled interruptions.

As mentioned in Sect. 3.1, the psychometric data about subjects' perceived stress during each task is measured by means of a self-report questionnaire with 7-point Likert-type scale. Thus, the neuroimaging data is complemented by psychometric data as supplementary evidence. On this basis, the research checks the consistency between psychometric evidence and neuroimaging results.

4 Discussion and Contribution

Traditionally, Information Systems research focuses on perceived stress to investigate the mechanism underlying mental stress. However, mental stress itself is an on-going process, such that the retrospective perception of stress can not represent mental stress per se. That is to say, the measurement of perceived stress by means of traditional methods cannot equivalently reflect mental stress occurring in real time. Instead, this research utilizes neuroscience techniques like fMRI and EEG to capture subjects' real-time neural activities evoked by stimuli, which puts forward a paradigm for measuring on-going mental stress in response to external interruptions like ICT-enabled interruptions.

This research investigates individuals' neural response related to mental stress during task execution in the presence of ICT-enabled interruptions. The results of this research will enrich literature on Human-Computer Interaction as well as NeuroIS. Moreover, this research examines the effect of different types of ICT-enabled interruptions on individuals' mental stress. The expected research findings will benefit both individuals and organizations to advance the understanding of ICTs adoption in either daily life or work. Ultimately, this research will provide guidance to alleviate individuals' metal stress during task execution with adaptive ICTs.

Acknowledgements This research is supported by Singapore Ministry of Education Academic Research Fund Tier 1, R-253-000-120-112.

References

1. Sproull, L., & Kiesler, S. (1986). Reducing social context cues: Electronic mail in organizational communication. *Management Science, 32*(11), 1492–1512.
2. Kessler, M. (2007). Fridays go from casual to e-mail free. *USA Today*.
3. Jackson, T., Dawson, R., & Wilson, D. (2001). The cost of email interruption. *Journal of Systems and Information Technology, 5*(1), 81–92.
4. Thompson, C. (2005). Meet the life hackers. *The New York Times*.
5. Benbunan-fich, R., & Truman, G. E. (2009). Multitasking with laptops during meetings. *Communications of the ACM, 52*(2), 139–141.
6. Carr, N. (2010). *The shallows: What the internet is doing to our brains*. New York: W. W. Norton & Company.
7. Vonnegut, B., & Youtube, O. (2011). Growing up digital, wired for distraction. *The New York Times*, 1–8.
8. Ragu-Nathan, T. S., Tarafdar, M., Ragu-Nathan, B. S., & Tu, Q. (2008). The consequences of technostress for end users in organizations: Conceptual development and empirical validation. *Information Systems Research, 19*(4), 417–433.
9. Galluch, P. S., Grover, V., & Thatcher, J. B. (2015). Interrupting the workplace: Examining stressors in an information technology context. *Journal of the Association for Information Systems, 16*(1), 1–47.
10. Jenkins, J. L., Anderson, B. B., Vance, A., Kirwan, C. B., & Eargle, D. (2016). More harm than good? How messages that interrupt can make us vulnerable. *Information Systems Research, 27*(4), 880–896.
11. Chen, D., & Vertegaal, R. (2004). Using mental load for managing interruptions in physiologically attentive user interfaces. In *CHI'04 Extended Abstracts on Human Factors in Computing Systems* (pp. 1513–1516). ACM.
12. Züger, M., & Fritz, T. (2015). Interruptibility of software developers and its prediction using psycho-physiological sensors. In *Proceedings of the 33rd Annual ACM Conference on Human Factors in Computing Systems* (pp. 2981–2990). ACM.
13. Chen, D., Hart, J., & Vertegaal, R. (2007). Towards a physiological model of user interruptability. In *IFIP Conference on Human-Computer Interaction* (pp. 439–451). Berlin, Heidelberg: Springer.
14. Coraggio, L. (1990). Deleterious effects of intermittent interruptions on the task performance of knowledge workers: A laboratory investigation. *Reproduction*.
15. Fisher, C. D. (1998). Effects of external and internal interruptions on boredom at work: Two studies. *Journal of Organizational Behavior, 19*(5), 503–522.
16. Wang, X., Ye, S., & Teo, H. H. (2014). Effects of interruptions on creative thinking. In *Proceedings of ICIS 2014* (pp. 1–10).

17. Tellegen, A., & Atkinson, G. (1974). Openness to absorbing and self-altering experiences ('absorption'), a trait related to hypnotic susceptibility. *Journal of Abnormal Psychology, 83*(3), 268–277.
18. Jett, Q. R., & George, J. M. (2003). Work interrupted: A closer look at the role of interruptions in organization life. *Academy of Management Review, 28*(3), 494–507.
19. Okhuysen, G. A. (2001). Structuring change: Familiarity and formal interventions in problem-solving groups. *Academy of Management Journal, 44*(4), 794–808.
20. Gefen, D., & Straub, D. W. (2004). Consumer trust in B2C e-Commerce and the importance of social presence: Experiments in e-Products and e-Services. *Omega, 32*(6), 407–424.
21. Chun, M. M. (2000). Contextual cueing of visual attention. *Trends in Cognitive Sciences, 4*(5), 170–178.
22. Carver, C. S., & Scheier, M. F. (1990). Origins and functions of positive and negative affect: A control-process view. *Psychological Review, 97*(1), 19–35.
23. Mullarkey, S., Jackson, P. R., Wall, T. D., Wilson, J. R., & Grey, S. (1997). The impact of technology characteristics and job control on worker mental health. *Journal of Organizational Behavior, 18,* 471–489.
24. Rohleder, N., Wolf, J. M., Maldonado, E. F., & Kirschbaum, C. (2006). The psychosocial stress-induced increase in salivary alpha-amylase is independent of saliva flow rate. *Psychophysiology, 43*(6), 645–652.
25. Beehr, T. A., Jex, S. M., Stacy, B. A., & Murray, M. A. (2000). Work stressors and coworker support as predictors of individual strain and job performance. *Journal of Organizational Behavior, 21*(4), 391–405.
26. Fenlason, K. J., & Beehr, T. A. (1994). Social support and occupational stress: Effects of talking to others. *Journal of Organizational Behavior, 15*(2), 157–175.
27. Meyer, D. E., & Kieras, D. E. (1997). A computational theory of executive cognitive processes and multiple-task performance: Part 1. Basic mechanisms. *Psychological review, 104*(1), 3–65.
28. Aziz, M. (2003). Organizational stress: A review and critique of theory, research, and applications (Book). *Vikalpa: The Journal for Decision Makers, 28*(1), 141.
29. Smith, T. W. (2006). Personality as risk and resilience in physical health. *Current Directions in Psychological Science, 15*(5), 227–231.
30. Ayyagari, R., Grover, V., & Purvis, R. (2011). Technostress: Technological antecedents and implications. *MIS Quarterly, 35*(4), 831.
31. Cooper, C. L. (1998). *Job stress*. Oxford: Oxford University Press.
32. Edwards, J. R. (1996). An examination of competing versions of the person-environment fit approach to stress. *Academy of Management Journal, 39*(2), 292–339.
33. Dedovic, K., Renwick, R., Mahani, N. K., Engert, V., Lupien, S. J., & Pruessner, J. C. (2005). The Montreal imaging stress task: Using functional imaging to investigate the effects of perceiving and processing psychosocial stress in the human brain. *Journal of Psychiatry and Neuroscience, 30*(5), 319–325.
34. Kern, S., Oakes, T. R., Stone, C. K., McAuliff, E. M., Kirschbaum, C., & Davidson, R. J. (2008). Glucose metabolic changes in the prefrontal cortex are associated with HPA axis response to a psychosocial stressor. *Psychoneuroendocrinology, 33,* 517–529.
35. Pruessner, J. C. (2004). Dopamine release in response to a psychological stress in humans and its relationship to early life maternal care: A positron emission tomography study using [11C] raclopride. *Journal of Neuroscience, 24*(11), 2825–2831.
36. Pruessner, J. C., Dedovic, K., Khalili-Mahani, N., Engert, V., Pruessner, M., Buss, C., et al. (2008). Deactivation of the limbic system during acute psychosocial stress: Evidence from positron emission tomography and functional magnetic resonance imaging studies. *Biological Psychiatry, 63*(2), 234–240.
37. Wang, J., Korczykowski, M., Rao, H., Fan, Y., Pluta, J., Gur, R. C., et al. (2007). Gender difference in neural response to psychological stress. *Social Cognitive and Affective Neuroscience,* 2(3), 227–239.
38. Davidson, R. J., Jackson, D. C., & Kalin, N. H. (2000). Emotion, plasticity, context and regulation: Perspectives from affective neuroscience. *Psychological Bulletin, 126*(6), 890–906.

39. Dedovic, K., Duchesne, A., Andrews, J., Engert, V., & Pruessner, J. C. (2009). The brain and the stress axis: The neural correlates of cortisol regulation in response to stress. *NeuroImage, 47*(3), 864–871 (Elsevier Inc.).
40. Seo, S.-H., & Lee, J.-T. (2010). *Stress and EEG*. INTECH Open Access Publisher.
41. Ryu, K., & Myung, R. (2005). Evaluation of mental workload with a combined measure based on physiological indices during a dual task of tracking and mental arithmetic. *International Journal of Industrial Ergonomics, 35,* 991–1009.
42. Harmony, T., Fernández, T., Silva, J., Bernal, J., Díaz-Comas, L., Reyes, A., et al. (1996). EEG delta activity: An indicator of attention to internal processing during performance of mental tasks. *International Journal of Psychophysiology, 24,* 161–171.
43. Papousek, I., & Schulter, G. (2002). Covariations of EEG asymmetries and emotional states indicate that activity at frontopolar locations is particularly affected by state factors. *Psychophysiology, 39*(3), 350–360.
44. Desmond, J. E., & Glover, G. H. (2002). Estimating sample size in functional MRI (fMRI) neuroimaging studies: Statistical power analyses. *Journal of Neuroscience Methods, 118*(2), 115–128.

User Performance in the Face of IT Interruptions: The Role of Executive Functions

Seyedmohammadmahdi Mirhoseini, Khaled Hassanein, Milena Head and Scott Watter

Abstract Information systems (IS) research has studied the consequences of IT interruption on user performance. However, our knowledge thus far of the cognitive mechanisms involved in processing different interruption types is limited. In response to this research gap, the present research-in-progress paper proposes that IT intrusions (unnecessary interruptions) and IT interventions (relevant interruptions) impose different types of load on users' cognitive resources. The study employs a self-regulation framework and borrows from the literature on executive functions (EFs), which are a set of general-purpose cognitive processes that control thought and actions. The moderating role of individuals' differences in terms of three EF capabilities as well as the effect of EF loads on task performance are hypothesized. A three-factor (Interruption Frequency × Interruption Type × Executive Function Capability) mixed-design experiment using electroencephalography is proposed to test the generated hypotheses.

Keywords IT interruptions · Executive functions · Self-regulation · User performance · Electroencephalography

---

S. Mirhoseini (✉) · K. Hassanein · M. Head
DeGroote School of Business, McMaster University, Hamilton, Canada
e-mail: mirhos1@mcmaster.ca

K. Hassanein
e-mail: hassank@mcmaster.ca

M. Head
e-mail: headm@mcmaster.ca

S. Watter
Department of Psychology, Neuroscience & Behaviour, McMaster University, Hamilton, Canada
e-mail: Watter@mcmaster.ca

F. D. Davis et al. (eds.), *Information Systems and Neuroscience*,
Lecture Notes in Information Systems and Organisation 32,
https://doi.org/10.1007/978-3-030-28144-1_5

1 Introduction

While information technology enables users to be online and accessible in both the work environment and daily life, it can negatively affect their performance [1]. An online quiz with 6000 respondents showed that 71% of users are frequently interrupted [2] and that various IT devices push four to six notifications every working hour [3]. While there are potential advantages to IT interruptions (e.g., providing timely information [4]), negative consequences of such interruptions include stress [5, 6] and reduced productivity [7].

Information Systems research has mostly studied how the characteristics of interruptions affect performance [8]; however, the cognitive mechanisms by which IT interruptions are processed are not yet explained. IS research has examined how interruption characteristics such as task complexity [9], congruence [10] and interrupting task modality [11] influence task performance [12], job performance [10] behavior [13], and affect [14]. A number of studies have used cognitive load construct as a mediator between interruption characteristics and performance, but the underlying cognitive mechanisms remain unclear [10]. The available published literature on interruptions informs us how performance is affected by interruptions; however, it is not yet clear which brain functions are involved in handling interruptions and how these functions affect behavior and performance.

Understanding how our brain works in response to IT interruptions is crucial because it helps researchers predict users' performance and behavior based on their cognitive capabilities. Executive functions (EFs)—a set of general purpose control mechanisms the brain uses to coordinate thought and action [15]—play a crucial role in various stages of processing an interruption. Depending on the type of interruption, different executive functions are involved in handling the main and interrupting tasks. Moreover, research findings suggest that people differ in their EF capabilities [16]. This individual difference factor may help to explain why people respond differently to interruptions.

Employing a NeuroIS approach, the present research-in-progress paper investigates the cognitive mechanisms used by the brain to handle interruptions. More specifically, our study pursues three goals: (1) Explaining how two categories of IT interruptions (i.e., intrusions and interventions) impose different types of cognitive load (i.e., three EF loads); (2) investigating how individuals' executive function capacity (EFC) moderates this link; and (3) examining the effect of three EF loads on users' performance of primary and interrupting tasks.

2 Literature Review

Executive functions (EFs) are a set of general purpose cognitive processes that coordinate human thought and action by modulating the operation of several cognitive sub-processes [15]. They are the means by which humans self-regulate their behavior

[17]. Although there are several categorizations of EFs [18], it is widely accepted that the three main EFs are: monitoring and updating of working memory representations (*updating*), inhibition of prepotent responses (*inhibition*), and shifting of mental sets (*shifting*) [15, 17, 19]. These three EFs are basic functions set at a lower level of cognition compared to the other higher level EFs such as "planning" [15]. Executive functions play a vital role in self-regulation, which is defined as an individual's goal-directed behavior [20].

The *updating* function refers to the active monitoring, coding, and manipulating of relevant information to the task at hand [15]. This function, which is closely linked to the concept of working memory [21], is responsible for effective shielding and controlled retrieval of information [19]. The *inhibition* function deals with an individual's ability to deliberately override a dominant, automatic or prepotent action, if necessary [15]. Because active inhibition is a mechanism by which the brain conserves its resources from other habitual behavior, it presents in the form of "*do not do X & Y*" [17]. *Shifting* is the EF that handles switching between tasks, i.e., disentangling cognitive engagement from one task and switching the attentional resources of the brain to a second task [15]. In contrast to *updating* and *inhibition* EFs that are necessary for keeping the focus and attention on the current task, *shifting* indicates how flexible and efficient an individual's brain is in switching from one task or sub-goal to another [17].

Research findings suggest that people differ in their EF capabilities [16]. For instance, one user may be high in *inhibition* (i.e., overriding an automatic response), while having low *shifting* (i.e., task-switching) capability. Another user may exhibit low *inhibition* (i.e., they are easily distracted by interruptions) but have high *shifting* capacity between tasks.

IT interruptions are defined as "perceived, IT-based external events with a range of content that captures cognitive attention and breaks the continuity of an individual's primary task activities" [8]. There are several taxonomies that classify IT interruptions based on different factors such as initiating entity (system or user-generated), relevance, and structure (e.g., actionable or informational) [8]. In the present work, we adopt IT intrusion-versus-intervention classification [8], which distinguishes between relevant and irrelevant IT interruptions. Interventions are IT interruptions that users perceive as being relevant or important to their job and thus prefer to address. On the contrary, intrusions are IT interruptions that are either not relevant to the users' ongoing task or are not perceived as necessary.

Thus far, IS research on IT interruptions has mostly dealt with examining how different IT interruption characteristics (e.g., the timing or relevance of the interruption) can influence users' cognition, affect, behavior, and performance. Attentional inhibition was studied as a mechanism by which aging affects technostress [22]. Another study investigated the effect of working memory capacity on the number of errors following interruptions [23].

3 Research Model

To address the research objectives above, we propose the research model of Fig. 1.

3.1 *Effect of IT Interruptions on EF Load*

As the frequency of IT intrusions increases, users need to ignore distractions and keep their focus on the ongoing task. The *inhibition* EF is responsible for overriding automatic, dominant, or prepotent responses to such intrusions [15]. Therefore, an increase in the number of IT intrusions imposes cognitive load on *inhibition* EF. Thus:

H1: The frequency of IT intrusions is positively associated with the inhibition load.

Updating EF is responsible for manipulating, storing, and retrieving task-related information. Addressing IT interventions requires users to load interrupting task information in the memory, which imposes a load on *updating* EF. Hence:

H2: The frequency of IT interventions is positively associated with the updating load.

Attending and responding to IT interventions means that users disentangle their attention from the main task and focus on the interrupting task, which imposes a load on *shifting* EF. Therefore:

H3: The frequency of IT interventions is positively associated with the shifting load.

3.2 *Moderation Effect of EFC*

Users with high executive functioning capability (EFC) experience less EF load compared to those with low EFC because the former group has more effective EF and is able to maintain a relative balance between available and required resources

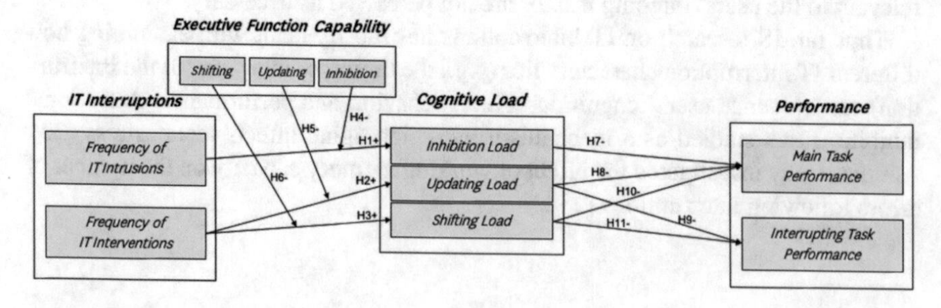

Fig. 1 Proposed research model

for performing the task [22]. Neuroimaging experiments using fMRI and EEG report low neural activation for users with high cognitive capacity [24, 25]. For instance, an EEG experiment compared the brain's alpha power between individuals with high and low cognitive capabilities while solving two well-defined and ill-defined tasks [26]. The results showed a significant difference of alpha power between the two groups, especially over the parieto-occipital areas. Since alpha power is inversely related to cognitive load [27], the results suggest that high-IQ individuals use less mental effort to perform the same task compared to low-IQ subjects. The efficiency theory justifies this phenomenon by attributing the extra cognitive load observed in individuals with low cognitive ability to the inefficient use of mental resources [28]. Therefore, we expect that individuals with higher *inhibition*, *updating*, and *shifting* EFCs experience less inhibition, updating, and shifting EF loads, respectively. Thus:

H4: Inhibition EFC negatively moderates the positive relationship between the frequency of IT intrusions and inhibition load.
H5: Updating EFC negatively moderates the positive relation between frequency of IT intrusions and updating load.
H6: Shifting EFC negatively moderates the positive relationship between frequency of IT intrusions and shifting load.

3.3 Effect of EF Load on Performance

The framework proposed by Hofmann et al. [17] links a number of self-regulatory mechanisms to each EF. We will use this framework to explain how EFs affect users' performance (accuracy and efficiency) in the face of receiving interruptions. Effective inhibition EF is associated with the successful self-regulatory mechanism of actively overriding behavioral responses to habits or stimuli that are not relevant to users' goals [17]. As inhibition EF increases, the available resources for ignoring irrelevant stimuli diminishes, resulting in less effective identification of irrelevant interruptions. Attending to IT intrusions consumes extra time and effort, which reduces users' performance on the main task. Hence:

H7: Inhibition load negatively influence performance on the main task.

Updating is linked to the active representation of goals and standards of the current task [29]; *updating* is also linked to passive inhibitory control (by protecting the goals and standards from interference [17, 30]) and to top-down direction of attentional resources [17]. Therefore, satisfactory performance on both the main and interrupting tasks depends on the availability of updating EF resources. Thus:

H8: The updating load negatively influences performance on the main task.
H9: The updating load negatively influences performance on the interrupting task.

The *shifting* EF is crucial in switching back and forth from the primary task to the interrupting task. Through the self-regulating mechanisms of (1) disentangling attention from the primary task and (2) switching attentional resources to the interrupting task, the *shifting* EF determines how effectively and efficiently one can redirect his/her attention from the main task to the interrupting task. Therefore, as the *shifting* EF increases, more time and attentional resources are needed to focus on the main and interrupting tasks. Therefore:

H10: Shifting load negatively influences performance on the main task.
H11: Shifting load negatively influences performance on the interrupting task.

4 Methodology

4.1 Experimental Design

A three-factor (Frequency of Interruption × Interruption Type × Executive Function Capability) mixed-design experiment is formulated to test the hypotheses in our research model (see Table 1). The Frequency and Type factors are within-subject and have three (Low, Medium, High) and two (Intrusions, Interventions) levels, respectively. The EFC is a between-subject factor with 8 levels (three dimensions each with low/high levels). A power analysis conducted using G*Power [31] yielded a sample size of 65 participants. The following criteria were used: eight groups, a moderate effect size (eta squared = 0.06), six number of measurements (three frequency levels and two interruption types), and a moderate correlation among repeated measures (0.25).

Table 1 Experimental factors

Factor		Type	Selection/Manipulation	Level
Frequency of interruptions		Within	Low frequency of interruptions (5)	1
			Medium frequency of interruptions (10)	2
			High frequency of interruptions (15)	3
Relevance		Within	Intrusions (participant instructed to ignore)	1
			Interventions (participant instructed to respond)	2
EFC	Inhibition	Between	Score on Antisaccade, Stop-Signal, and Stroop tasks	1, 2
	Updating		Score on Keep Track, Tone Monitoring, and Letter Memory tasks	1, 2
	Shifting		Score on Plus-Minus, Number-Letter, and Local Global tasks	1, 2

4.2 Experimental Task

The experimental task must (1) impose an extensive load on users' working memory and (2) require users to focus their attention on the screen so that interruptions distract them, thus representing a real-world user-IT interaction. We chose a variation of the information search task Memory [22, 32], since it maximizes the load on different elements of working memory and meets our criteria.

In the Memory task, subjects need to identify matching pairs of objects or symbols by flipping cards. In the more challenging version of this game, users must perform simple mathematical operations and remember the items behind each card [22] while being interrupted by secondary tasks. Instead of pairing symbols, subjects must pair a mathematical operation with its answer (e.g., 12×4 pairs with 48). This task imposes extensive load not only on the short-term memory system (memorizing numbers) [33] but on the central executive unit of the working memory (performing mathematical operations) [34]. The interruptions appear at random locations (four places) during the game and include simple multiple-choice math questions. As in the real world, some interruptions are irrelevant and must be ignored, while others need to be addressed immediately. Two tones at 1 and 5 kHz are used to distinguish between intrusions and interventions. Participants are asked to ignore interruptions with 1 kHz tone and respond to the ones with 5 kHz. The task is developed in MATLAB 2019a using the *App designer* tool. A screenshot of the task interface is presented in Fig. 2.

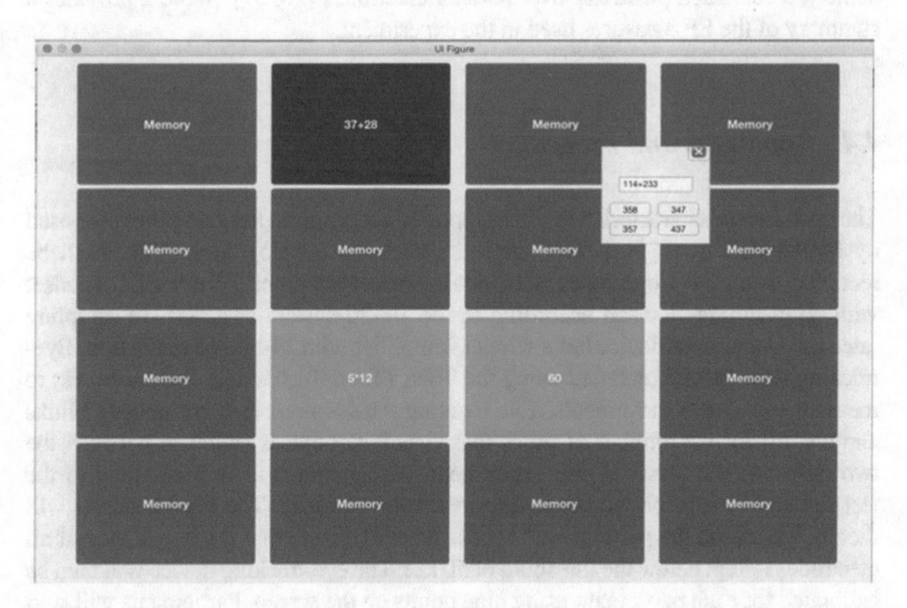

Fig. 2 Task interface

Table 2 Executive function measures

EF load	EEG measure	References
Inhibition load	Alpha event-related synchronization (alpha power) increases as individuals inhibit a cognitive or motor response	[38–40]
Updating load	Theta event-related synchronization (theta power) increases with encoding new information into working memory	[27, 41]
Shifting load	The amplitude difference between switch cues (shifting from the main task to the interrupting task) and repeat cues (staying on the main task) reflects the cognitive set-shifting process	[36]

4.3 Measurement

Executive functions capability is measured as the participants' combined score on nine computerized tests (three tests for each EFC dimension) [15]. Subjects perform all nine tasks before the main tasks. The number of clicks on the memory cards and time on the main task are used as the main task performance. The number of errors and time on multiple choice questions capture interrupting task performance. Three EEG measures are used to capture EF load. The inhibition load is measured by the alpha band (8–13 Hz) event-related-synchronization (ERS) proposed by [35]. The updating load is captured by the strength of the theta band oscillations (4–8 Hz) [35]. The shifting load is measured as the amplitude of the posterior switch positivity—namely, a sustained positivity over parietal electrodes [36, 37]. Table 2 provides a summary of the EF measures used in the experiment.

4.4 Apparatus and Procedure

The instruments' characteristics will be reported according to the guidelines proposed by Müller-Putz et al. [42]. The subjects' electroencephalography (EEG) will be recorded using the Cognionics (Cognionics Inc., CA) quick-20 dry EEG headset with 20 electrodes located according to the 10–20 system at a 500 Hz sampling rate. The Cognionics device has a wireless amplifier with 24-bit AD resolution. Eye-tracking data will be recorded using the Tobii (Tobii Technology AB) at 60 Hz to measure users' eye movements. Eye-tracking data is used to both identify blinks and control for the number of times that visual attention is switched between the two tasks. At the outset of the experiment, participants will be welcomed to the test site and their informed consent form will be obtained. The EEG headsets will then be placed on the participants' heads. We will ensure that the impedance of all electrodes is kept below the threshold of 50 kΩ. The eye-tracking device will then be calibrated for each participant using nine points on the screen. Participants will start the experiment by watching an introductory video explaining the experimental task. They will then be invited to answer a short demographic questionnaire followed by nine executive function tasks. Their score on these nine tasks will determine their EFC value on three dimensions. Finally, they will perform six memory tasks (six repeated measure conditions) in random order and their score will be presented to them afterwards.

5 Expected Contributions

This study is expected to make several contributions. First, it explains how distinct executive functions loads (i.e., *inhibition*, *updating*, and *shifting*) are imposed by IT interruptions. Although cognitive load has been used previously to study the consequences of IT interruptions [10], this is the first study, to the best of our knowledge, that attempts to open the cognitive load box and identify the cognitive mechanisms of handling interruptions. Second, the research introduces an individual difference factor (EFC) that affects users' performance and behavior in the face of IT interruptions. The literature on interruption management systems (IMSs), which are IT artifacts designed to improve users' performance by filtering irrelevant interruptions, will benefit from this research. These systems prioritize interruptions based on their characteristics to identify opportune moments (i.e., the period when users' attention can be interrupted with a minimal adverse effect on their performance) [43, 44]. Nonetheless, existing IMS systems have not yet considered cognitive differences among users. A third anticipated contribution of this study is suggesting new ways to improve users' performance in the face of IT interruptions. Specifically, current IMSs mainly help users by blocking distractions and delivering interruptions at opportune moments [45]. Such IMS features work as supporting pillars to users' inhibition executive function (i.e., overriding an automatic response) that plays the main role in blocking distractions. However, other main EFs (e.g., *updating* and *shifting*) are not supported by current IMSs.

References

1. Baethge, A., & Rigotti, T. (2013). Interruptions to workflow: Their relationship with irritation and satisfaction with performance, and the mediating roles of time pressure and mental demands. *Work & Stress, 27*(1), 43–63.
2. Murphy, M. (2016). Interruptions at work are killing your productivity.
3. Rissler, R., Nadj, M., Adam, M., & Maedche, A. (2017). Towards an integrative theoretical framework of IT-mediated interruptions.
4. Jett, Q. R., & George, J. M. (2003). Work interrupted: A closer look at the role of interruptions in organizational life. *Academy of Management Review, 28*(3), 494–507.
5. Galluch, P. S., Grover, V., & Thatcher, J. B. (2015). Interrupting the workplace: Examining stressors in an information technology context. *Journal of the Association for Information Systems, 16*(1), 1.
6. Riedl, R., Kindermann, H., Auinger, A., & Javor, A. (2012). Technostress from a neurobiological perspective. *Business & Information Systems Engineering, 4*(2), 61–69.

7. Spira, J. B., & Feintuch, J. B. (2005). The cost of not paying attention: How interruptions impact knowledge worker productivity. *Report from Basex*.
8. Addas, S., & Pinsonneault, A. (2015). The many faces of information technology interruptions: A taxonomy and preliminary investigation of their performance effects. *Information Systems Journal, 25*(3), 231–273.
9. Speier, C., Vessey, I., & Valacich, J. S. (2003). The effects of interruptions, task complexity, and information presentation on computer-supported decision-making performance. *Decision Sciences, 34*(4), 771–797.
10. Addas, S., & Pinsonneault, A. (2018). E-mail interruptions and individual performance: Is there a silver lining? *Management Information Systems Quarterly, 42*(2), 381–405.
11. Lu, S. A., Wickens, C. D., Prinet, J. C., Hutchins, S. D., Sarter, N., & Sebok, A. (2013). Supporting interruption management and multimodal interface design: Three meta-analyses of task performance as a function of interrupting task modality. *Human Factors, 55*(4), 697–724.
12. Gupta, A., Li, H., & Sharda, R. (2013). Should I send this message? Understanding the impact of interruptions, social hierarchy and perceived task complexity on user performance and perceived workload. *Decision Support Systems, 55*(1), 135–145.
13. Jenkins, J. L., Anderson, B. B., Vance, A., Kirwan, C. B., & Eargle, D. (2016). More harm than good? How messages that interrupt can make us vulnerable. *Information Systems Research, 27*(4), 880–896.
14. Barley, S. R., Meyerson, D. E., & Grodal, S. (2011). E-mail as a source and symbol of stress. *Organization Science, 22*(4), 887–906.
15. Miyake, A., Friedman, N. P., Emerson, M. J., Witzki, A. H., Howerter, A., & Wager, T. D. (2000). The unity and diversity of executive functions and their contributions to complex "frontal lobe" tasks: A latent variable analysis. *Cognitive Psychology, 41*(1), 49–100.
16. Del Missier, F., Mäntylä, T., & Bruine de Bruin, W. (2010). Executive functions in decision making: An individual differences approach. *Thinking & Reasoning, 16*(2), 69–97.
17. Hofmann, W., Schmeichel, B. J., & Baddeley, A. D. (2012). Executive functions and self-regulation. *Trends in Cognitive Sciences, 16*(3), 174–180.
18. Toplak, M. E., West, R. F., & Stanovich, K. E. (2013). Practitioner review: Do performance-based measures and ratings of executive function assess the same construct? *Journal of Child Psychology and Psychiatry, 54*(2), 131–143.
19. Miyake, A., & Friedman, N. P. (2012). The nature and organization of individual differences in executive functions: Four general conclusions. *Current Directions in Psychological Science, 21*(1), 8–14.
20. Carver, C. S., & Scheier, M. F. (2004). Self-regulation of action and affect. In *Handbook of self-regulation: Research, theory, and applications* (pp. 13–39).
21. Smith, E. E., & Jonides, J. (1997). Working memory: A view from neuroimaging. *Cognitive Psychology, 33*(1), 5–42.
22. Tams, S., Thatcher, J. B., & Grover, V. (2018). Concentration, competence, confidence, and capture: An experimental study of age, interruption-based technostress, and task performance. *Journal of the Association for Information Systems, 19,* 9.
23. Foroughi, C. K., Malihi, P., & Boehm-Davis, D. A. (2016). Working memory capacity and errors following interruptions. *Journal of Applied Research in Memory and Cognition, 5*(4), 410–414.
24. Jann, K., Koenig, T., Dierks, T., Boesch, C., & Federspiel, A. (2010). Association of individual resting state EEG alpha frequency and cerebral blood flow. *Neuroimage, 51*(1), 365–372.
25. Neuper, C., Grabner, R. H., Fink, A., & Neubauer, A. C. (2005). Long-term stability and consistency of EEG event-related (de-) synchronization across different cognitive tasks. *Clinical Neurophysiology, 116*(7), 1681–1694.
26. Jaušovec, N. (1996). Differences in EEG alpha activity related to giftedness. *Intelligence, 23*(3), 159–173.
27. Klimesch, W. (1999). EEG alpha and theta oscillations reflect cognitive and memory performance: A review and analysis. *Brain Research Reviews, 29*(2–3), 169–195.

28. Jaušovec, N., & Jaušovec, K. (2001). Differences in EEG current density related to intelligence. *Cognitive Brain Research, 12*(1), 55–60.
29. Kane, M. J., Bleckley, M. K., Conway, A. R., & Engle, R. W. (2001). A controlled-attention view of working-memory capacity. *Journal of Experimental Psychology: General, 130*(2), 169.
30. Dreisbach, G., & Haider, H. (2009). How task representations guide attention: Further evidence for the shielding function of task sets. *Journal of Experimental Psychology. Learning, Memory, and Cognition, 35*(2), 477.
31. Faul, F., Erdfelder, E., Buchner, A., & Lang, A. G. (2009). Statistical power analyses using G* Power 3.1: Tests for correlation and regression analyses. *Behavior Research Methods, 41*(4), 1149–1160.
32. Washburn, D. A., Gulledge, J. P., James, F., & Rumbaugh, D. M. (2007). A species difference in visuospatial working memory: Does language link "what" with "where"? *International Journal of Comparative Psychology, 20,* 1.
33. Zook, N. A., Davalos, D. B., DeLosh, E. L., & Davis, H. P.(2004). Working memory, inhibition, and fluid intelligence as predictors of performance on Tower of Hanoi and London tasks. *Brain and Cognition, 56*(3), 286–292.
34. Baddeley, A. (2007). *Working memory, thought, and action* (Vol. 45). Oxford: OUP Oxford.
35. Roux, F., & Uhlhaas, P. J. (2014). Working memory and neural oscillations: Alpha–gamma versus theta–gamma codes for distinct WM information? *Trends in Cognitive Sciences, 18*(1), 16–25.
36. Lange, F., Lange, C., Joop, M., Seer, C., Dengler, R., Kopp, B., & Petri, S. (2016). Neural correlates of cognitive set shifting in amyotrophic lateral sclerosis. *Clinical Neurophysiology, 127*(12), 3537–3545.
37. Elchlepp, H., Best, M., Lavric, A., & Monsell, S. (2017). Shifting attention between visual dimensions as a source of switch costs. *Psychological Science, 28*(4), 470–481.
38. Klimesch, W., Sauseng, P., & Hanslmayr, S. (2007). EEG alpha oscillations: The inhibition—timing hypothesis. *Brain Research Reviews, 53*(1), 63–88.
39. Hummel, F., Andres, F., Altenmüller, E., Dichgans, J., & Gerloff, C. (2002). Inhibitory control of acquired motor programmes in the human brain. *Brain, 125*(2), 404–420.
40. Klimesch, W., Doppelmayr, M., Schwaiger, J., Auinger, P., & Winkler, T. (1999). Paradoxical'alpha synchronization in a memory task. *Cognitive Brain Research, 7*(4), 493–501.
41. Klimesch, W., Schimke, H., & Schwaiger, J. (1994). Episodic and semantic memory: An analysis in the EEG theta and alpha band. *Electroencephalography and Clinical Neurophysiology, 91*(6), 428–441.
42. Müller-Putz, G. R., Riedl, R., & Wriessnegger, S. C. (2015). Electroencephalography (EEG) as a research tool in the information systems discipline: Foundations, measurement, and applications. *CAIS, 37*, 46.
43. Mehrotra, A., Pejovic, V., Vermeulen, J., Hendley, R., & Musolesi, M. (2016). My phone and me: Understanding people's receptivity to mobile notifications. In *Proceedings of the 2016 CHI Conference on Human Factors in Computing Systems*. ACM.
44. Fischer, J. E., Greenhalgh, C., & Benford, S. (2011). Investigating episodes of mobile phone activity as indicators of opportune moments to deliver notifications. In *Proceedings of the 13th International Conference on Human Computer Interaction with Mobile Devices and Services*. ACM.
45. Shrot, T., Rosenfeld, A., Golbeck, J., & Kraus, S. (2014). Crisp: An interruption management algorithm based on collaborative filtering. In *Proceedings of the SIGCHI Conference on Human Factors in Computing Systems*. ACM.

Investigating the Role of Mind Wandering in Computer-Supported Collaborative Work: A Proposal for an EEG Study

Michael Klesel, Frederike M. Oschinsky, Bjoern Niehaves, René Riedl and Gernot R. Müller-Putz

Abstract Mind wandering is a mental activity that allows us to easefully escape from current situations and tasks. Being the opposite of goal-directed thinking, existing research suggests that mind wandering is an important antecedent of creativity and innovation behavior. Moreover, there is initial evidence that technology characteristics may influence mind wandering. Despite a growing academic interest in mind wandering, there is only limited research that provides insights into the relationship between technology characteristics and mind wandering. We seek to address this research gap by proposing a research model that investigates whether technology supported collaborative work has an impact on the degree of mind wandering. In this research-in-progress paper, we describe the use of self-report measures and neurophysiological measures (specifically, Electroencephalography, EEG) to study mind wandering in an Information Systems research context. Ultimately, our research seeks to inform design science research in order to enhance creativity and innovation behavior.

M. Klesel (✉) · F. M. Oschinsky · B. Niehaves
University of Siegen, Siegen, Germany
e-mail: michael.klesel@uni-siegen.de

F. M. Oschinsky
e-mail: frederike.oschinsky@uni-siegen.de

B. Niehaves
e-mail: bjoern.niehaves@uni-siegen.de

M. Klesel
University of Twente, Enschede, The Netherlands

R. Riedl
University of Applied Sciences Upper Austria, Steyr, Austria
e-mail: rene.riedl@jku.at

Johannes Kepler University, Linz, Austria

G. R. Müller-Putz
Graz University of Technology, Graz, Austria
e-mail: gernot.mueller@tugraz.at

F. D. Davis et al. (eds.), *Information Systems and Neuroscience*,
Lecture Notes in Information Systems and Organisation 32,
https://doi.org/10.1007/978-3-030-28144-1_6

Keywords Mind wandering · Technology use · Distraction · Experimental research · Creativity · Collaborative systems · EEG

1 Introduction

Research has found great potential in conceptualizing and investigating the role of daydreaming [1] and mind wandering [2]. Several studies demonstrate that mind wandering is related to positive outcomes such as creativity [2–4]. At the same time, research also found negative consequences, particularly reduced performance [5–7]. A major reason for the increasing interest in mind wandering relates to the fact that it allows us to derive thoughts and significant meaning without external influence (e.g., [8]). More specifically, the mental retreat from stressful and unpleasant situations allows for future planning and sense making of past events. Hence, the relevance of mind wandering has increased in our fast-paced times, where the perceived level of stress increases, among other reasons, due to the blurring of boundaries between the private and the business domain (e.g., [9, 10]).

Since many employees rely on the intensive use of information technology (IT), mind wandering is also relevant for Information Systems (IS) research [11–13]. Studies show that mind wandering has an impact on pivotal IS constructs, such as task performance and knowledge retention [12, 13]. Furthermore, research demonstrates that the degree of mind wandering varies among the use of hedonic and utilitarian systems [11]. Despite the increasing interest in technology-related mind wandering, or technology-induced mind wandering, current literature leaves important questions unanswered. Specifically, there is only limited research that investigates how technology can be designed in order to stimulate or inhibit mind wandering (c.f. strategy 1 [14]). Consequently, design and development of neuroadaptive systems is hampered [15].

Against this background, we propose a research model whose long-term goal is to shed light on how to design technology to influence users' mind wandering states. In the present research-in-progress paper, we investigate whether IT supported collaborative work has an impact on mind wandering and consequently on creativity. In addition to self-report measures, we suggest to use EEG to measure users' mind wandering state.

The remainder is structured as follows: First, we briefly describe the extant literature on mind wandering and highlight current shortcomings. Second, we propose a research model that seeks to address the identified gap. Third, we describe our experimental procedure and conclude with final remarks on potential contributions of this research.

2 Theoretical Background

Literature from psychology shows that unconstrained mental processes are rather the norm than the exception. Specifically, evidence indicates that between one third and one half of our daily mental activity is unrelated to the external environment and off-task [16]. This finding instigated research on the exploration of the mind's capacity to wander, yielding in a new stream of research on mind wandering [2, 4, 16]. Mind wandering is commonly understood as "a shift of executive control away from a primary task to the processing of personal goals (p. 946)" [17].

Evidence from psychology and neuroscience illustrates that mind wandering occurs aimlessly during the resting state, in non-demanding circumstances and predominantly during task-free activities [18–23]. The state of decoupled attention is characterized by thinking exclusively about internal notions and feelings and by the temporal inability to process external information [24]. Due to this inability, it is often perceived as cumbersome [4, 24]. Mind wandering implies a lack of awareness and is a cause of poor performance, errors, disruption, disengagement, and carelessness [5–7]. Also, mind wandering is enhanced by stress, unhappiness and substance abuse, and consequently by states which are negatively connoted themselves [25–27]. However, research also studied mind wandering's positive outcomes. In essence, it was found that mind wandering positively affects creativity and innovative thinking [12, 28, 29], and it may lead to an increased ability to solve problems [13, 16]. In summary, evidence shows that mind wandering has negative and positive effects.

First attempts have been made to investigate the relationship between technology use and mind wandering. For that purpose, technology-related mind wandering has been defined as "task-unrelated thought which occurs spontaneously and the content is related to the aspects of computer systems" ([13], p. 4). A few IS studies provide empirical insights into mind wandering. Wati et al., who introduced the concept of mind wandering into IS research, demonstrate that user performance is influenced by an individual's focus ability and mind wandering [12]. At a later stage, the same research team studied the content of thought during mind wandering in technology-related and non-technology-related settings [13]. Their study indicates that mind wandering moderates the relationship between on-task thought and creativity; it was also found that mind wandering has a significant impact on task performance and knowledge retention. While these studies primarily focused on the moderating role of mind wandering, others provided initial evidence that depending on what kind of technology is used (e.g., hedonic or utilitarian systems) there are different levels of mind wandering [11]. Despite the valuable first research efforts in the IS discipline on mind wandering, IS research would benefit from a better understanding of *why* and *how* an IT artifact influences the degree of mind wandering, because these insights can inform design decisions of systems and applications, including the development of neuroadaptive systems.

Methodologically, previous literature proposed and used various measurement instruments to study mind wandering. Table 1 provides and overview.

Table 1 Overview measurement instruments

Type of measure		References
Self-report	Experience sampling	[30–33]
	Online	[11, 34]
Psychophysiological	Eye Tracking (ET)	[35–37]
	Skin Conductance Response (SCR)	[38]
	Electrocardiogram (ECG)	[39–41]
Brain imaging tools	Functional Magnetic Resonance Imaging (fMRI)	[31, 42]
	Electroencephalography (EEG)	[5, 41, 43–45]
	Magnetoencephalography (MEG)	[46, 47]

In summary, mind wandering is an important psychological construct with relevance in the IS domain, and this construct, according to prior studies, has neurobiological correlates, both on the autonomous nervous system level (ET, SCR, ECG) and the central nervous system level (fMRI, EEG, MEG). Yet, there is a major research gap when it comes to the relationship between technology characteristics and their impact on the degree of mind wandering as well as potential outcome variables. Against this background, we propose a research model that seeks to shed further light on these relationships.

3 Research Model

Our research model is shown in Fig. 1 and explained in the following.

We build on previous literature that repeatedly suggested that there is a strong relationship between mind wandering and creativity [2–4]. Consequently, we assume that this relationship is also given in the context of an IT supported task. Moreover, previous literature argued that un-demanding tasks and cognitive ease increase the degree of mind wandering (e.g., [29]). Stan and Christoff tellingly describe the state of

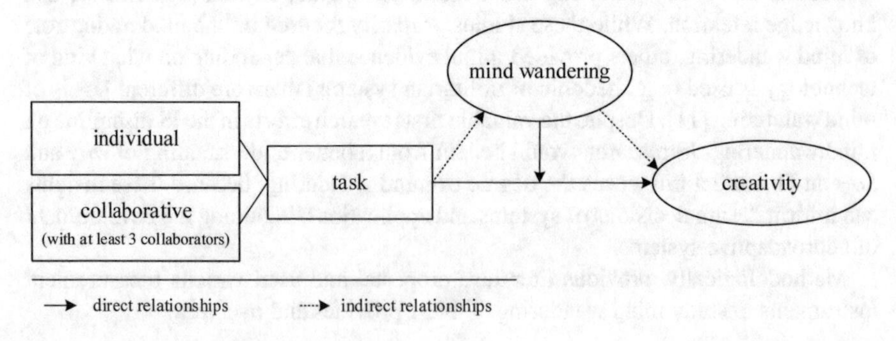

Fig. 1 Proposed research model

mind wandering, they write: “we would find ourselves transitioning gently between thoughts, rather than being pressured toward them” ([48], p. 45). The logic of our model with respect to the task is built on the fact that in a collaborative task with at least three persons, there is a higher probability to fall in a state of mind wandering if compared to a state in which two persons work jointly on a task or in which one person works alone. This reasoning is based on the fact that a collaborative task can be spread on the shoulders of a number of persons, which increases the probability for cognitive ease, an antecedent of mind wandering [48]. Note that cognitive ease is less likely in two-person interactions because the two individuals are in a direct communication process, which is not the case in groups of three or more people (because it is possible that two persons interact, while the other person(s) may drift into a mind wandering state). Consequently, we argue that collaborative tasks (except those including two persons only) result in a higher degree of mind wandering compared to individual tasks in which a person has to complete a task alone.

4 Methodology

4.1 Experimental Design

Based on our research model, we use a between-subjects design in which we manipulate task execution (either individual or group of three people). We acquire data from healthy students and employees from a middle-size university in a German-speaking country to conduct the experiments. Each participant receives financial compensation.

4.2 Measurement Instruments

Self-reported measures. As part of the experimental procedure, we use a probe-caught methodology to sample mind wandering [49, 50]. After the experimental procedure, we furthermore use established scales to measure creativity by means of the Creative Achievement Questionnaire [51] and mind wandering as suggested in previous literature [11, 12, 52].

Objective measures. Besides self-reported measures, we also use EEG to study mind wandering. While there are several instruments to collect neurophysiological data on mind wandering (c.f. Table 1), we use EEG because it has been successfully used most often in this domain [39–41]. For an introduction into EEG from an IS perspective, please see Müller-Putz et al. [53]. Time-frequency analyses, spectral features as well as more sophisticated methods like connectivity measures will be derived from the EEG. Comparison of different rounds, e.g. round 1 with round 4 and others can be used to derive evidence of neural correlates. The theta/beta ratio is used to measure mind wandering episodes [44, 45].

4.3 Experimental Task

The investigator welcomes the participants, presents them the informed consent procedure and fits them with the EEG cap and scalp electrodes. Next, s/he will ask them to accomplish a 20-min creative task. In line with previous literature, we use the title task [3, 54] to measure divergent thinking capabilities. We schedule 4 min and 1-min break per round, resulting in a total of 20 min. The test is specifically designed to foster creative ideas. It requires coming up with alternative titles for widely known books or movies. We chose two books and two movies well known to German speaking people. The participants are asked to either independently come up with as many titles as possible (the 'individual' group) or to perform the same task two further participant (the 'collaborative' group). They are assured that their answers will neither be graded nor publicized. To ensure that the collaboration does not cause confounding effects, only one subject is part of the 'collaborative' group while the other two group members are actually researchers. Importantly, in the 'collaborative' group only the EEG of the actual subject is measured. Moreover, the experiment is designed so that the two researchers start intensive interaction in the collaboration task so that the actual subject is likely to fall in a state of mind wandering. To operationalize the individual and collaborative tasks, we use Google Docs (as instance for computer-supported collaborative work) which allows both individual and synchronous work in collaboration. An example of the collaborative task with three collaborators is shown in Fig. 2.

After completing the task, the participants are asked to fill out a short questionnaire assessing their general creativity and their self-reported degree of mind wandering,

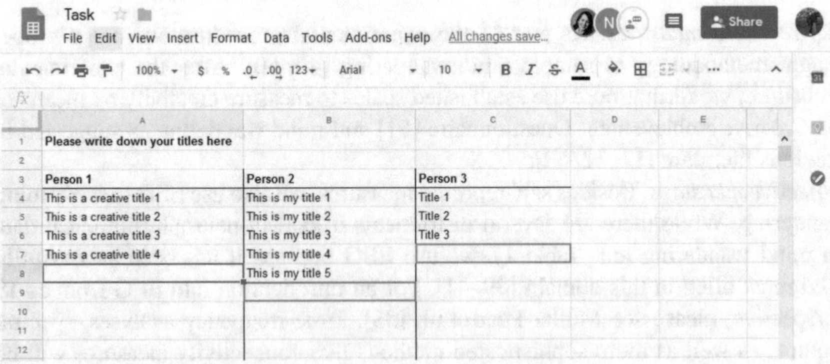

Fig. 2 Experimental task (collaboration)

along with demographic questions. Finally, they are thanked, debriefed and given the compensation for participation.

4.4 Data Analysis

Analysis of variance (ANOVA) is used to investigate the rate of mind wandering across both groups for the self-reported measures. Repeated measures ANOVA is used to investigate differences based on the thought probes. We use linear modeling with interaction terms to investigate the mediating effect of mind wandering on creativity.

5 Outlook

Based on our proposed research model, we expect important contributions for IS research and NeuroIS alike. For IS research, this is one of the first studies that investigates how a class of systems, i.e. group collaboration systems, can influence mind wandering. In addition, it highlights that computer-supported collaborative work has the potential to enhance creativity, a process which is eventually mediated by mind wandering (see Fig. 1). If follows that the insight derived by our study is predominantly relevant for jobs that require a high degree of creativity (i.e., knowledge work), our results can inform workplace design for creative jobs. With regards to NeuroIS, this study provides new insights into the neural (EEG) correlates of a cognition-related construct, namely mind wandering [15]. Ultimately, our research could have significant potential to inform future studies on EEG-based correlates in mind wandering and consequently to conduct studies that seek to develop neuroadaptive systems [14, 15, 55]. We envision a system which can, based on the users' current state, influence his or her mind wandering state in order to affect important outcome variables, particularly creativity.

References

1. Merlo, K. L., Wiegand, K. E., Shaughnessy, S. P., Kuykendall, L. E., & Weiss, H. M. (2019). A qualitative study of daydreaming episodes at work. *Journal of Business and Psychology, 10,* 389–415.
2. Fox, K. C. R., & Christoff, K. (Eds.). (2018). *The Oxford handbook of spontaneous thought. Mind-wandering, creativity, and dreaming*. New York, NY: Oxford University Press.
3. Agnoli, S., Vanucci, M., Pelagatti, C., & Corazza, G. E. (2018). Exploring the link between mind wandering, mindfulness, and creativity: A multidimensional approach. *Creativity Research Journal, 30,* 41–53.

4. Smeekens, B. A., & Kane, M. J. (2016). Working memory capacity, mind wandering, and creative cognition: An individual-differences investigation into the benefits of controlled versus spontaneous thought. *Psychology of Aesthetics, Creativity, and the Arts, 10,* 389–415.
5. Baldwin, C. L., Roberts, D. M., Barragan, D., Lee, J. D., Lerner, N., & Higgins, J. S. (2017). Detecting and quantifying mind wandering during simulated driving. *Frontiers in Human Neuroscience, 11,* 406.
6. Drescher, L. H., van den Bussche, E., & Desender, K. (2018). Absence without leave or leave without absence: Examining the interrelations among mind wandering, metacognition and cognitive control. *PLOS ONE, 13,* e0191639.
7. Zhang, Y., Kumada, T., & Xu, J. (2017). Relationship between workload and mind-wandering in simulated driving. *PLoS ONE, 12,* e0176962.
8. Morewedge, C. K., & Kupor, D. M. (2018). When the absence of reasoning breeds meaning: Metacognitive appraisals of spontaneous thought. In K. C. R. Fox & K. Christoff (Eds.), *The Oxford handbook of spontaneous thought. Mind-wandering, creativity, and dreaming* (pp. 35–46). New York, NY: Oxford University Press.
9. Addas, S., & Pinsonneault, A. (2018). E-mail interruptions and individual performance: Is there a silver lining? *MIS Quarterly, 42,* 381–405.
10. Riedl, R. (2013). On the biology of technostress. *ACM SIGMIS Database, 44,* 18–55.
11. Oschinsky, F. M., Klesel, M., Ressel, N., & Niehaves, B. (2019). Where are your thoughts? On the relationship between technology use and mind wandering. In *Proceedings of the 52nd Hawaii International Conference on System Sciences*, Honolulu, Hi, USA.
12. Wati, Y., Koh, C., & Davis, F. (2014). Can you increase your performance in a technology-driven society full of distractions? In *Proceedings of the Thirty Fifth International Conference on Information Systems*, Auckland, New Zealand.
13. Sullivan, Y., Davis, F., & Koh, C. (2015). Exploring mind wandering in a technological setting. In *Proceedings of the Thirty Sixth International Conference on Information Systems*. Fort Worth, United States of America.
14. Vom Brocke, J., Riedl, R., & Léger, P.-M. (2013). Application strategies for neuroscience in information systems design science research. *Journal of Computer Information Systems, 53,* 1–13.
15. Riedl, R., & Léger, P.-M. (2016). *Fundamentals of NeuroIS. Information systems and the brain.* Berlin, Heidelberg: Springer.
16. Smallwood, J., & Schooler, J. W. (2015). The science of mind wandering: Empirically navigating the stream of consciousness. *Annual Review of Psychology, 66,* 487–518.
17. Smallwood, J., & Schooler, J. W. (2006). The restless mind. *Psychological Bulletin, 132,* 946–958.
18. Giambra, L. M. (1995). A laboratory method for investigating influences on switching attention to task-unrelated imagery and thought. *Consciousness and Cognition, 4,* 1–21.
19. Schooler, J. W., Smallwood, J., Christoff, K., Handy, T. C., Reichle, E. D., & Sayette, M. A. (2011). Meta-awareness, perceptual decoupling and the wandering mind. *Trends in Cognitive Sciences, 15,* 319–326.
20. Posner, M. I., & Petersen, S. E. (1990). The attention system of the human brain. *Annual Review of Neuroscience, 13,* 25–42.
21. Schooler, J. W. (2002). Re-representing consciousness: Dissociations between experience and meta-consciousness. *Trends in Cognitive Sciences, 6,* 339–344.
22. Buckner, R. L., & Vincent, J. L. (2007). Unrest at rest: Default activity and spontaneous network correlations. *NeuroImage, 37,* 1091–1099.
23. Smith, S. M., Fox, P. T., Miller, K. L., Glahn, D. C., Fox, P. M., Mackay, C. E., et al. (2009). Correspondence of the brain's functional architecture during activation and rest. *Proceedings of the National Academy of Sciences of the United States of America, 106,* 13040–13045.
24. Smallwood, J., Fishman, D. J., & Schooler, J. W. (2007). Counting the cost of an absent mind: Mind wandering as an underrecognized influence on educational performance. *Psychonomic Bulletin & Review, 14*, 230–236.

25. Epel, E. S., Puterman, E., Lin, J., Blackburn, E., Lazaro, A., & Mendes, W. B. (2013). Wandering minds and aging cells. *Clinical Psychological Science, 1,* 75–83.
26. Sayette, M. A., Dimoff, J. D., Levine, J. M., Moreland, R. L., & Votruba-Drzal, E. (2012). The effects of alcohol and dosage-set on risk-seeking behavior in groups and individuals. *Psychology of Addictive Behaviors/Journal of the Society of Psychologists in Addictive Behaviors, 26,* 194–200.
27. Smallwood, J., O'Connor, R. C., Sudbery, M. V., & Obonsawin, M. (2007). Mind-wandering and dysphoria. *Cognition and Emotion, 21,* 816–842.
28. Mooneyham, B. W., & Schooler, J. W. (2013). The costs and benefits of mind-wandering: A review. *Canadian Journal of Experimental Psychology [Revue Canadienne De Psychologie Experimentale], 67,* 11–18.
29. Baird, B., Smallwood, J., Mrazek, M. D., Kam, J. W. Y., Franklin, M. S., & Schooler, J. W. (2012). Inspired by distraction: Mind wandering facilitates creative incubation. *Psychological Science, 23,* 1117–1122.
30. Smith, G. K., Mills, C., Paxton, A., & Christoff, K. (2018). Mind-wandering rates fluctuate across the day: Evidence from an experience-sampling study. *Cognitive Research: Principles and Implications, 3,* 1–20.
31. Christoff, K., Gordon, A. M., Smallwood, J., Smith, R., & Schooler, J. W. (2009). Experience sampling during fMRI reveals default network and executive system contributions to mind wandering. *Proceedings of the National Academy of Sciences of the United States of America, 106,* 8719–8724.
32. Stawarczyk, D., Majerus, S., Maj, M., van der Linden, M., & D'Argembeau, A. (2011). Mind-wandering: Phenomenology and function as assessed with a novel experience sampling method. *Acta Psychologica, 136,* 370–381.
33. Song, X., & Wang, X. (2012). Mind wandering in Chinese daily lives. An experience sampling study. *PLOS ONE, 7,* e44423.
34. Killingsworth, M. A., & Gilbert, D. T. (2010). A wandering mind is an unhappy mind. *Science, 330,* 932.
35. Franklin, M. S., Broadway, J. M., Mrazek, M. D., Smallwood, J., & Schooler, J. W. (2013). Window to the wandering mind: Pupillometry of spontaneous thought while reading. *Quarterly Journal of Experimental Psychology, 66,* 2289–2294.
36. Smilek, D., Carriere, J. S. A., & Cheyne, J. A. (2010). Out of mind, out of sight: Eye blinking as indicator and embodiment of mind wandering. *Psychological Science, 21,* 786–789.
37. Uzzaman, S., & Joordens, S. (2011). The eyes know what you are thinking: Eye movements as an objective measure of mind wandering. *Consciousness and Cognition, 20,* 1882–1886.
38. Blanchard, N., Bixler, R., Joyce, T., & D'Mello, S. (2014). Automated physiological-based detection of mind wandering during learning. In S. Trausan-Matu, K. E. Boyer, M. Crosby, & K. Panourgia (Eds.), *12th International Conference on Intelligent Tutoring Systems, ITS 2014* (8474, pp. 55–60). Cham: Springer.
39. Ottaviani, C., Shapiro, D., & Couyoumdjian, A. (2013). Flexibility as the key for somatic health: From mind wandering to perseverative cognition. *Biological Psychology, 94,* 38–43.
40. Ottaviani, C., & Couyoumdjian, A. (2013). Pros and cons of a wandering mind: A prospective study. *Frontiers in Psychology, 4,* 524.
41. Conrad, C., & Newman, A. (2019). Measuring the impact of mind wandering in real time using an auditory evoked potential. In F. D. Davis, R. Riedl, J. Vom Brocke, P.-M. Léger, & A. B. Randolph (Eds.), *Information systems and neuroscience* (pp. 37–45).
42. Mason, M. F., Norton, M. I., van Horn, J. D., Wegner, D. M., Grafton, S. T., & Macrae, C. N. (2007). Wandering minds: The default network and stimulus-independent thought. *Science, 315,* 393–395.
43. Smallwood, J., Beach, E., Schooler, J. W., & Handy, T. C. (2008). Going AWOL in the brain: Mind wandering reduces cortical analysis of external events. *Journal of Cognitive Neuroscience, 20,* 458–469.
44. Braboszcz, C., & Delorme, A. (2011). Lost in thoughts: Neural markers of low alertness during mind wandering. *NeuroImage, 54,* 3040–3047.

45. van Son, D., de Blasio, F. M., Fogarty, J. S., Angelidis, A., Barry, R. J., & Putman, P. (2019). Frontal EEG theta/beta ratio during mind wandering episodes. *Biological Psychology, 140,* 19–27.
46. Zhigalov, A., Heinilä, E., Parviainen, T., Parkkonen, L., & Hyvärinen, A. (2019). Decoding attentional states for neurofeedback: Mindfulness vs. wandering thoughts. *NeuroImage, 185,* 565–574.
47. Marzetti, L., Di Lanzo, C., Zappasodi, F., Chella, F., Raffone, A., & Pizzella, V. (2014). Magnetoencephalographic alpha band connectivity reveals differential default mode network interactions during focused attention and open monitoring meditation. *Frontiers in Human Neuroscience, 8,* 832.
48. Stand, D., & Christoff, K. (2018). The mind wanders with ease: Low motivational intensity is an essential quality of mind wandering. In K. C. R. Fox & K. Christoff (Eds.), The Oxford handbook of spontaneous thought. Mind-wandering, creativity, and dreaming (pp. 47–54). New York, NY: Oxford University Press.
49. Seli, P., Risko, E. F., Smilek, D., & Schacter, D. L. (2016). Mind-wandering with and without intention. *Trends in Cognitive Sciences, 20,* 605–617.
50. Choi, H., Geden, M., & Feng, J. (2017). More visual mind wandering occurrence during visual task performance: Modality of the concurrent task affects how the mind wanders. *PLoS ONE, 12,* e0189667.
51. Carson, S. H., Peterson, J. B., & Higgins, D. M. (2005). Reliability, validity, and factor structure of the creative achievement questionnaire. *Creativity Research Journal, 17,* 37–50.
52. Mrazek, M. D., Phillips, D. T., Franklin, M. S., Broadway, J. M., & Schooler, J. W. (2013). Young and restless: Validation of the mind-wandering questionnaire (MWQ) reveals disruptive impact of mind-wandering for youth. *Frontiers in Psychology, 4,* 1–7.
53. Müller-Putz, G. R., Riedl, R., & Wriessnegger, S. C. (2015). Electroencephalography (EEG) as a research tool in the information systems discipline: Foundations, measurement, and applications. *CAIS, 37.*
54. Guilford, J. P. (Eds.). (1968). *Intelligence, creativity and their educational implications.* Robert R. Knapp.
55. Loos, P., Riedl, R., Müller-Putz, G. R., Vom Brocke, J., Davis, F. D., Banker, R. D., & Léger, P.-M. (2010). NeuroIS: Neuroscientific approaches in the investigation and development of information systems. *Business & Information Systems Engineering, 2,* 395–401.

The Effect of Technology on Human Social Perception: A Multi-methods NeuroIS Pilot Investigation

Peter Walla and Sofija Lozovic

Abstract Effects of digital communication have been reported, but with only little physiological data backing. The purpose of this pilot study was to use a multi-methods approach to investigate in digital natives the effects of reading from a mobile device, listening to an audio recording and listening to an actual person present, who reads out loud. Self-reported pleasantness and arousal as conscious data, startle reflex modulation, skin conductance and heart rate as non-conscious data were recorded for each condition. The findings indicate that physiological arousal measures tend to match respective self-report measures both indicating higher arousal levels for social conditions. However, physiological valence measures do not match their corresponding self-report measures. Listening to an audio recording and listening to a real person reading were rated as more pleasant than reading alone. However, listening to a present person reading out loud resulted in the most negative subcortical raw affective responses in digital native's brains.

Keywords Self report versus objective data · Multi-methods · NeuroIS · Social neuroscience · Digital technology · Socialization · Digital overuse

1 Introduction

Technology has seeped deep into the core of human society and is becoming increasingly more integrated into our everyday lives. We spend a great deal of time using our devices—the recent Nielsen Audience Report [1] found that Americans spend between 4 and 13 h of each day looking at screens, with 65% of the population surpassing the 10 h mark. Over a decade later, a Pew Research Center survey found the same results [2]. Although technology can aid socialization, there is already evidence

P. Walla (✉) · S. Lozovic
CanBeLab, Department of Psychology, Webster Vienna Private University, Praterstrasse 23, 1020 Vienna, Austria
e-mail: peter.walla@webster.ac.at

P. Walla
School of Psychology, Newcastle University, Newcastle, Callaghan, Australia

F. D. Davis et al. (eds.), *Information Systems and Neuroscience*,
Lecture Notes in Information Systems and Organisation 32,
https://doi.org/10.1007/978-3-030-28144-1_7

that it has a negative impact on the quantity and quality of face-to-face communication [3]. Uhls et al. [4] found that preteens who did not engage with screens for five days better recognized nonverbal emotional cues in facial expressions shown in photos and videos, as compared to those who had access to screens for that time. As Montag and Walla [5] point out, genuine emotion recognition depends on facial and other nonverbal cues, and it is possible that not using our ability to read them could lead to us gradually losing our ability to do so, whereby we would also lose the ability to comprehend and even detect emotions, both our own and other people's.

The purpose of this experimental study was to investigate how people (with a focus on digital natives) respond to different ways of taking in information under varying levels of social interaction from reading alone (*reading* condition) over listening to a recorded voice (*audio* condition) to listening to a real person reading out loud (*social* condition). Crucially, no person was present in both the reading condition and the audio condition, while in the *social* condition a person was present both visually and acoustically by sitting face-to-face with the participant. The audio condition was included to investigate whether a person being "semi-present" makes a difference in valence and arousal responses during socialization. It was hypothesised that digital natives respond more negatively to the social condition as a result of digital overuse.

2 Materials and Methods

2.1 Participants

Seventeen participants were recruited. All were students between the ages of 20 and 30 (mean age = 22.47; SD = 2.58). This age group was chosen, as it includes the generation that Prensky [6] calls "digital natives". Only 5 of the participants who volunteered were male. All had normal or corrected to normal vision. The participants received no compensation. This study was approved by the ethics committee of Webster University in Saint Louis (Missouri).

2.2 Measures

Five different measures were taken. Self-reported measures included *valence* [on a scale from 1 to 9 (1 = negative; 9 = positive)] and *arousal* [on a scale of 1–9 (1 = low; 9 = high)]. Both of those are understood as representing responses to affective impact on conscious levels. Physiological measures included *startle reflex modulation* (see Walla and Koller [7]) sensitive to valence, *skin Conductance* (see Braithwaite et al. [8]) sensitive to arousal and h*eart rate* (sensitive to valence and arousal). Those measures are understood as representing responses to affective impact on non-conscious levels. Consequently, this multi-methods approach could be used to compare both valence and arousal aspects of affective responses to the three selected independent variables with respect to conscious and non-conscious processing.

2.3 *Stimuli*

Three counterbalanced experimental conditions were introduced in the study. Each consisted of a different story depicting made-up discoveries about the universe. The participants were initially told that the stories were legitimate news articles about real scientific discoveries. (1) *Audio Condition*: The participant listens to a pre-recorded audio track of a research team member reading a story without a research team member being present; (2) *Reading Condition*: The participant reads a written story on his/her own phone in his/her own head without a research team member being present; (3) *Social Condition*: The participant watches and listens as a research team member reads him/her a story face-to-face.

2.4 *Procedure*

2.4.1 Lab Experiment

After giving informed consent participants were seated in a comfortable chair where he/she had all sensors for collecting physiological data attached to them. Each of the conditions lasted between 7 and 10 min. During each condition, six startle probes were presented at an interval of 40–70 s to minimize habituation to the stimulus. The startle stimulus was a 50 ms burst of acoustic white noise at a sound pressure level of 105 dB delivered through headphones. During each condition, physiological data were recorded with a NeXus 10 wireless recording device (from Mind Media BV) and Bio-trace + software was used for data processing. Finally, after each condition, the participants were asked to fill out a self-report questionnaire on their perceived valence and arousal. Self-reported valence and arousal were measured using two nine-point likert scales (1 = negative valence/low arousal; 9 = positive valence/high arousal).

2.4.2 Analysis

All data was subject to analysis of variance (ANOVA) for each of the dependent variables with SPSS. Greenhouse-Geisser correction was applied to the repeated measures tests. Contrasts were calculated to look at each possible pair of condition combinations (Bonferroni-corrections were made in case of multiple comparisons). Effect sizes and observed power were also calculated for a better interpretation of statistical findings.

3 Results

3.1 Self-reported Valence

Mean self-reported valence for the *audio condition* was 7.12 (SD = 1.45). In the *reading condition* it was 6.06 (SD = 1.60) and in the *social condition* it was equal to that of the audio condition at 7.12 (SD = 1.50). ANOVA revealed a significant condition main effect on self-reported valence ($F = 5.355$, $p = 0.012$, $\eta 2 = 0.251$). Pairwise comparisons found that there was a significant difference between the *audio condition* and the *reading condition* ($p = 0.045$) and an almost significant difference between the *social condition* and the *reading condition* ($p = 0.057$). There was no significant difference between the *audio condition* and the *social condition* ($p = 1.000$).

3.2 Self-reported Arousal

Mean self-reported arousal for the *audio condition* was 5.47 (SD = 1.42), for the *reading condition* it was 4.59 (SD = 1.73) and for the *social condition* it was 5.71 (SD = 1.61). ANOVA did not find a significant condition main effect (very strong trend though) on self-reported arousal ($F = 3.004$, $p = 0.076$, $\eta 2 = 0.158$). Pairwise comparisons found that there was no significant difference between the *audio condition* and the *reading condition* ($p = 0.195$), no significant difference between the *social condition* and the *reading condition* ($p = 0.271$), and, as expected, no difference between the *audio condition* and the *social condition* ($p = 1.000$).

3.3 Startle Reflex Modulation

Mean eye blink response for the *audio condition* was 57.96 μV (SD = 31.05). For the *reading condition* it was 49.54 μV (SD = 28.34) and for the *social condition* it was highest at 63.58 μV (SD = 37.07). ANOVA found a significant main effect ($F = 4.273$, $p = 0.029$, $\eta 2 = 0.211$). Pairwise comparisons found that there was no significant difference between the *audio condition* and the *reading condition* ($p = 0.157$), a very strong trend, but not a significant difference between the *social condition* and the *reading condition* ($p = 0.077$), and no difference between the *audio condition* and the *social condition* ($p = 0.726$).

3.4 Skin Conductance

Mean skin conductance for the *audio condition* was 6.49 μS (SD = 2.87). In the *reading condition* it was 6.37 μS (SD = 3.03). The highest skin conductance was observed in the *social condition* with a mean of 7.05 μS (SD = 2.81). ANOVA found no significant main effect on skin conductance (F = 2.152, $p = 0.135$, $\eta 2 = 0.119$). Accordingly, pairwise comparisons found that there was no difference between the *audio condition* and the *reading condition* ($p = 1.000$), no significant difference between the *social condition* and the *reading condition* ($p = 0.246$), and no significant difference between the *audio condition* and the *social condition* ($p = 0.436$).

3.5 Heart Rate

Mean heart rate for the *audio condition* was 77.90 bpm (SD = 6.31), for the *reading condition* it was 81.78 bpm (SD = 9.76) and it was 82.68 bpm (SD = 10.02) for the social condition. ANOVA found a significant main effect on heart rate (F = 3.848, $p = 0.043$, $\eta 2 = 0.194$). Pairwise comparisons found that there was a statistically significant difference between the *audio condition* and the *reading condition* ($p = 0.031$), no difference between the *social condition* and the *reading condition* ($p = 1.000$), and only a strong trend, but no significant difference between the *audio condition* and the *social condition* ($p = 0.070$). See Table 1 that summarizes all numbers and Fig. 1 displaying respective bar diagrams.

Table 1 Mean values and standard deviations of self-reported valence, self-reported arousal, eye blink response, skin conductance and heart rate presented by the three conditions being tested: audio, reading and social

Measure	Audio		Reading		Social	
	Mean	SD	Mean	SD	Mean	SD
Self-reported valence	7.12	1.45	6.06	1.6	7.12	1.50
Self-reported arousal	5.47	1.45	4.59	1.73	5.71	1.61
Eye blink response	57.96	31.05	49.54	28.34	63.58	37.07
Skin conductance	6.49	2.87	6.37	3.03	7.05	2.81
Heart rate	77.90	6.31	81.78	9.76	82.68	10.02

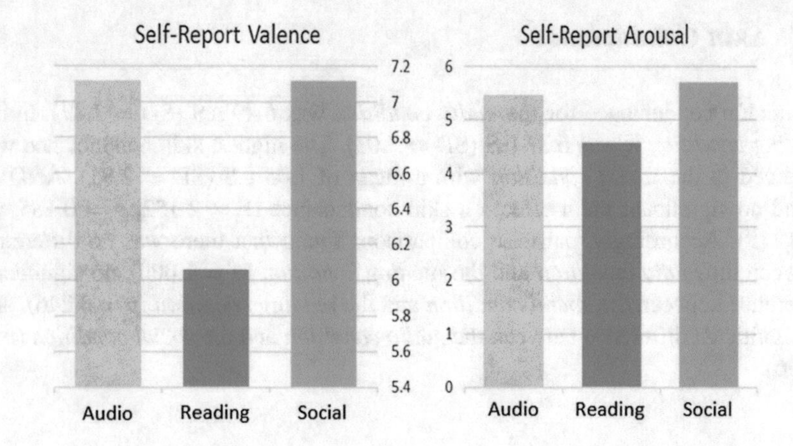

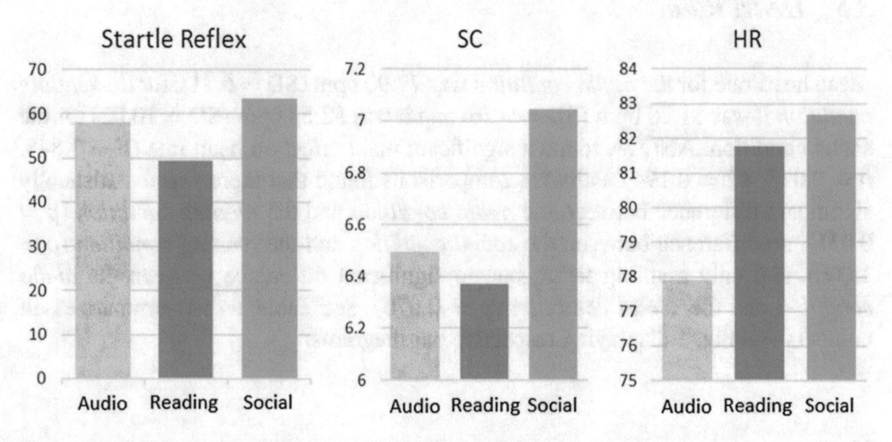

Fig. 1 Bar diagrams showing mean values of self-reported valence, self-reported arousal, eye blink response, skin conductance and heart rate presented by the three conditions being tested: audio, reading and social

4 Discussion

Technology has provided us with a whole new dimension of socialization, which allows us to communicate quickly and efficiently no matter where we, or the parties we're communicating with, are located. Paradoxically, however, the price of this new and convenient means of socialization might come at the cost of precisely our ability to socialize. Montag and Walla [5] caution that by choosing to spend our time staring at screens instead of looking into each other's eyes when we communicate, we may, as predicted by the use it or lose it principle, end up losing our complex socializing abilities because we do not utilize them.

By using various dependent variables sensitive to both conscious and non-conscious processes we aimed to examine whether people respond more negatively when taking in information via a mobile device than when taking in information face-to-face from another person. The study provides evidence indicating that self-reported valence and startle reflex modulation differ between conditions where there was a human presence (the audio and social conditions) and where there was no human presence (the reading condition). A strong trend in support of the hypothesis was detected in self-reported arousal, although statistical significance was missing. There was only a small difference between mean skin conductance values in each of the conditions. Heart rate was the only measure where a significant difference was found between the *reading* and *audio* condition with the *reading condition* surpassing that of the *audio condition*. Additionally, a strong trend was observed in this measure where the *social condition* elicited a higher level of arousal than the *audio condition*.

Most interestingly, the two measures that are sensitive to valence revealed a discrepancy between what participants say when asked versus how participants' brains responded. Startle reflex modulation results indicated that the *reading condition* was experienced as most pleasant, followed by the *audio condition*, and then by the *social condition*. These results contradicted those of the self-reported valence measure, in which the *reading condition* was rated as less pleasant than the *audio* and *social conditions*, with the latter two being rated as equally pleasant. Various previous research has demonstrated similar discrepancies between self-reported pleasantness and valence measured via startle reflex modulation (see Refs. [9–24]). The present study could be viewed as supporting the idea that young people are intimidated by in-person communication and are shifting more and more towards communicating via devices. This, however, can in this study only be seen in the non-conscious data set that is reflective of deep inside affective processing. Conscious self-report in this study on the other hand reflects most positive experience during the actual person present. One could argue that the conscious mind is more influenced by cognitive knowledge about the benefits of face-to-face contact, but that the non-conscious mind tells the truth (only speculation). It is, however, also possible that the participants found being face-to-face with a stranger, where they could more easily feel self-conscious, to be more unpleasant than being on their own, focusing on the task of reading. Further research that controls for such factors is necessary to determine whether taking in information via a mobile device is, as this study suggests, experienced as more pleasant than taking in information face-to-face.

References

1. The Nielsen Company. (2017). *The Nielsen total audience report: Q1 2017* (The Nielsen total audience series). US. Retrieved from http://3xyemy1let2g2jeg0pea76v1.wpengine.netdna-cdn.com/wp-content/uploads/NLSN.pdf.
2. Anderson, M. (2015). *6 facts about Americans and their smartphones*. Retrieved May 8, 2018, from http://www.pewresearch.org/fact-tank/2015/04/01/6-facts-about-americans-and-their-smartphones/.

3. Drago, E. (2015). The effect of technology on face-to-face communication. *Elon Journal of Undergraduate Research in Communications*, *6*(1).
4. Uhls, Y. T., Michikyan, M., Morris, J., Garcia, D., Small, G. W., Zgourou, E., et al. (2014). Five days at outdoor education camp without screens improves preteen skills with nonverbal emotion cues. *Computers in Human Behavior, 39*, 387–392. https://doi.org/10.1016/j.chb.2014.05.036.
5. Montag, C., & Walla, P. (2016). Carpe diem instead of losing your social mind: Beyond digital addiction and why we all suffer from digital overuse. *Cogent Psychology, 3*(1), 1157281. https://doi.org/10.1080/23311908.2016.1157281.
6. Prensky, M. (2001). *Digital natives, digital immigrants.*
7. Walla, P., & Koller, M. (2015). Emotion is not what you think it is: Startle reflex modulation (SRM) as a measure of affective processing in NeuroIs. In *NeuroIs conference proceedings*, Springer. http://dx.doi.org/10.1007/978-3-319-18702-0_24.
8. Braithwaite, J. J., Watson, D. G., Jones, R., & Rowe, M. (2013). A guide for analysing electrodermal activity (EDA) and skin conductance responses (SCRs) for psychological experiments. *Psychophysiology, 49*(1), 1017–1034.
9. Bosshard, S., Bourke, J. D., Kunaharan, S., Koller, M., & Walla, P. (2016). Established liked versus disliked brands: Brain activity, implicit associations and explicit responses. *Cogent Psychology, 3,* 1. https://doi.org/10.1080/23311908.2016.1176691.
10. Geiser, M., & Walla, P. (2011). Objective measures of emotion during virtual walks through urban neighbour-hoods. *Applied Sciences, 1*(1): 1–11. http://dx.doi.org/10.3390/app1010001.
11. Grahl A., Greiner, U., & Walla, P. (2012). Bottle shape elicits gender-specific emotion: A startle reflex modulation study. Psychology, *7*: 548–554. http://dx.doi.org/10.4236/psych.2012.37081.
12. Koller, M., & Walla, P. (2015). Towards alternative ways to measure attitudes related to consumption: Introducing startle reflex modulation. *Journal of Agricultural and Food Industrial Organization, 13*(1), 83–88.
13. Koller, M., & Walla, P. (2012). Measuring affective information processing in information systems and consumer research—Introducing startle reflex modulation. In *ICIS Proceedings, Breakthrough ideas, full paper in conference proceedings*, Orlando 2012. http://aisel.aisnet.org/icis2012/proceedings/BreakthroughIdeas/1/.
14. Kunaharan, S., Halpin, S., Sitharthan, T., Bosshard, S., & Walla, P. (2017). Conscious and non-conscious measures of emotion: Do they vary with frequency of pornography use? *Applied Sciences, 7*, 493, 17 pp.
15. Lyons, G. S., Walla, P., & Arthur-Kelly, M. (2013). Toward improved ways of knowing children with profound multiple disabilities (PMD): Introducing startle reflex modulation. *Developmental Neurorehabilitation, 16*(5), 340–344. https://doi.org/10.3109/17518423.2012.737039.
16. Mavratzakis, A., Herbert, C., & Walla, P. (2016). Emotional facial expressions evoke faster orienting responses, but weaker emotional responses at neural and behavioural levels compared to scenes: A simultaneous EEG and facial EMG study. *Neuroimage, 124,* 931–946.
17. Mavratzakis, A., Molloy, E., & Walla, P. (2013). Modulation of the startle reflex during brief and sustained exposure to emotional pictures. *Psychology, 4,* 389–395. https://doi.org/10.4236/psych.2013.44056.
18. Walla, P. (2018). Affective processing guides behavior and emotions communicate feelings: Towards a guideline for the NeuroIS community. In F. Davis, R. Riedl, J. vom Brocke, P. M. Léger, & A. Randolph (Eds.), *Information systems and neuroscience* (Vol. 25). Cham: Springer.
19. Walla, P., Koller, M., Brenner, G., & Bosshard, S. (2017). Evaluative conditioning of established brands: Implicit measures reveal other effects than explicit measures. *Journal of Neuroscience, Psychology and Economics, 10*(1), 24–41.
20. Walla, P., & Schweiger, M. (2017). Samsung versus apple: Smartphones and their conscious and non-conscious affective impact. In Full Conference Paper in Information Systems and Neuroscience. Volume 16 of the series Lecture Notes in Information Systems and Organisation (pp 73–82).

21. Walla, P., Koller, M., Brenner, G.., & Bosshard, S. (2016). Evaluative conditioning of brand attitude—Comparing explicit and implicit measures. In Conference paper accepted for the 2016 European Marketing Academy conference in Oslo.
22. Walla, P., Koller, M., & Meier, J. (2014). Consumer neuroscience to inform consumers—Physiological methods to identify attitude formation related to over-consumption and environmental damage. Frontiers in Human Neuroscience, 20 May 2014. http://dx.doi.org/10.3389/fnhum.2014.00304.
23. Walla, P., Rosser, L., Scharfenberger, J., Duregger, C., & Bosshard, S. (2013). Emotion ownership: different effects on explicit ratings and implicit responses. *Psychology, 3,* 213–216. https://doi.org/10.4236/psych.2013.43A032.
24. Walla, P., Brenner, G., & Koller, M. (2011). Objective measures of emotion related to brand attitude: A new way to quantify emotion-related aspects relevant to marketing. *PloS ONE, 6*(11), e26782. https://doi.org/10.1371/journal.pone.0026782.

Intelligent Invocation: Towards Designing Context-Aware User Assistance Systems Based on Real-Time Eye Tracking Data Analysis

Christian Peukert, Jessica Lechner, Jella Pfeiffer and Christof Weinhardt

Abstract Recently introduced virtual and augmented reality devices such as the HTC Vive Pro Eye or Microsoft's HoloLens 2 come with integrated eye tracking technology. Eye tracking technology is thus closer to the consumer market than ever before. Should these systems make the leap into the end consumer market, possibilities also evolve to use the data to feed intelligent user assistance systems in real time. One application could be to detect phases in consumers' decision making processes based on eye tracking data, which, in turn, can be used to offer context-aware assistance to consumers. By analyzing eye tracking data from an experiment in a virtual reality shopping environment, we test existing approaches to detect decision phases and evaluate their applicability for an intelligent invocation of real-time user assistance. Furthermore, we propose a new approach, called on-the-fly-detection, since we conclude that existing approaches are not suitable for real-time phase detection.

Keywords User assistance systems · Decision phases · NeuroIS · Context-aware systems · Eye tracking · Virtual reality

C. Peukert (✉) · C. Weinhardt
Karlsruhe Institute of Technology (KIT), Institute of Information Systems and Marketing, Karlsruhe, Germany
e-mail: christian.peukert@kit.edu

C. Weinhardt
e-mail: christof.weinhardt@kit.edu

J. Lechner
Karlsruhe Institute of Technology (KIT), Karlsruhe, Germany
e-mail: jessica.lechner@student.kit.edu

J. Pfeiffer
Department of Economics and Business Studies, Justus Liebig University Giessen, Giessen, Germany
e-mail: jella.pfeiffer@wirtschaft.uni-giessen.de

F. D. Davis et al. (eds.), *Information Systems and Neuroscience*,
Lecture Notes in Information Systems and Organisation 32,
https://doi.org/10.1007/978-3-030-28144-1_8

1 Introduction

Eye tracking (ET) technology has recently enjoyed an upswing in attention. Not only the acquisitions of technology giants in the field of ET technology (e.g., Apple bought SMI or Oculus acquired The Eye Tribe) have contributed to this, but also the introduction of new products that are already equipped with ET technology (HTC Vive Pro Eye, Microsoft's HoloLens 2). The symbiosis of ET and virtual reality (VR) technology is obvious: Real-time gaze detection allows to only render the currently looked at regions in highest quality (foveated rendering) [1], which in turn saves computational resources. Furthermore, advanced interactions can be offered by system designers (gaze selection), and avatars can be modeled more naturally by transferring the real eye movements to the virtual representation. We thus more and more experience a shift from ET, that solely serves as a post hoc diagnosis tool, to real-time ET data processing and usage due to the possibility to unobtrusively observe gaze behavior in VR [2].

In literature more than 20 years ago, researchers have tried to detect different phases in customers' decision making based on ET data [3] measured in real-life contexts and in recent years this idea was further pursued [4–6]. The knowledge in which decision phase a user is currently in could be of great interest for system designers who could use this information for the intelligent invocation of assistance systems [7]. In particular, a high potential is stated to so-called advanced user assistance systems (UAS) which promise to be context-aware and adaptive [8]. Depending on the decision phase (context-awareness), an assistance system might proactively offer suitable decision aids to the user (adaptive), e.g., a comparison matrix as soon as someone starts comparing products with each other (evaluation phase), or a filter function, that helps to eliminate products, which reduces the overall complexity right from the start (orientation phase). Such a system would, therefore, take over the selection of the aid, i.e., *how*, and the time of use, i.e., *when* they shall be supported, adaptively for a user.

Within this article, we, therefore, aim to detect phases in consumers' decision-making, which can potentially be used by advanced UAS to support consumers during their shopping in regular stores. In this context, we assume that image processing and object recognition can be performed in real time for practical applications. This article is closely related to the field of NeuroIS, since the information systems (IS) research problem, i.e., the identification of the right moment in time to assist consumers in their decision-making process (intelligent invocation), is solved using the neurophysiological tool of ET [9].

2 Context-Aware Assistance Systems Meet Eye Tracking Phase Detection

In order to build a UAS with intelligent invocation based on ET, it must first be ensured that a phase detection can be performed in real time. Therefore, we start to examine existing approaches for their suitability for real-time phase detection.

Today, UAS are almost ubiquitous in e-commerce settings trying to increase a user's task performance by addressing the known problems of an enormous product range and information overload [10, 11]. A UAS can have many facets: e.g., recommendation agents [10], interactive decision aids [12], or chatbots. According to Mädche et al. [8], UAS can be further specified along the dimensions *intelligence* and *interactivity*, defining a system that at least employs features of one dimension as an advanced UAS rather than a basic one. Advanced UAS primarily differentiate themselves by the fact that they are context-aware and adaptive. Especially context-awareness can be related to the intelligent time of automatic invocation. Whereas in the area of notifications much research has already been done focusing on the best time of invocation (e.g., Bailey and Konstan [13]), empirical studies on intelligent invocation in the context of UAS are relatively scarce [7]. However, user assistance is by no means limited to an e-commerce context and assistance systems for regular stores are also in the focus of studies (e.g., mobile auto-ID technology [14]).

Most commonly, studies that support a phase theory build upon at least two distinct phases, namely an *orientation* and an *evaluation* phase [5]. The orientation phase is thereby characterized by acquiring an initial overview of available products, whereas the evaluation phase primarily consists of paired comparisons between products. Besides, several studies argue that a subsequent *verification* phase can take place, in which the product choice is verified again [3, 4]. In related literature, different approaches are proposed on how a phase detection based on ET data might be operationalized [3, 4, 15]. Within this work-in-progress article, we will examine the approaches by Russo and Leclerc [3] (R&L) and Gidlöf et al. [4] (G) more closely, who both build upon three phases (orientation, evaluation, verification).

According to Russo and Leclerc [3], an initial orientation phase consists of consecutive dwells on products up to the first re-dwell (the latter already counts to the second phase). As an indicator for the beginning of the third phase, a verbal announcement made by the participant is used from which the last re-dwell (going backward in time) is determined, which, in turn, defines the end of the second phase [3]. In contrast, Gidlöf et al. [4] solely rely on *dwells on the chosen product* to determine the different phases: The orientation phase includes all dwells up to the *first* dwell on the chosen product, which, however, already counts to the second phase [4]. The second phase, in turn, ends with the *last* dwell on the selected product (counting to the second phase), meaning that the last phase only consists of dwells not directed to the chosen product.

Even though the strict phase distinction is partly criticized, we consider it a good starting point to investigate points in time for an intelligent invocation. However, since both prior described approaches either require knowledge about the outcome of the decision process (product choice) or can only be determined through post hoc analysis (last re-dwell before an announcement), we conclude that they are not applicable for real-time phase detection. We, therefore, propose an approach to detect the previously mentioned phases in real time and in the following reference to it as *On-the-fly-detection* (OFD). Similar to the approach by Russo and Leclerc [3], we argue that a re-dwell can be used as an indicator for the start of the evaluation phase since the evaluation phase is characterized by paired product comparisons between alternatives [3, 16]. However, to address the criticism expressed by Gidlöf et al. [4] that a re-dwell on any product can also occur by chance, we define the delineation rule more strictly and demand an X-Y-X sequence of transitions, i.e., a direct pair-wise comparison, as begin of the evaluation phase (re-dwell belongs to the second phase). For the distinction between the evaluation and verification phase, we preliminarily suggest using the time at which any product is first placed in the shopping cart. This point in time is meaningful because then a person has made an initial decision (evaluation phase is certainly finished) but still has the chance (if they want) to verify their choice. Of course, it may be possible that verification starts slightly in advance, but we have also decided to use this event because of the possible real-time detection (assuming that image processing and object recognition can be performed in real time in the future).

In order to test the three approaches with our existing data set, we needed to adapt the approaches slightly. In our experiment, participants did not announce their choice verbally, which is why the (R&L) approach cannot be tested one-to-one. Following Gidlöf et al. [4]—who encountered the same problem within their study—we will also use the last dwell on the chosen product as a decision rule for the (R&L) approach. Unlike Gidlöf et al. [4], we consider the last dwell on the chosen product as part of the verification phase, as a result of the experimental setup, because when placing a product in the shopping cart, a glance at the product happens nearly automatically. Furthermore, a dwell including the way to and the process of putting the product into the shopping cart is conceptually no longer an evaluation. Within the new OFD approach, a special case can occur that the event *product enters the shopping cart* takes place precisely during a dwell. In order to do justice to this we distinguish in OFD(A), the dwell will be counted to phase two or OFD(B), the dwell belongs to phase three. This subdivision is only essential for the following analysis, but not for an intelligent invocation, as the invocation would exclusively depend on the event. Figure 1 illustrates the operationalization of the approaches.

3 Experimental Design and Procedure

The dataset (n = 129) originates from an experiment in which students performed several consecutive choice tasks in front of a virtual supermarket shelf [17]. The experimental design followed a choice-based conjoint analysis (CBC) [18], and the specific task was to always pick the product (muesli package) out of a set of 24 products which they would most likely buy in reality. The interaction possibilities

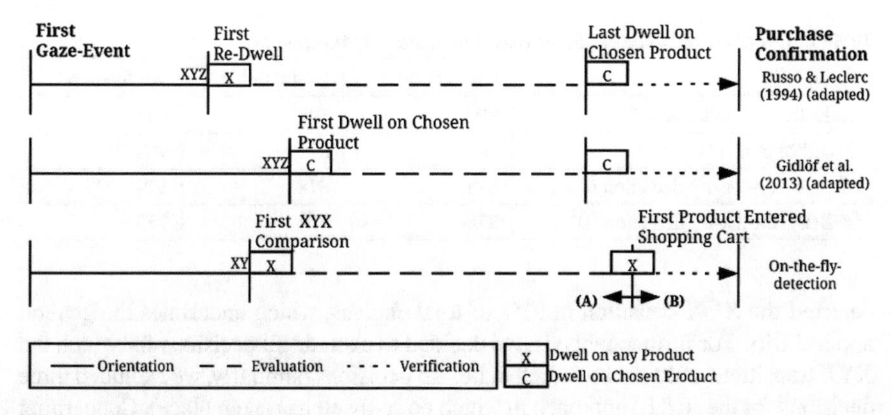

Fig. 1 Comparison of phase definitions

were close to reality: participants could take products from the shelf, view them from all sides and in the end they had to put the chosen product in a virtual shopping cart. Similar to many e-commerce websites, a screen then appeared on which they had to confirm the purchase finally. In each task, the displayed products, as well as the prices, changed according to the CBC design (in total we had six price levels and 40 different products). For a description in more detail, we refer to Peukert et al. [17].

To record participants' eye movements, we used an HTC Vive head-mounted display with an integrated SMI eye tracker (250 Hz). We applied a velocity-based algorithm [consecutive eye movements with a velocity below (above) 50°/s were determined to be a fixation (saccade)] for fixation determination. We only considered data of the participant's dominant eye and only included fixations with a minimum fixation duration of 100 ms [19]. Fixations were automatically annotated to predefined areas of interest (AOIs). For this article, the product package as a whole, as well as the related price tag, were regarded as one interrelated AOI. As a next step, we aggregated consecutive fixations on the same AOI to dwells.

4 Preliminary Results

For the preliminary data analysis, we only consider the average values across all tasks irrespective of the order and only refer to the most interesting results. In order to be able to evaluate the applicability of the approaches for real-time detection of the decision phases, it is initially important to determine whether it is generally possible to reliably detect the different phases based on the approaches. Table 1 provides an overview of the presence of the phases for the four tested approaches.

Overall, the table shows that in almost all cases the approaches detect an orientation and evaluation phase. In the approach by Gidlöf et al. [4] [(G) approach] it could have happened that by chance the first dwell was already directed to the chosen product (1/24 chance) which means that the first phase is skipped. The OFD approach

Table 1 Overview presence of phases (based on a total of 903 choices)

	Orientation	Evaluation	Verification
R&L: Russo and Leclerc [3]	900	898	899
G: Gidlöf et al. [4]	869	890	902
OFD(A): on-the-fly-detection (A)	878	878	149
OFD(B): on-the-fly-detection (B)	878	874	753

detected the XYX transition in 97% of total choices, which underlines the general applicability. For further analysis, we decided to exclude all decisions for which the XYX transition could not be detected, i.e., 25 decisions (similarly, we excluded three decisions for the (R&L) approach in which no re-dwell has taken place). Concerning the detection of the verification phase, the result of OFD(A) is especially noticeable, since only in 16.5% of the cases a verification phase is detected. Accordingly, following the same argumentation as for shifting the last dwell to the verification phase in the (R&L) and (G) approach, OFD(B) shall better be applied in the future.

Further, we were interested in key indicators for the phases (Table 2) to investigate to what extent they differ between the phases as well as between the approaches. With regard to phase durations, in all approaches, the evaluation stage dominates the decision-making process (similar to Russo and Leclerc [3]). The orientation phase of (R&L) is the shortest followed by OFD and (G), whereby OFD's mean orientation phase duration (7.5 s) could be sufficiently long for designing a meaningful intelligent invocation at the begin of the evaluation phase. Consistent with the literature [3, 4, 20], we find shorter dwells within the orientation phase compared to the evaluation phase, which supports the theory of an initial scanning of products. For the verification phase, we find extremely long dwell durations, which can be most probably traced back to the experimental design (participants had to grab the product, turn away from the shelf and finally place it in the cart). Regarding the number of different fixated products per phase, it can be seen that the number is the highest in the evaluation phase, followed by the orientation phase, whereby the proportion of different products fixated in relation to the phase duration is much higher during orientation. Many participants only fixated one product (the chosen product) during the verification phase (note that the OFD(A) value is only based on 149 decisions, in which primarily participants are included who have fixated several other products).

5 Limitations and Future Research

Within this work-in-progress paper, we preliminarily analyzed data retrieved from a preference measurement study in order to evaluate whether phases in decision making can be detected in VR shopping environments in real time. Since the primary objective of the underlying study was different, the experimental setting was not precisely tailored to the purpose of detecting phases. As mentioned in the prior

Table 2 Measures differentiated by phase; values represent the mean (standard deviation)

	Orientation phase			Evaluation phase				Verification phase			
	R&L	G	OFD(A)/(B)	R&L	G	OFD(A)	OFD(B)	R&L	G	OFD(A)	OFD(B)
Avg. phase duration [s]	4.11 (2.83)	10.00 (13.61)	7.50 (7.16)	46.82 (34.27)	40.91 (31.53)	46.79 (33.82)		5.66 (4.03)	5.68 (4.02)	3.35 (4.33)	
Avg. dwell duration [s]	0.42 (0.13)	0.51 (0.16)	0.47 (0.13)	0.69 (0.20)	0.70 (0.20)	0.75 (0.20)	0.71 (0.20)	2.10 (1.33)	2.11 (1.33)	0.57 (0.42)	2.14 (1.44)
Number of different products fixated	6.44 (3.34)	9.48 (6.15)	8.77 (5.40)	19.71 (4.73)	17.74 (5.91)	18.76 (5.56)	18.82 (5.47)	1.36 (1.62)	1.36 (1.62)	2.87 (4.26)	1.41 (2.10)

section, the fact that consumers had to turn away from the shelf to confirm the purchase, could have influenced consumer behavior and thus the results. In future research, hence, a study shall be designed that focuses exclusively on phase detection. In addition to observing gaze behavior throughout a decision task, a further post hoc analysis shall be conducted to test if the proposed phases are a reliable proxy for the consumers' actual executed decision process. Therefore, for instance, participants could retrospectively state in which decision phase they were currently in (e.g., retrospective think aloud based on a video recording [21]), or the decision making process could be coded by independent judges with respect to the decision phases (e.g., thematic coding).

Further, the analysis reported within this paper is preliminary and additional analyses need to be conducted to fully understand the phenomenon. This includes, for example, that at both—task—and respondent-level—it needs to be examined whether patterns can be identified, e.g., whether it is possible to determine clusters of participants that are similar to each other. Besides, so far, we have not assigned any relevance to the task sequence and have simply averaged across all tasks. Especially related to the verification phase it can be interesting for future research to consider people who have fixated more than one product during this phase, which might indicate that they did verify their selection to a greater extent. Furthermore, it should not be ignored that the standard deviation of most of the measures is quite high (e.g., the standard deviation of the average phase duration of the OFD approach is 7.16 s) and therefore the meaningfulness of these indicators for the real-time phase detection needs to be investigated. Although the large variance between the time of invocation would underline the adaptivity of the UAS, it remains to be tested whether the point in time is considered useful by all users.

Given the assumption that such context-aware UAS are available in the near future, the key question then arises as to whether consumers accept these systems. From a technological perspective, an implementation in a VR shopping environment is already possible today, but users nonetheless have to be willing to share their gaze and interaction data to feed the UAS. Therefore, questions regarding privacy concerns need to be addressed in future research. In the same vein, it could be interesting to investigate to what extent explanations influence the acceptance of context-aware UAS [22].

Finally, within this article, we followed the idea to use decision phases [3, 4] to trigger an intelligent invocation. However, further approaches shall be pursued in the future. For instance, other ET measures such as the saccade length or fixation duration could serve as additional indicators.

6 Conclusion

Within this article, we introduced the idea to design context-aware UAS based on the real-time analysis of ET data especially focusing on intelligent invocation. Based on the theory of decision phases in consumer decision-making, we propose an approach for on-the-fly-detection of decision phases which can potentially be used for intelligent invocations. In the future, such advanced UAS could help to reduce consumers' experienced cognitive load when doing their shopping by ensuring that the UAS steps in at the right time and also provides adequate support for the respective context. Such context-aware UAS could represent an entirely new shopping experience and thus be of great interest to practitioners. It remains to be seen whether such systems will gain acceptance in the future; nevertheless, it is important to start doing research in this area at an early stage.

References

1. Patney, A., Salvi, M., Kim, J., Kaplanyan, A., Wyman, C., Benty, N., et al. (2016). Towards foveated rendering for gaze-tracked virtual reality. *ACM Transactions on Graphics, 35,* 1–12.
2. Meißner, M., Pfeiffer, J., Pfeiffer, T., & Oppewal, H. (2019). Combining virtual reality and mobile eye tracking to provide a naturalistic experimental environment for shopper research. *Journal of Business Research, 100,* 445–458.
3. Russo, J. E., & Leclerc, F. (1994). An eye-fixation analysis of choice processes for consumer nondurables. *Journal of Consumer Research, 21,* 274–290.
4. Gidlöf, K., Wallin, A., Dewhurst, R., & Holmqvist, K. (2013). Using eye tracking to trace a cognitive process: Gaze behaviour during decision making in a natural environment. *Journal of Eye Movement Research, 6,* 1–14.
5. Glaholt, M. G., & Reingold, E. M. (2011). Eye movement monitoring as a process tracing methodology in decision making research. *Journal of Neuroscience, Psychology, 4,* 125–146.
6. Pfeiffer, J., Meißner, M., Brandstätter, E., Riedl, R., Decker, R., & Rothlauf, F. (2014). On the influence of context-based complexity on information search patterns: An individual perspective. *Journal of Neuroscience, Psychology, and Economics, 7,* 103–124.
7. Friemel, C., Morana, S., Pfeiffer, J., & Maedche, A. (2017). On the role of users' cognitive-affective states for user assistance invocation. *Lecture Notes in Information Systems and Organisation, 25,* 37–46.
8. Maedche, A., Morana, S., Schacht, S., Werth, D., & Krumeich, J. (2016). Advanced user assistance systems. *Business and Information Systems Engineering, 58,* 367–370.
9. vom Brocke, J., & Liang, T. (2014). Guidelines for neuroscience studies in information systems research. *Journal of Management Information Systems, 30,* 211–233.
10. Xiao, B., & Benbasat, I. (2007). E-Commerce product recommendation agents: Use, characteristics, and impact. *MIS Quarterly, 31,* 137–209.
11. Wang, W., & Benbasat, I. (2005). Trust in and adoption of online recommendation agents. *Journal of the Association for Information Systems, 6,* 72–101.
12. Groissberger, T., & Riedl, R. (2017). Do online shops support customers' decision strategies by interactive information management tools? Results of an empirical analysis. *Electronic Commerce Research and Applications, 26,* 131–151.
13. Bailey, B. P., & Konstan, J. A. (2006). On the need for attention-aware systems: Measuring effects of interruption on task performance, error rate, and affective state. *Computers in Human Behavior, 22,* 685–708.
14. Venkatesh, V., Aloysius, J. A., Hoehle, H., Burton, S., & Walton, S. M. (2017). Design and evaluation of auto-ID enabled shopping assistance artifacts in customers' mobile phones: Two retail store laboratory experiments. *MIS Quarterly, 41,* 83–113.
15. Reutskaja, E., Nagel, R., Camerer, C. F., & Rangel, A. (2011). Dynamics in consumer choice under time pressure: An eye-tracking study. *American Economic Review, 101,* 900–926.
16. Orquin, J. L., & Mueller Loose, S. (2013). Attention and choice: A review on eye movements in decision making. *Acta Psychologica, 144,* 190–206.

17. Peukert, C., Pfeiffer, J., Meißner, M., Pfeiffer, T., & Weinhardt, C. (2019). Shopping in virtual reality stores: The influence of immersion on system adoption. *Journal of Management Information Systems, 36,* 755–788.
18. The CBC system for choice-based conjoint analysis. (2013). Sawtooth Software Inc. *Sawtooth Software Technology Paper Series, 8,* 1–27.
19. Holmqvist, K., Nyström, M., Andersson, R., Dewhurst, R., Jarodzka, H., & Van de Weijer, J. (2011). *Eye tracking: A comprehensive guide to methods and measures*. Oxford: Oxford University Press.
20. Gloeckner, A., & Herbold, A.-K. (2011). An eye-tracking study on information processing in risky decisions: Evidence for compensatory strategies based on automatic processes. *Journal of Behavioral Decision Making, 24,* 71–98.
21. Guan, Z., Lee, S., Cuddihy, E., Ramey, J. (2006). The validity of the stimulated retrospective think-aloud method as measured by eye tracking. In *Proceedings of the SIGCHI Conference on Human Factors in Computing Systems*, *CHI 2006* (pp. 1253–1262). ACM: New York.
22. Gregor, S., & Benbasat, I. (1999). Explanations from intelligent systems: Theoretical foundations and implications for practice. *MIS Quarterly, 23,* 497–530.

Designing Self-presence in Immersive Virtual Reality to Improve Cognitive Performance—A Research Proposal

Katharina Jahn, Bastian Kordyaka, Caroline Ressing, Kristina Roeding and Bjoern Niehaves

Abstract With the increasing availability of immersive virtual reality (IVR) technologies, new opportunities to change individuals' behavior become possible. Notably, recent research showed that by creating a full-body ownership illusion of a virtual avatar looking similar to Einstein, users' cognitive performance can be enhanced. However, although research is quite consistent in reporting that visuomotor synchrony in IVR achieved with body tracking suffices to elicit body ownership illusions that change behavior, it is still unclear whether strengthening these visuomotor illusions with additional technological design elements, such as visuotactile feedback, can contribute to increase desired outcomes even more. In this research in progress paper, we aim to conduct a 2 (physical feedback: low vs. high) × 2 (avatar design: normal vs. high intelligence) between-subjects experiment in IVR to test this assumption. In addition to subjective measures, we use heart rate and electrodermal activity to assess the strength of self-presence induced through the illusions.

Keywords Body ownership illusions · Heart rate · Electrodermal activity · Cognitive performance · Physical feedback

K. Jahn (✉) · B. Kordyaka · C. Ressing · K. Roeding · B. Niehaves
University of Siegen, Chair of Information Systems, Siegen, Germany
e-mail: katharina.jahn@uni-siegen.de

B. Kordyaka
e-mail: bastian.kordyaka@uni-siegen.de

C. Ressing
e-mail: caroline.ressing@uni-siegen.de

K. Roeding
e-mail: kristina.roeding@uni-siegen.de

B. Niehaves
e-mail: bjoern.niehaves@uni-siegen.de

F. D. Davis et al. (eds.), *Information Systems and Neuroscience*,
Lecture Notes in Information Systems and Organisation 32,
https://doi.org/10.1007/978-3-030-28144-1_9

1 Introduction

With the ability to present user's visual, auditory, and tactile senses with completely virtual content, Immersive Virtual Reality (IVR) provides new opportunities to represent the bodily self of users. IVR describes a set of technologies that, by enclosing the user with head-mounted displays (HMD) or cage systems heightens sensory immersion. Sensory immersion is a characteristic of the technology, which is high when users are separated into a technology from the real world and their real movements are matched to the virtual environment [1]. In contrast to this technological viewpoint, the sense of telepresence describes the psychological perception of the "illusion of being in a distant place" or "being there" of the individual, [2, p. 438], which should arise in individuals when technology provides a high degree of sensory immersion.

In IVR, full-body ownership illusions can be created by combining HMDs with full-body tracking, creating a high degree of self-presence [2]. Self-presence relates to the "Illusion [of] inhabiting the virtual body" [2, p. 438], when interacting with a virtual body in an environment. Self-presence elicited through body ownership illusions arises when the users' real movements are tracked in real-time and then transferred to a virtual body in the IVR. As a result, the movements of the users' virtual body are displayed in synchrony to the users' real body movements (visuomotor synchrony). This synchrony is sufficient for individuals to experience self-presence [3]. However, when design elements such as visuotactile or visuomotor synchrony are disrupted, self-presence can be diminished [4].

Research already showed, that self-presence created by full-body ownership illusions offer many opportunities to enhance desired behavioral and cognitive outcomes when working alone or interacting with other people. As an example, individuals embodied in a virtual body with dark skin drum differently [5] and show decreased racial bias and prejudice [6, 7] compared to individuals in a virtual body with white skin. Additionally, individuals embodied in the body of Sigmund Freud show different cognitive processing of problems [8]. Furthermore, full-body ownership illusions can even change male users' cognitive performance if they are embodied in an avatar that is associated with high intelligence [9].

Whereas a main factor to elicit full-body ownership illusions with sufficient strength seems to be first person perspective, the strength of body ownership illusions is dependent upon multiple factors. Research has indicated that the strength of body ownership illusions is related to questionnaire items, but can also be measured by biophysiological variables, for example through skin conductance response or heart rate in reaction to a threat [4, 10, 11]. However, whether increasing the effectivity of the body ownership illusions through specific design elements to enhance the cognitive or behavioral outcomes induced through a specific avatar design, is still unclear. Therefore, we want to investigate the following research question to contribute to close this research gap:

RQ: How can the interaction between users and virtual avatars be designed to increase users' self-presence and cognitive performance in immersive virtual realities?

To answer our research question, we plan to conduct a 2 (physical feedback: low vs. high) × 2 (avatar design: normal vs. high intelligence) between-subjects experiment.

2 Background and Research Model

In this section, we develop our hypotheses based on literature on the antecedents and outcomes of self-presence through full-body ownership illusions. Our research model is displayed in Fig. 1, which we explain in the following paragraphs.

2.1 Full-Body Ownership Illusions and Effects on the Self

Rooted in the classical rubber hand illusion experiment [12], in which a rubber hand is touched in synchrony with the individuals' real hand, subsequently arising a sense of ownership over the rubber hand, full-body ownership illusions elicit a sense of ownership over a complete body [3, 13]. When IVR is used with body tracking, these illusions can create a quite realistic experience of having another body.

From a theoretical point of view, self-presence initiated through body ownership illusions constitutes a passive form of perspective taking [14, 15], in which, rather than imagining to be in the shoes of another person, users can directly experience owning another body [9, 16]. As a consequence, if full-body ownership illusions arise for avatars with specific design elements (e.g. skin color or similarity to a person with competencies in a specific area), individuals cognitive processing and behavior can be influenced [6, 17]. It is assumed that this process occurs by activating existing resources of the individual previously not accessible through this form of perspective-

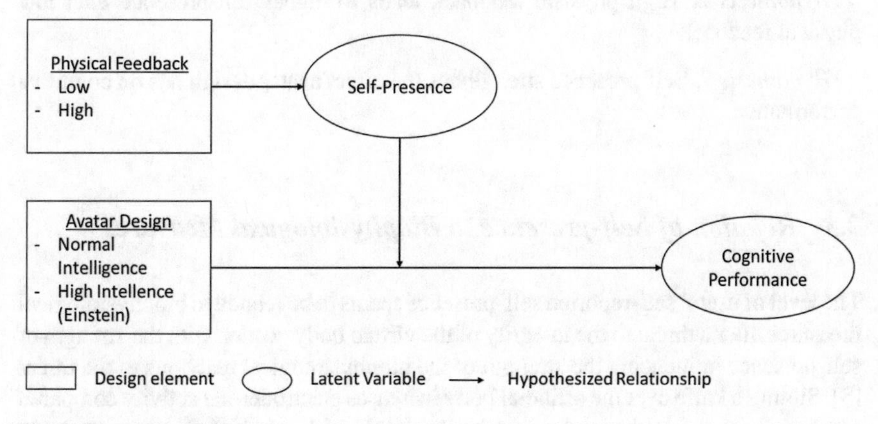

Fig. 1 Research model

taking [8, 9]. For example, when individuals were embodied in a virtual avatar of Sigmund Freud when they counselled themselves, they showed more positive mood changes than when they were embodied in a body-scanned version of themselves [8]. Additionally, individuals who are embodied in an avatar of Einstein show higher performance outcomes in a cognitive task than when they are embodied in a regular unknown body with which they most likely connect lower intelligence levels [9]. We therefore hypothesize that:

Hypothesis 1. Being embodied in a virtual body that is associated with high intelligence leads to higher cognitive performance than being embodied in a virtual body that is associated with normal intelligence.

2.2 Strength of Self-presence

Previous research on virtual arms has indicated that self-presence can be induced by synchronous visuomotor stimulation, even when tactile stimulation is absent [10]. Comparing the effects of visuomotor and visuotactile interaction has shown that the disruption of visuotactile synchrony leads to a lower body ownership illusion [4]. Thus, we suspect that sustaining congruence for visual stimuli coming in contact to the body and touch that is subsequently felt is highly important for keeping the level of self-presence high. This should be especially important in situations in which users have to interact with their hand's multiple times in fine granularity, as this is the case with many virtual reality applications. However, when users' bodies are fully tracked, including their fingers, physical feedback can be incomplete after interaction with virtual objects if no feedback mechanism is implemented in addition to the tracking device. Therefore, we assume that self-presence is higher when physical feedback is presented, and, that, this strengthened self-presence leads to an increased effect of avatar design on cognitive performance.

Hypothesis 2. High physical feedback leads to higher self-presence than low physical feedback.

Hypothesis 3. Self-presence strengthens the effect avatar design has on cognitive performance.

2.3 Relation of Self-presence to Biophysiological Measures

The level of users' self-reported self-presence seems to be related to biophysiological measures after a threat to the integrity of the virtual body occurs, with the strength of self-presence influencing the strength of the biophysiological reactions to the threat [3]. Sliding a knife over the artificial body increases electrodermal activity compared to a spoon or asynchronous physical feedback [18] and a knife sliding over the body in a condition of first person perspective with synchronous physical feedback results in higher electrodermal activity than a third person perspective or asynchronous physical feedback [19]. In both related studies, these differences were also reflected by the questionnaire items for self-presence. However, other research indicated that synchronous and asynchronous physical feedback is not necessarily reflected by a change in skin conductance response [4]. To gain more insights into these effects, we hypothesize:

Hypothesis 4. Higher levels of self-presence are reflected by an increase in electrodermal activity after the presentation of a threat to the virtual body.

Another biophysiological measure that has been shown to be related to self-presence is heart rate deceleration. After seeing a woman slapping the face of a virtual body from a first person perspective, heart rate deceleration increased compared to a third person perspective, which was also related to the questionnaire items for self-presence [20]. Additionally, heart rate deceleration is positively related to self-reported self-presence in a questionnaire after the legs of the virtual body were visually separated [3]. Thus, we hypothesize:

Hypothesis 5. Higher levels of self-presence are reflected by an increase in heart rate deceleration after the presentation of a threat to the virtual body.

3 Method

3.1 Participants and Design

We will recruit at least 128 male participants to take part in our experiment and use a 2 (physical feedback: low vs. high) × 2 (avatar design: normal vs. high intelligence) between-subjects design to test our hypotheses.

3.2 Materials and Measures

IVR: A HTC Vive HMD will be used to display the virtual environment, which will be designed with Unity 3D. Full-body tracking will be implemented with five HTC Vive trackers (2 for hands, 2 for feet, 1 for hip) and hand-tracking will be implemented by using Hi5 VR Gloves. Avatars are created using Adobe Fuse.

Electrodermal Activity. We will use electrodermal activity (EDA) as a biophysiological measure for self-presence. In line with previous research in the area of body ownership illusions, EDA will be measured in the 6 s baseline period and in 2–8 s period after the threat [4]. The latency window during which a response will be assumed to be elicited by the stimulus will be based on frequency distributions of response latencies to simple stimuli (1–4 s) [21].

Heart Rate. We will use the Polar H7 belt to measure participants' heart rate deceleration. In line with previous research, we will measure the mean heart rate for a baseline period of six seconds before and six seconds after the presentation of a threat [4, 22]. As dependent variable for our data analysis, the base measure will be subtracted from the threat measure.

Tower of London Task. This task assesses the level of cognitive performance and is implemented similar to Banakou et al. [9] in which three differently colored beads on three chopsticks are displayed at descending height. Within three moves, the beads have to manipulate from a predetermined starting position to another set of pins to match the position of the beads in the model. As in Banakou, a point-based algorithm will be used to evaluate the performance (similar to Krikorian et al. [23]).

Questionnaire. We will use the five questions adapted from Banakou et al. [9] to assess self-presence (body ownership) and agency.

3.3 Design Elements

Physical feedback. Physical feedback will be designed by providing feedback in form of vibrations through the IVR gloves. Thus, when individuals in the high physical feedback condition touch objects, the gloves will vibrate. For individuals in the low physical feedback condition, this vibration will be missing.

Avatar design. Avatar design will be operationalized by either using a normal-looking male avatar (normal intelligence condition) or an avatar looking similar to Einstein (high intelligence condition).

3.4 Procedure

Apart from the physical feedback conditions, the threat to the virtual body, and the psychophysiological measurement, the overall procedure is adapted from Banakou et al. [9]. Participants will be told that they will take part in a study investigating the effects of virtual reality on user experience. They will be invited to the laboratory at two time points: during their first visit participants will sign informed consent, complete measures for self-esteem as well as cognitive ability, and complete the premeasure of the tower of London task. One week later, the IVR session takes place. First, participants are lead into a changing room to put on the HRV belt. Next, the experimenter attaches the electrodes for EDA measurement to the inside of the middle and index finger. Afterwards, participants will get instructions on how to put on HTC Vive Trackers and Hi5 VR Gloves. Subsequently, they will put on the HTC Vive HMD and will see a virtual environment which consists of a room with a mirror, a chair, and a virtual body (which either looks like a human or like Einstein, according to the condition) from a first person perspective. When looking in the mirror, participants can see the virtual body mirrored, thus, in a third person perspective. Participants are then asked to get accustomed to the virtual body by moving their body parts and to look around in the virtual room.

To engage participants into being in the virtual environment, and to make the physical feedback conditions salient, participants will be asked to complete a task in which they have to locate numbers in the room and sort them in ascending order using their hands. In the high physical feedback condition, participants will receive physical feedback when touching the numbers, whereas this feedback will be missing for participants in the low physical feedback condition.

In the next part of the experiment, participants will be seated on a chair and asked to answer the virtually presented questionnaire regarding self-presence (body ownership) and telepresence. After they have finished answering the questionnaire, participants will be told that they have the chance to play a game with a box-shaped robot. In this game, participants will be asked to put their right hand on a virtual pad which is tantalized to them by the box-shaped robot. Then, the robot will pull out a knife and starts to stab the knife quickly in the space between the fingers of the participants. This serves as a threat for the virtual body. We chose a game in which the virtual body is not actually hurt because we wanted to refrain from permanently damaging the virtual body, as we expected that this might interfere with the intelligence salience of the Einstein body (participants could remember their experience as threatening rather than as being embodied in the body of an intelligent person). Afterwards, participants will take off the HMD and do the post measure of the tower of London task. Finally, participants will be thanked and debriefed.

4 Discussion

With our results, we aim to gain insights into the working mechanisms through which body ownership illusions affect cognitive performance. First, our research contributes to the literature indicating that self-presence in the form of body ownership illusions can be measured by biophysiological variables [3, 4] by delivering a more practice-oriented view on physical feedback. Second, we aim to contribute to literature indicating that visuotactile feedback can indeed strengthen self-presence [4]. Third, by testing whether strengthening self-presence can increase cognitive performance, we contribute to practice increasing the knowledge on how immersive virtual reality can be designed to shape behavioral and cognitive outcomes in a beneficial way [8, 9, 24].

References

1. Slater, M., & Wilbur, S. (1997). A framework for immersive virtual environments (FIVE): Speculations on the role of presence in virtual environments. *Presence: Teleoperators and Virtual Environments, 6,* 603–616. https://doi.org/10.1162/pres.1997.6.6.603.

2. Schultze, U. (2010). Embodiment and presence in virtual worlds: a review. *Journal of Information Technology., 25,* 434–449. https://doi.org/10.1057/jit.2010.25.
3. Maselli, A., & Slater, M. (2013). The building blocks of the full body ownership illusion. *Front Hum Neurosci., 7,* 83. https://doi.org/10.3389/fnhum.2013.00083.
4. Kokkinara, E., & Slater, M. (2014). Measuring the effects through time of the influence of visuomotor and visuotactile synchronous stimulation on a virtual body ownership illusion. *Perception, 43,* 43–58. https://doi.org/10.1068/p7545.
5. Kilteni, K., Bergstrom, I., & Slater, M. (2013). Drumming in immersive virtual reality: The body shapes the way we play. *IEEE Transactions on Visualization and Computer Graphics, 19,* 597–605.
6. Banakou, D., Hanumanthu, P. D., & Slater, M. (2016). Virtual embodiment of white people in a black virtual body leads to a sustained reduction in their implicit racial bias. *Frontiers in human neuroscience, 10,* 601. https://doi.org/10.3389/fnhum.2016.00601.
7. Hasler, B. S., Spanlang, B., & Slater, M. (2017). Virtual race transformation reverses racial in group bias. *PLoS ONE, 12,* e0174965. https://doi.org/10.1371/journal.pone.0174965.
8. Osimo, S. A., Pizarro, R., Spanlang, B., & Slater, M. (2015). Conversations between self and self as Sigmund Freud-A virtual body ownership paradigm for self counselling. *Sci Rep., 5,* 13899. https://doi.org/10.1038/srep13899.
9. Banakou, D., Kishore, S., & Slater, M. (2018). Virtually being einstein results in an improvement in cognitive task performance and a decrease in age bias. *Frontiers in Psychology, 9,* 917. https://doi.org/10.3389/fpsyg.2018.00917.
10. Sanchez-Vives, M. V., Spanlang, B., Frisoli, A., Bergamasco, M., & Slater, M. (2010). Virtual hand illusion induced by visuomotor correlations. *PLoS ONE, 5,* e10381. https://doi.org/10.1371/journal.pone.0010381.
11. Tieri, G., Tidoni, E., Pavone, E. F., & Aglioti, S. M. (2015). Body visual discontinuity affects feeling of ownership and skin conductance responses. *Scientific Reports, 5,* 17139. https://doi.org/10.1038/srep17139.
12. Botvinick, M., & Cohen, J. (1998). Rubber hands 'feel' touch that eyes see. *Nature, 391,* 756. https://doi.org/10.1038/35784.
13. Kilteni, K., Maselli, A., Kording, K. P., & Slater, M. (2015). Over my fake body: body ownership illusions for studying the multisensory basis of own-body perception. *Frontiers in Human Neuroscience., 9,* 141. https://doi.org/10.3389/fnhum.2015.00141.
14. Davis, M. H. (1980). *A multidimensional approach to individual differences in empathy*.
15. Regan, D. T., & Totten, J. (1975). Empathy and attribution: Turning observers into actors. *Journal of Personality and Social Psychology, 32,* 850–856.
16. Oh, S. Y., Bailenson, J., Weisz, E., & Zaki, J. (2016). Virtually old: Embodied perspective taking and the reduction of ageism under threat. *Computers in Human Behavior, 60,* 398–410. https://doi.org/10.1016/j.chb.2016.02.007.
17. Maister, L., Slater, M., Sanchez-Vives, M. V., & Tsakiris, M. (2015). Changing bodies changes minds: Owning another body affects social cognition. *Trends in Cognitive Sciences., 19,* 6–12. https://doi.org/10.1016/j.tics.2014.11.001.
18. Petkova, V. I., & Ehrsson, H. H. (2008). If I were you: Perceptual illusion of body swapping. *PLoS ONE., 3*(1–9), e3832. https://doi.org/10.1371/journal.pone.0003832.
19. Petkova, V. I., Khoshnevis, M., & Ehrsson, H. H. (2011). The perspective matters! Multisensory integration in ego-centric reference frames determines full-body ownership. *Frontiers in Psychology, 2,* 1–7. https://doi.org/10.3389/fpsyg.2011.00035.
20. Slater, M., Spanlang, B., Sanchez-Vives, M. V., & Blanke, O. (2010). First person experience of body transfer in virtual reality. *PLoS ONE, 5,* e10564. https://doi.org/10.1371/journal.pone.0010564.
21. Cacioppo, J. T., Tassinary, L. G., Berntson, G. G.: (2007). *Handbook of psychophysiology*. Cambridge University Press, Cambridge; New York. https://doi.org/10.13140/2.1.2871.1369.
22. Pollatos, O., Herbert, B. M., Matthias, E., & Schandry, R. (2007). Heart rate response after emotional picture presentation is modulated by interoceptive awareness. *International Journal of Psychophysiology, 63,* 117–124. https://doi.org/10.1016/j.ijpsycho.2006.09.003.

23. Krikorian, R., Bartik, J., & Gay, N. (1994). Tower of London procedure: A standard method and developmental data. *Journal of Clinical and Experimental Neuropsychology, 16,* 840–850. https://doi.org/10.1080/01688639408402697.
24. Ott, M., Freina, L. (2015). A literature review on immersive virtual reality in education: State of the art and perspectives. In: *Conference proceedings of »eLearning and Software for Education«* (pp. 133–141) (eLSE).

Using fMRI to Measure Stimulus Generalization of Software Notification to Security Warnings

Brock Kirwan, Bonnie Anderson, David Eargle, Jeffrey Jenkins and Anthony Vance

Abstract This paper examines how habituation to frequent software notifications may carry over to infrequent security warnings. This general process—known as stimulus generalization or simply generalization—is a well-established phenomenon in neurobiology that has clear implications for information security. Because software user interface guidelines call for visual consistency, software notifications and security warnings have a similar look and feel. Consequently, through frequent exposure to notifications, people may become habituated to security warnings they have never seen before. The objective of this paper to propose an fMRI experimental design to measure the extent to which this occurs. We also propose testing security warning designs that are resistant to generalization of habituation effects.

Keywords Security warnings · Habituation · Generalization · fMRI · Mouse cursor tracking · NeuroIS

B. Kirwan (✉) · B. Anderson · J. Jenkins
Brigham Young University, Provo, UT, USA
e-mail: kirwan@byu.edu

B. Anderson
e-mail: bonnie_anderson@byu.edu

J. Jenkins
e-mail: jeffrey_jenkins@byu.edu

D. Eargle
University of Colorado, Boulder, CO, USA
e-mail: dave@daveeargle.com

A. Vance
Temple University, Philadelphia, PA, USA
e-mail: anthony@vance.name

F. D. Davis et al. (eds.), *Information Systems and Neuroscience*,
Lecture Notes in Information Systems and Organisation 32,
https://doi.org/10.1007/978-3-030-28144-1_10

1 Introduction

In neurobiology's habituation theory, *stimulus generalization*—or simply *generalization*—occurs when the effects of habituation to one stimulus generalize, or carry over, to other novel stimuli that are similar in appearance [1, 2]. Applied to the domain of human–computer interaction, generalization suggests that users not only habituate to individual security warnings, but also to whole classes of user interface dialogs (e.g., notifications, alerts, confirmations, etc.—hereafter referred to collectively as "notifications" for brevity) that share a similar look and feel (see Fig. 1). If true, then the threat and potential impact of habituation is much broader than previous work has suggested [3–5], as users may already be deeply habituated to a security warning that they have never seen before.

Building on prior research [6], we outline an experiment using fMRI and, mouse cursor tracking to (1) measure the extent to which a non-clicking mode of interaction for security warning designs can reduce the occurrence of generalization and (2) which mode of interaction is the most effective in reducing the occurrence of generalization.

2 Literature Review

2.1 Habituation and Generalization to Security Warnings

Although habituation to security warnings is well known and has been examined in a number of studies [4, 7–9], the phenomenon of generalization is less well recognized. West noted that "Security messages often resemble other messages dialogs. As a result, security messages may not stand out in importance and users often learn to disregard them" [10, p. 39]. Böhme and Köpsell observed that users' automatic response to notifications "seems to spill over from moderately relevant topics (e.g., EULAs) to more critical ones (online safety and privacy)" [11, p. 2406]. However, neither of these studies empirically examined this effect.

Similarly, researchers have observed that habituation to a single warning in one context can carry over to a different context. For example, Sunshine et al. [12]

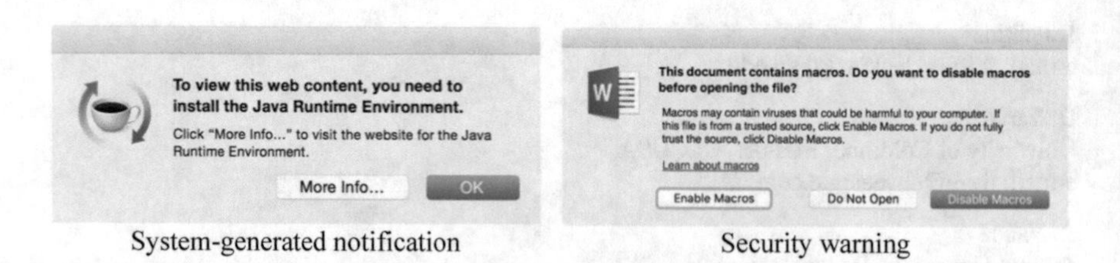

System-generated notification Security warning

Fig. 1 A notification and security warning. Note the similarities in UI and mode of interaction

observed that users who correctly identified the risks of an SSL warning in a library context inappropriately identified these same risks in a banking context. Likewise, Amer et al. [13] found that users who habituated to exception notifications in one context were habituated to a different through visually identical exception notification in a different context. However, in each of these cases, users habituated to the same type of security warning or notification. As a result, it is unclear to what extent software notifications generalize to security warnings.

2.2 Hypotheses

Extending a pilot study that examined the generalization of habituation using Amazon Mechanical Turk, we hypothesize that users' habituation to security messages will generalize to security warning messages. When users repeatedly see software notifications, the brain creates a mental model of these notifications. Rather than giving attention to future exposures to the notifications or similar looking warnings, the brain increasingly relies on this mental model. As a result, users' responses to future warnings decrease (i.e., habituate) in response to repeated exposures of notifications [14]. In summary, we predict:

H1: Security warnings which are designed with a distinctive look will be more resistant to generalization (as measured by both fMRI activation and mouse cursor movements), and the greater the difference in the look, the lower the amount of generalization.

In addition to habituating to the visual features of notifications, participants may also form high-level memory representations (or schemas) for how to interact with notifications and warnings. According to schema theory [15], schemas can represent general knowledge about objects, situations, or sequences of actions. Similar to habituation, which conserves attentional resources to increase efficiency, the development of schemas for sequences of actions improves behavioral efficiency. Unfortunately, behavioral schemas developed in one situation may generalize to another situation, leading to inappropriate responses. In order to test variations of interaction models, we will develop alternatives to what we call the "click to dismiss" model for interacting with warnings, such as a swipe or slider bar (Fig. 2).

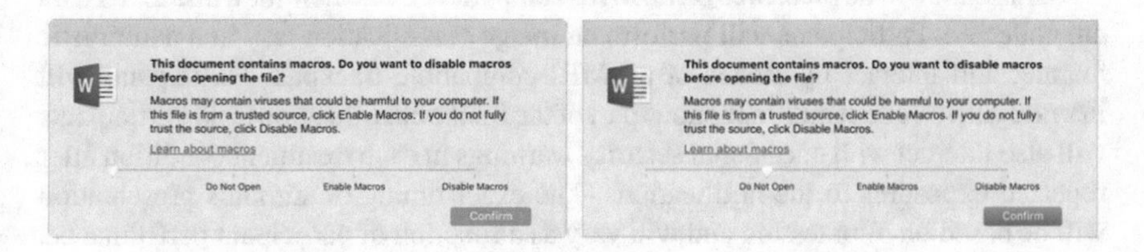

Fig. 2 Warning with slider to evaluate an alternative mode of interaction

By changing the mode of interaction, we predict that users will be able to break out of their schemas and make more considered responses to warnings. Again, we will examine both mouse cursor movements and fMRI activation in response to notifications, compared to warnings that have a distinctive interaction paradigm. In summary, we predict:

H2: Security warning messages which are designed with a change the mode of interaction will be more resistant to generalization (as measured by fMRI activation and mouse cursor movements).

3 Experimental Design

3.1 Methodology

We plan to use fMRI and mouse cursor tracking tools simultaneously while participants receive repeated exposures to notifications and occasional exposures to warnings. Following our pilot work, we will use mouse cursor tracking to behaviorally demonstrate generalization of notification habituation to warnings. We will use fMRI to confirm habituation and generalization in neural activity. fMRI gives us a sensitive measure of neurocognitive processes that are otherwise difficult to directly observe. MRI data will be collected with a Siemens 3T Tim-Trio scanner and mouse cursor-tracking data will be collected with a custom-built MRI-compatible touchpad.

3.2 Task

We will expose participants to (1) repeated notifications while they perform a simple classification task in the MRI scanner. Additionally, participants will encounter (2) standard security warnings, (3) security warning designs that vary visually from the look and feel of the notifications as in our pilot study, and (4) security warning designs that vary in the mode of interaction from notifications (see Fig. 2). Finally, (5) novel software images will be displayed to rule out fatigue.

All stimuli will be presented on an MRI-compatible LCD monitor while fMRI data are collected. Participants will perform an image classification task in a naturalistic manner and interact by means of an MRI-compatible trackpad. Participants will interact with frequent notifications as part of the image classification task. Participants will also interact with occasional security warnings in each treatment condition after repeated exposures to the notifications. The exact timing of stimulus presentation will be based on pilot testing and will vary as a function of participant performance. Timing information for the analysis of fMRI time course data will be determined by both stimulus presentation time and mouse cursor movement onset times and latencies. Standard structural and functional MRI scanning parameters will be used.

In addition to exploratory whole-brain analyses, a priori anatomical regions of interest will be examined, including the visual cortex and ventral visual pathway, medial temporal lobe, and motor control regions such as the prefrontal cortex and basal ganglia. The analysis of each of these regions allows for an examination of generalization at different levels of processing. First, the visual cortex and ventral visual pathway are involved in object perception [see 16 for review]. Second, the medial temporal lobe is involved in memory specificity, or the detection of differences between similar stimuli [17]. Lastly, motor control regions will allow us to examine the change in motoric scripts and schemas.

3.3 Analysis

We will examine established behavioral and neural indices of habituation, including faster response times and decreased neural activation to repeated stimuli. If generalization occurs, we would expect these same responses to security warnings as well. To isolate these effects from effects due to fatigue, we will compare behavioral and neural responses to warnings and notifications against a novel stimulus that should result in full recovery of responses if the participant is not fatigued.

4 Anticipated Contributions

We anticipate that our findings of this study will complement and extend previous work that examined habituation to individual warnings. With the proposed experimental design, we intend to examine how to mitigate the effects of the generalization of habituation to frequent software notifications. Specifically, our anticipated contributions are:

1. Determine how the effects of habituation to frequent notifications and warnings generalize to novel warnings.
2. Measure the if changing the mode of interaction can reduce generalization.
3. Test warning designs with distinctive modes of interaction to determine the highest resistance to generalization.

5 Future Research

In this paper, we theoretically explain how changing the mode of interaction will contradict existing metal schemas and thereby make a warning more resistant to generalization. We propose running the above experiment to empirically test our hypotheses related to decreasing generalization.

In addition, the mode of interaction can improve security behaviors through several other mechanisms that can be examined in future research. First, the mode of interaction can be used to improve comprehension. For example, Bravo-Lillo et al. [4] had people highlight key text-components in the warning before they were allowed to reject it, improving comprehension of the security risk. The mode of interaction can also help users understand the consequence of the action. For instance, one could have a person perform an action that imitates the potential danger of ignoring the warning before allowing them to reject it (e.g., dragging a password to a hacker icon, disabling a lock that makes a computer public, etc.).

Second, the mode of interaction can influence the amount of work required to behave non-securely, and thereby make alternative, more secure behaviors, more appealing. It is often easier to engage in risky behavior than secure behavior, resulting in poor security decisions. For example, ignoring an SSL warning often requires less effort than try to find an alternative website to address the user's need that is secure [18]. While much research has focused on making security more usable [19], less research has focused on making non-secure behaviors less usable. However, the mode of interaction can be used to accomplish this objective and thereby improve secure behaviors. For example, for SSL warnings on Chrome, you must click on "Advanced" before finding an option to dismiss the warning. This extra work can help deter non-secure behavior and promote easier secure behaviors.

Acknowledgements This research was funded by NSF Grant #CNS-1931108.

References

1. Rankin, C. H., et al. (2009). Habituation revisited: An updated and revised description of the behavioral characteristics of habituation. *Neurobiology of Learning and Memory, 92*(2), 135–138.
2. Thompson, R. F., & Spencer, W. A. (1966). Habituation: A model phenomenon for the study of neuronal substrates of behavior. *Psychological Review, 73*(1), 16–43.
3. Anderson, B. B., Kirwan, C. B., Jenkins, J. L., Eargle, D., Howard, S., & Vance, A. (2015). How polymorphic warnings reduce habituation in the brain: Insights from an fMRI study. In *Proceedings of the 33rd Annual ACM Conference on Human Factors in Computing Systems* (pp. 2883–2892). ACM: Seoul, Republic of Korea.
4. Bravo-Lillo, C., Komanduri, S., Cranor, L. F., Reeder, R. W., Sleeper, M., Downs, J., et al. (2013). Your attention please: Designing security-decision UIs to make genuine risks harder to ignore. In *Proceedings of the Ninth Symposium on Usable Privacy and Security* (pp. 1–12). ACM: Newcastle, United Kingdom.
5. Egelman, S., Cranor, L. F., & Hong, J. (2008). You've been warned: An empirical study of the effectiveness of web browser phishing warnings. In *Proceedings of the SIGCHI Conference on Human Factors in Computing Systems* (pp. 1065–1074). ACM: Florence, Italy.
6. Anderson, B. B., Vance, A., Jenkins, J. L., Kirwan, C. B., & Bjornn, D. (2017). It all blurs together: How the effects of habituation generalize across system notifications and security warnings. *Information Systems and Neuroscience* (pp. 43–49). Cham: Springer.

7. Bravo-Lillo, C., Cranor, L., Komanduri, S., Schechter, S., & Sleeper, M. (2014). Harder to ignore? Revisiting pop-up fatigue and approaches to prevent it. In *10th Symposium on Usable Privacy and Security (SOUPS 2014)*. USENIX Association.
8. Brustoloni, J. C., & Villamarín-Salomón, R. (2007). Improving security decisions with polymorphic and audited dialogs. In *Proceedings of the Third Symposium on Usable Privacy and Security (SOUPS 2007)*. New York, NY, USA: ACM.
9. Vance, A., et al. (2018). Tuning out security warnings: a longitudinal examination of habituation through fMRI, eye tracking, and field experiments. *MIS Quarterly, 42*(2), 355–380.
10. West, R. (2008). The psychology of security. *Communications of the ACM, 51*(4), 34–40.
11. Böhme, R., & Köpsell, S. (2012). *Trained to accept? A field experiment on consent dialogs*. In *Proceedings of the ACM Conference on Human Factors in Computing Systems (CHI)*. Atlanta: ACM.
12. Sunshine, J., Egelman, S., Almuhimedi, H., Atri, N., & Cranor, L. F. (2009). Crying wolf: An empirical study of SSL warning effectiveness. In *SSYM'09 Proceedings of the 18th Conference on USENIX Security Symposium*. Montreal, Canada.
13. Amer, T. S., & Maris, J.-M. B. (2007). Signal words and signal icons in application control and information technology exception messages—Hazard matching and habituation effects. *Journal of Information Systems, 21*(2), 1–25.
14. Groves, P. M., & Thompson, R. F. (1970). Habituation: A dual-process theory. *Psychological Review, 77,* 419–450.
15. Rumelhart, D. E. (1980). Schemata: the building blocks of cognition. In R. J. Spiro (Ed.), *Theoretical issues in reading comprehension*. Hillsdale, NJ: Lawrence Erlbaum.
16. Grill-Spector, K. (2003). The neural basis of object perception. *Current Opinion in Neurobiology, 13*(2), 159–166.
17. Kirwan, C. B., & Stark, C. E. L. (2007). Overcoming interference: An fMRI investigation of pattern separation in the medial temporal lobe. *Learning and Memory, 14*(9), 625–633.
18. Adams, A., & Sasse, M. A. (1999). Users are not the enemy. *Communications of the ACM, 42*(12), 40–46.
19. Balfanz, D., et al. (2004). In search of usable security: Five lessons from the field, IEEE. *IEEE Security and Privacy, 2*(5), 19–24.

Do We Protect What We Own?: A Proposed Neurophysiological Exploration of Workplace Information Protection Motivation

Shan Xiao, Merrill Warkentin, Eric Walden and Allen C. Johnston

Abstract Part-time and temporary employees and contractors become a major cybersecurity threat for organizations due to the ephemeral nature of their engagement. Compared with full-time employees, they may be less commited to the welfare of the organization and, therefore, less willing to engage in security recommendations to protect it. Perceived psychological ownership is an important factor that shapes employees' security behaviors. The endowment effect also explains employees' tendencies to overvalue information that belongs to them, and conversely, extend fewer protections to information that they view as belonging to others. Thus, employees may be more motivated to safeguard their own information than organizational information. From a principle-agent perspective, this study investigates how three types of employees perceive organizational and personal information, and how different employees make decisions about protecting their own versus organizational information.

Keywords Full-time and temporary employees · Contractors · Agency theory · Psychological ownership · Endowment effect · IS security · fMRI

S. Xiao · M. Warkentin (✉)
Mississippi State University, Mississippi State, MS, USA
e-mail: m.warkentin@msstate.edu

S. Xiao
e-mail: sx45@msstate.edu

E. Walden
Texas Tech University, Lubbock, TX, USA
e-mail: Eric.walden@ttu.edu

A. C. Johnston
University of Alabama, Tuscaloosa, AL, USA
e-mail: ajohnston@cba.ua.edu

F. D. Davis et al. (eds.), *Information Systems and Neuroscience*,
Lecture Notes in Information Systems and Organisation 32,
https://doi.org/10.1007/978-3-030-28144-1_11

1 Introduction

As Internet connectivity steadily grows, cybersecurity remains a key concern for organizations. A recent industry report [1] indicates that 77% of companies, globally, rank cybersecurity as a top priority, with vulnerabilities related to organizational insiders serving as one of the most pressing concerns. Organizational insiders with authorized access and practical knowledge of business processes pose a major threat to organizations [2], and while previous research has investigated insider threats through various approaches, one under-examined aspect of that research is the impact employment status has on insiders' security-related behaviors. In fact, among insiders, temporary employees and contractors are responsible for a large portion of data breaches [3].

D'Arcy and Herath [4] suggested that employees' positions and the characteristics of their positions in the organization are associated with their intention to comply with security policies. Employment status—including full-time, part-time, temporary, and contractual—significantly affects employees' attitudes and behaviors [5]. In particular, temporary employees or contractors may lack the necessary commitment to their organizations to engage in protective security behaviors, while contractors are self-interested as part of an opportunistic workforce [6, 7]. Indeed, Sharma and Warkentin [8] found that permanent employees have a higher level of organizational commitment and maintain a greater intention to undertake recommended security actions than temporary employees.

Tied closely to the temporary or permanent nature of one's employment with a company, psychological ownership and endowment bias are other influencing factors that shape security compliance behaviors [9, 10]. Psychological ownership refers to a mental state in which individuals perceive objects as their own [11], while endowment bias explains how individuals estimate prices for their own and others' items. Employees are more likely to engage in security behaviors when experiencing a sense of ownership of the assets in jeopardy, but, how they engage in those security behaviors may be dependent upon the extent to which they believe they have been endowed with the assets. In this study, by using agency theory as a theoretical foundation and drawing upon psychological ownership and the endowment effect, we argue that employees with three types of employment status perceive their ownership of organizational information differently and may overvalue their own information relative to organizational information. Hence, this study attempts to answer the following questions:

RQ1: How does employment status influence how one thinks about the safeguarding of organizational information?
RQ2: How does psychological ownership affect how one thinks about the safeguarding of organizational information?
RQ3: How does endowment bias affect how one thinks about safeguarding organizational information relative to personal information?

2 Theoretical Background

2.1 Agency Theory

Agency theory advocates a contractual relationship in which a principal entrusts work to an agent who is authorized to perform that work [12]. Agents undertake actions on behalf of the principal. The concept of agency relationships stemmed from the economics discipline, where research expanded the literature of risk sharing to explain an agency problem in which different attitudes exist between cooperating parties [13]. This theory has been widely studied in organizational context.

According to agency theory, two primary problems may take place in an agency relationship. First, an agent and principal are self-interested parties with conflicting goals. Naturally, agents prefer to devote efforts toward maximizing their own outcomes rather than achieving the principal's goals. It also becomes challenging and costly for the principal to observe and monitor the agent's behavior. In other words, a principal might not be able to verify how an agent acts in certain circumstance. Secondly, principals and agents possess different attitudes toward the same risk. As a result, both parties may engage in dissimilar behaviors with respect to the perception of risk. This perspective on the principal and agent relationship has been applied to a variety of transactional exchanges which have occurred in socio-economic systems in which information asymmetry, bounded rationality, and opportunism exist [14].

Agency theory has informed a range of relationships in the business management context. The owners of a firm (e.g. shareholders of a publicly-traded firm) hire agents (e.g. CEO) to manage the firm on their behalf. One particularly interesting business organizational form in this context is the family-owned firm in which some managers are also members of the family (and may be owners, thus principals), whereas other managers are agents without the same ownership perspective [15, 16].

IS researchers have leveraged agency theory in security settings to help explain employment relationships with organizations and the challenges those relationships pose with regard to security outcomes. For instance, Herath and Rao [17] argued that the phenomenon in which employees do not comply with information security policies could be explored through an agency lens. Employees have divergent views of security policies and engage in actions based on personal preferences. For example, it is often recommended by organizations that their employees change their passwords, yet the employees continue to use their old passwords for the sake of convenience. Thus, to encourage employees to take the recommended security actions, organizations often turn to incentive or disincentive mechanisms. In a similar vein, Chen et al. [18] suggested organizations should develop reward structures tailored to their employees' compliance behaviors, since employees rationally behave in their preferred way regardless of organizational security policies. Both studies extended agency theory within an information security context.

In our study, we follow the basic assumptions of agency theory. Employees with distinct employment status, including full-time and temporary employees, as well as contractors, have an agency relationship with their organizations. With regard to information security, organizations attempt to motivate employees to act in accordance with their security policies, but despite their efforts to promote compliance behaviors, employees might not have the same interest in following their organizations' recommendations. If the organizations' objectives are either not relevant to or consistent with an employee's personal goals, an interest-conflict dynamic will arise. Employees may also think differently about their own information than organizational information and may form different safeguarding intentions, accordingly. Given the disparity of goals that may exist between and organization and its employees, it is not surprising that employees are likely to act in a manner which accomplishes their own goals rather than those of their organizations. Furthermore, employees' behaviors may be unobservable for organizations. This inability to monitor their employees' compliance reactions concerning security policies is commonly described by organizations. Hence, there is indeed a principal-agent problem relating to information security in organizations.

2.2 Psychological Ownership

As defined earlier, psychological ownership refers to a mental state in which individuals perceive an object as if it were their own [11]. Such ownership is applicable not only for physical objects (e.g., houses or cars) but also for non-physical objects, such as ideas [19]. Psychological ownership was originally developed in the psychology discipline and has since been expanded to study organizational behaviors in management [20–22].

The establishment of psychological ownership is driven by efficacy, self-identity, and a sense of place [21]. The sense of ownership allows individuals to experience efficacy when they feel capable of achieving desired outcomes and altering an environment. Also, individuals may define or express themselves to others through the ownership of an object, such as customized personal computers. Finally, feelings of ownership fulfill individuals' needs for having a sense of place. The place is not limited to a physical location, but may also be instantiated as an extension of the self through perceptions of affiliation. These basic human needs explain why psychological ownership exists.

Psychological ownership involves both cognitive and affective processes [21]. The cognitive aspect represents perceptions related to cognizance, beliefs, and opinions regarding the ownership of an object. This cognition is also formed by an emotional connection with that object. On the other hand, the affective component may be induced by a third party in such a way that the party threatens an individual's sense of ownership [11]. These two processes conceptually separate psychological ownership from legal ownership [23]. Since we study the perceived ownership of organizational and personal data, we consider psychological ownership consists of both cognitive and affective elements.

Researchers have stated that psychological ownership causes ethical and responsible behaviors in organizations [11, 24]. Employees may be more willing to engage in recommended behaviors when experiencing a sense of ownership. The influence of psychological ownership has been considered by IS researchers in security settings. Anderson and Agarwal [9] found that this psychological state positively affected security-related behavioral intentions for home users. Similarly, Thompson et al. [25] concluded that psychological ownership of home computers and mobile devices significantly influenced security intentions. Menard et al. [10] suggested ownership as an antecedent of protective behaviors and that a sense of belonging is impacted by cultural dimensions. Finally, Yoo et al. [26] demonstrated that psychological ownership played a role in security training effectiveness as well as security compliance intentions.

The literature of psychological ownership within a security context is promising, but limited. In our context, we argue that three types of employees have different attitudes toward organizational and personal data. Depending on their employment status, employees may be more motivated to protect their own information than that of the organization. The stronger the sense of ownership, the more likely an employee engages in protective behaviors. Furthermore, employment status may influence employees' perceptions by virtue of different attachments to the organization [5, 27]. In such a manner, employees with high organizational commitment might be more likely to perceive the organization's information as their own, to a certain degree.

2.3 *Endowment Effect*

Economists suggest that individuals tend to overvalue the goods they own, attributing this to an endowment effect. The endowment effect demonstrates the discrepancy between willingness to pay and willingness to accept [28]. In general, individuals value items that they are endowed with at a higher level than identical items, which they do not own. Furthermore, individuals are more reluctant to lose or give up an object than they are to acquire it. The effect could occur for tangible (e.g. mug) and intangible (e.g. intellectual property) artifacts [29]. Empirical evidence presented by previous research attests to the endowment effect [30], with the majority of this research occurring in the areas of marketing, psychology, and organizational behavior.

Goes [31] asserted that the endowment effect, as a cognitive bias, describes how individuals make decisions "through a process of 'building' their preferences (p, iv)." IS research has been informed by the research findings from behavioral economics. For example, Renaud et al. [32] argued that individuals are unwilling to change the method they use to select or create new passwords, owing to the endowment effect, even when a better alternative is offered. The ownership of a password created by an individual activates the endowment effect. The individual feels endowed with that password creation process and is resistant to give up the old password creation process in exchange for a new one.

In this study, we apply the endowment effect to explore how employees make decisions about protective behaviors with respect to information security. According to the endowment effect, employees may overvalue their personal information relative to organizational information. When combined with psychological ownership theory, agency theory suggests that these employees may make biased decisions about protecting organizational information versus personal information.

2.4 Neuroscience Literature into Psychological Ownership

Previous neuroscience literature has suggested that cognitive processes related to the allocation of selective attention occur when individuals encounter familiar and self-relevant items, such as a name, face, and personal possession [33–36]. Turk and colleagues [37] further explored the attentional biases triggered by object ownership and observed different ERP responses in the brain when participants faced "my" or "their" objects like apples and socks. They found that perceived ownership had a relationship with increased attentional processing when the objects belonged to the self.

In general, prior studies have focused on brain activity when participants were exposed to self-relevant or self-owned objects. Within our context, we target the valuable information owned by organizations and employees. However, we not only seek to determine how employees process the two kinds of information, but also their reactions with respect to the information even when they're making the exact same decisions with the same value of information.

3 Research Design

To investigate these research questions, we will design a neurophysiological research protocol, to assess the processes that reflect the latent constructs inherent in these psychological states. Similar to Warkentin et al. [38], we will evaluate brain activation of subjects with functional magnetic resonance imaging (fMRI) while undergoing various treatments. We plan to recruit employees who are either full-time employees, temporary employees, or contractors. 20 participants who are 18 years or older will be grouped according to their employment status. Stimuli will be designed to be ecologically relevant in the sense that participants will easily perceive if the information belongs to the organization or to them. For example, a business report or customer information will be regarded as organizational information, whereas a participant's name or phone number will be regarded as personal information. The same order of stimuli will be represented to the participants and they will be required to decide whether they intend to protect the information or not. Brain activation will be recorded to represent the findings. Additional questions will be placed after each stimulus to collect data about participants' perceptions of psychological ownership (and perhaps also about the threats to that information).

Each subject will be placed in the position of data advisors. They will be told that they have been asked to examine different sorts of data that the company stores and prioritize the security spending on the data. We have identified sixty different types of data and validated in a pre-test with a different group that thirty types are personal and hence tend to be high in psychological ownership (e.g. your birthday, your address) and thirty types are business related and tend to be low in psychological ownership (e.g. the company's address, the company's revenue). Subjects are asked to rate how high of a budget priority each piece of information should be given. The ratings are based on percentiles and subjects are asked to assign roughly 15 to the bottom 25th percentile, 15 to the 25–50th percentile, and so forth. This process will help to ensure that subjects really have to contemplate how important each piece of data is.

Following Kim and Johnson [39], we propose that ownership will activate the ventromedial prefrontal cortex, ventral anterior cingulate cortex, and medial orbitofrontal cortex. Furthermore, we predict that activation in these areas will be a significant predictor of willingness to prioritize spending.

After rating each piece of information, subjects are then asked to rate each item on psychological ownership (e.g. to what degree is this data yours vs. the company's). We then compare the brain activation when prioritizing data they own to brain activation when prioritizing data the company owns. Details will be discussed and feedback from workshop participants will be requested.

4 Conclusion

This study explores how employees' reactions are different when safeguarding organizational information versus personal information. Employees' perceptions of ownership toward information may lead to different responses to security recommendations. Also, employment status may influence employees' decisions at the same time, including full-time and temporary employees, and contractors. A neuroIS technique enables direct observation of participants' brains. This might be the first attempt to investigate how full-time and temporary employees and contractors react to organizational and personal information security threats. The findings may benefit both researchers and practitioners in promoting security compliance behaviors.

References

1. Schwab, W., & Poujol, M. (2018). The state of industrial cybersecurity 2018. Available at https://ics.kaspersky.com/media/2018-Kaspersky-ICS-Whitepaper.pdf.
2. Johnston, A. C., & Warkentin, M. (2010). Fear appeals and information security behaviors: An empirical study. *MIS Quarterly, 34*(3), 549–566.
3. Goldman, J. (2014 January). Data breach roundup. Available at https://www.esecurityplanet.com/network-security/data-breach-roundup-january-2014.html.
4. D'Arcy, J., & Herath, T. (2011). A review and analysis of deterrence theory in the IS security literature: Making sense of the disparate findings. *European Journal of Information Systems, 20*(6), 643–658.
5. De Cuyper, N., & De Witte, H. (2007). Job insecurity in temporary versus permanent workers: Associations with attitudes, well-being, and behaviour. *Work and Stress, 21*(1), 65–84.
6. Theoharidou, M., Kokolakis, S., Karyda, M., & Kiountouzis, E. (2005). The insider threat to information systems and the effectiveness of ISO17799. *Computers and Security, 24*(6), 472–484.
7. Williamson, O. E. (1991). Comparative economic organization: The analysis of discrete structural alternatives. *Administrative Science Quarterly, 36*(2), 269–296.
8. Sharma, S., & Warkentin, M. (2018). Do I really belong? Impact of employment status on information security policy compliance. Computers and Security, forthcoming, (published online September 23, 2018 at https://www.sciencedirect.com/science/article/pii/S0167404818304024).
9. Anderson, C. L., & Agarwal, R. (2010). Practicing safe computing: A multimedia empirical examination of home computer user security behavioral intentions. *MIS Quarterly, 34*(3), 613–643.
10. Menard, P., Warkentin, M., & Lowry, P. B. (2018). The impact of collectivism and psychological ownership on protection motivation: A cross-cultural examination. *Computers & Security, 75,* 147–166.
11. Pierce, J. L., Kostova, T., & Dirks, K. T. (2003). The state of psychological ownership: Integrating and extending a century of research. *Review of General Psychology, 7*(1), 84–107.
12. Eisenhardt, K. M. (1989). Agency theory: An assessment and review. *Academy of Management Review, 14*(1), 57–74.
13. Jensen, M. C., & Meckling, W. H. (1976). Theory of the firm: Managerial behavior, agency costs and ownership structure. *Journal of Financial Economics, 3*(4), 305–360.
14. Milgrom, P. R., & Roberts, J. D. (1992). *Economics, organization and management*. Englewood Cliffs, NJ: Prentice-Hall.
15. Chrisman, J. J., Chua, J. H., & Litz, R. A. (2004). Comparing the agency costs of family and non–family firms: Conceptual issues and exploratory evidence. *Entrepreneurship Theory and Practice, 28*(4), 335–354.
16. Chrisman, J. J., Chua, J. H., Kellermanns, F. W., & Chang, E. P. (2007). Are family managers agents or stewards? An exploratory study in privately held family firms. *Journal of Business Research, 60*(10), 1030–1038.
17. Herath, T., & Rao, H. R. (2009). Encouraging information security behaviors in organizations: Role of penalties, pressures and perceived effectiveness. *Decision Support Systems, 47*(2), 154–165.
18. Chen, Y., Ramamurthy, K., & Wen, K. W. (2012). Organizations' information security policy compliance: Stick or carrot approach? *Journal of Management Information Systems, 29*(3), 157–188.
19. Dittmar, H. (1992). The social psychology of material possessions: To have is to be. Harvester Wheatsheaf and St. Martin"s Press.
20. Pierce, J. L., & Furo, C. A. (1990). Employee ownership: Implications for management. *Organizational Dynamics, 18*(3), 32–43.
21. Pierce, J. L., Kostova, T., & Dirks, K. T. (2001). Toward a theory of psychological ownership in organizations. *Academy of Management Review, 26*(2), 298–310.

22. Van Dyne, L., & Pierce, J. L. (2004). Psychological ownership and feelings of possession: Three field studies predicting employee attitudes and organizational citizenship behavior. *Journal of Organizational Behavior: The International Journal of Industrial, Occupational and Organizational Psychology and Behavior, 25*(4), 439–459.
23. Isaacs, S. (1993). *Social development in young children*. London: Routledge and Kegan Paul.
24. Avey, J. B., Avolio, B. J., Crossley, C. D., & Luthans, F. (2009). Psychological ownership: Theoretical extensions, measurement and relation to work outcomes. *Journal of Organizational Behavior: The International Journal of Industrial, Occupational and Organizational Psychology and Behavior, 30*(2), 173–191.
25. Thompson, N., McGill, T. J., & Wang, X. (2017). "Security begins at home": Determinants of home computer and mobile device security behavior. *Computers and Security, 70,* 376–391.
26. Yoo, C. W., Sanders, G. L., & Cerveny, R. P. (2018). Exploring the influence of flow and psychological ownership on security education, training and awareness effectiveness and security compliance. *Decision Support Systems, 108,* 107–118.
27. Chambel, M. J., & Castanheira, F. (2006). Different temporary work status, different behaviors in organization. *Journal of Business and Psychology, 20*(3), 351–367.
28. Kahneman, D., Knetsch, J. L., & Thaler, R. H. (1991). Anomalies: The endowment effect, loss aversion, and status quo bias. *Journal of Economic Perspectives, 5*(1), 193–206.
29. Horowitz, J. K., & McConnell, K. E. (2002). A review of WTA/WTP studies. *Journal of Environmental Economics and Management, 44*(3), 426–447.
30. Knetsch, J. L. (1992). Preferences and nonreversibility of indifference curves. *Journal of Economic Behavior and Organization, 17*(1), 131–139.
31. Goes, P. B. (2013). Editor's comments: information systems research and behavioral economics. *MIS Quarterly, 37*(3), iii–viii.
32. Renaud, K., Otondo, R., & Warkentin, M. (2019). This is the way 'I' create my passwords"… does the endowment effect deter people from changing the way they create their passwords? *Computers and Security, 82,* 241–260.
33. Gray, H. M., Ambady, N., Lowenthal, W. T., & Deldin, P. (2004). P300 as an index of attention to self-relevant stimuli. *Journal of Experimental Social Psychology, 40*(2), 216–224.
34. Miyakoshi, M., Nomura, M., & Ohira, H. (2007). An ERP study on self-relevant object recognition. *Brain and Cognition, 63*(2), 182–189.
35. Ninomiya, H., Onitsuka, T., Chen, C. H., Sato, E., & Tashiro, N. (1998). P300 in response to the subject's own face. *Psychiatry and Clinical Neurosciences, 52*(5), 519–522.
36. Fischler, I., Jin, Y. S., Boaz, T. L., Perry, N. W., Jr., & Childers, D. G. (1987). Brain potentials related to seeing one's own name. *Brain and Language, 30*(2), 245–262.
37. Turk, D. J., Van Bussel, K., Brebner, J. L., Toma, A. S., Krigolson, O., & Handy, T. C. (2011). When "it" becomes "mine": Attentional biases triggered by object ownership. *Journal of Cognitive Neuroscience, 23*(12), 3725–3733.
38. Warkentin, M., Walden, E. A., Johnston, A. C., & Straub, D. W. (2016). Neural correlates of protection motivation for secure IT behaviors: An fMRI examination. *Journal of the Association for Information Systems, 17*(3), 194–215.
39. Kim, K., & Johnson, M. K. (2015). Distinct neural networks support the mere ownership effect under different motivational contexts. *Social Neuroscience, 10*(4), 376–390.

Investigating Phishing Susceptibility—An Analysis of Neural Measures

Rohit Valecha, Adam Gonzalez, Jeffrey Mock, Edward J. Golob and H. Raghav Rao

Abstract Phishing is an attempt to acquire sensitive information from a user by malicious means. The losses due to phishing have exceeded a trillion dollars globally. In investigating phishing susceptibility, literature has largely examined structural and individual characteristics. Very little attention has been paid to neural measures within phishing contexts. In this paper, we explore the role of cognitive responses and correlated brain responses in phishing context. Such research is useful because a deeper understanding of persuasion techniques can inform the design of effective countermeasures for detecting and blocking phishing messages.

Keywords Phishing susceptibility · EEG · Neural measures

1 Introduction

Phishing is an act of deception that involves an attacker who generally masquerades as a legitimate institution to trick users into disclosing sensitive information that is later used in fraudulent activities [1, 2]. More than 100,000 Internet users around the world are subjected to phishing attacks daily [3]. In 2018, the FBI has reported that recent phishing attacks targeting business e-mails, have resulted in more than $12.5 billion in losses [4]. In most organizations, the remedy against phishing is training and education [5]. Most training programs treat an individual's susceptibility to phishing as a black box by viewing phishing knowledge and efficacy as the input, and phishing response as the output [6]. Furthermore, most training programs have mostly focused on email structure and individual characteristics in phishing detection [7, 8]. Relying on such practices to consciously look for indicators of phishing may not always be sufficient to help to "win" against the scammers. There is little research on neural measures in the context of phishing emails.

R. Valecha (✉) · A. Gonzalez · H. Raghav Rao
Department of Information Systems and Cyber Security, San Antonio, USA
e-mail: rohit.valecha@utsa.edu

A. Gonzalez · J. Mock · E. J. Golob
Department of Psychology, University of Texas at San Antonio, San Antonio, USA

F. D. Davis et al. (eds.), *Information Systems and Neuroscience*,
Lecture Notes in Information Systems and Organisation 32,
https://doi.org/10.1007/978-3-030-28144-1_12

Along this backdrop, in this paper, we argue that accounting for neural measures may enhance effectiveness in phishing detection and add value to interventions for organizations and their employees. This paper utilizes neurophysiological measures and signal detection theory [9] (both using EEG) to identify brain activity associated with susceptibility to phishing emails. It aims to investigate the neural measures to defend against phishing by analyzing a corpus of phishing and benign email messages to understand the use EEG-based biomarkers that index processing of emails for detection of phishing email messages. Using EEG and behavioral measures, we propose to examine how subjects process information in phishing situations. This study addresses two research questions: (1) What is the role of cognitive responses and correlated brain responses in phishing context? (2) How does this response differ between victims and non-victims in phishing context?

The significance for cybersecurity is that neural signals generated by the brain could be an untapped resource for additional countermeasures to phishing and other security issues in human-computer interactions. Such investigation means that people would not be asked to modify their behavior, such as focusing less on their current task and more on cybersecurity. The EEG measures, instead, are capable of indexing unconscious indicators of threat. Such threat signals may be present in the brain even when subjects do not consciously notice a phishing threat. The rest of the paper is organized as follows: First we discuss the literature on phishing susceptibility. Then we discuss the methodology consisting of data collection, experimental design and quantitative analysis. In the subsequent section, we conclude with practical implications.

2 Phishing Susceptibility

In phishing literature, suspicion has been used as a measure of susceptibility [10]. Suspicion is the degree of uncertainty one experiences when interacting with phishing emails that points to its detection. It is both necessary for detecting phishing, as well as a predictor of detection accuracy. Several studies have utilized detection accuracy for explaining individual susceptibility to email-based phishing [11]. Wang et al. (2016) identify that frequency estimates are generally more accurate than probability estimates [12]. Accordingly, this proposal utilizes frequency of (in)accurate detection (of phishing messages) as a measure of individual phishing susceptibility.

Prior literature on phishing susceptibility has focused on two main streams of research: visual cues and individual traits [8]. The first stream deals with structural characteristics of messages that can be used by individuals to signal deceptive content. Some of these include web page text, code, images, URLs, and link information [13, 14]. In the second stream, the literature has attributed phishing susceptibility to individual characteristics. There have been numerous individual characteristics such as age, gender, affiliation and department that have been examined in the literature [15–17]. Yet other contextual factors affecting phishing susceptibility are time [18], culture [19], work environment [20] and personality traits [21].

While most of the existing research has focused on textual elements of the messages (such as analysis of URLs, typos) or contextual elements (for example persuasive forces), this research investigates cognitive responses and correlated brain responses in the context of phishing detection. The literature investigating cognitive responses and correlated brain responses in phishing context is still in its infancy. A review of the literature shows only a couple of studies that investigates neural correlates involved in phishing tasks. Neupane and colleagues [22, 23] have investigated user's security performance and underlying neural activity by focusing on two security tasks: phish detection and response to malware warnings. In this endeavor, they have focused on several regions of the brain such as prefrontal, frontal and occipital cortex, frontal, orbitofrontal and temporal gyrus, and parietal lobule. To the best of our knowledge, this paper is one of the early attempts at investigating the role of neural measures to defend against phishing for investigating perceptions and behaviors of phishing susceptibility.

3 Methodology

3.1 Emails

Prior research has identified an approach in phishing detection research, which relies on participants' responses to phishing emails. Participants are asked how they would respond to a mix of phishing and benign emails [11, 24, 25], or alternately they are asked if the emails are genuine [26–28]. Following the prior literature, we provided a sample of 75 phishing emails along with 75 spam emails (randomly chosen based on the various combinations of structural and persuasion characteristics, see Fig. 1 for example stimulus) separated into 3 blocks of 50 emails to 24 participants. Each email was displayed for 6 seconds, with 10 ± 2 seconds between presentation of successive emails. A viewing duration of 6 seconds was chosen to allow the subject enough time to quickly read the email, which introduced some time pressure, and to also maximize the number of trials for EEG analysis. It is important to note that we chose spam emails to compare with phishing emails because spam emails utilize structural characteristics similar to phishing email, albeit for manipulating the recipient into buying various products. The major difference between a phish and a spam email is that a spam is unsolicited email with product ads while a phish is a malicious email intended to trick the recipient into disclosing personal information [29].

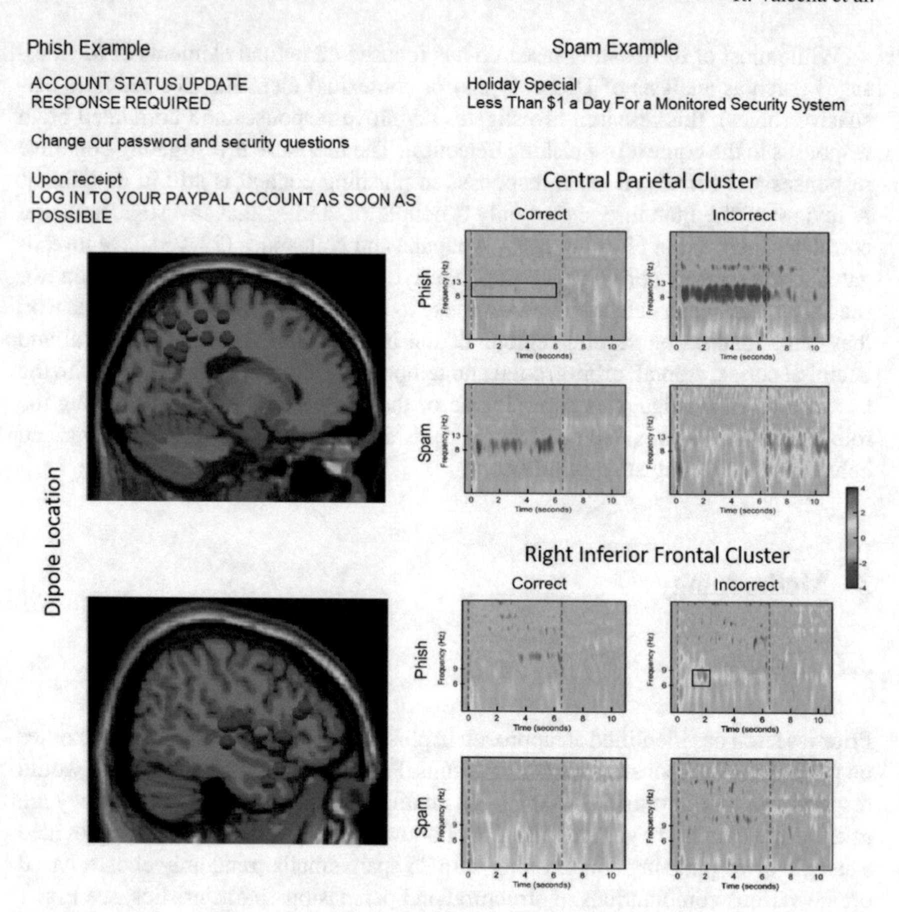

Fig. 1 Example email stimuli and experimental results. Top row shows examples of emails categorized as phish and spam. MRI images show the estimated location of neural sources for the central parietal and right inferior frontal ICA clusters. Time frequency analyses of the ICA components is a function of trial type (phish, spam) and subject decision (correct, incorrect)

3.2 *Subject Description and Data Acquisition*

Subjects consisted of 24 adults, between the ages of 18–23 (13 males, 11 females), who were students at a southern university in the US, and received course credit for participation. Each participant signed a consent form, and experiment protocols were followed in accordance with the Institutional Review Board. Subjects also filled out a survey that captured demographics information, such as age, gender, race, and situation context, such as security awareness, amount and type of security training, frequency of e-mail usage, prior experience.

3.3 Procedure

EEG data was recorded at 500 Hz using a 64 channel electrode cap (Ag/AgCl electrodes impedances $\leq$10 kΩ) acquired with Curry 7 Neuroimaging Suite (Compumedics Neuroscan, Charlotte, NC). Four electrodes were used to monitor eye movements, one above and one below the left eye and one lateral to each eye. Subjects were seated in an audiometric room in front of a computer monitor that presented all emails. A 4-button keypad was given to each participant with verbal instructions on what button to press for either spam or phish (counter balanced across participants). Each email was displayed for 6 seconds with a 6 seconds intertrial interval.

3.4 Data Processing

All post processing of EEG data was done offline using the EEGLAB toolbox for MATLAB (The Mathworks Inc., Natwick, MA). Post processing steps included (1) interpolation of bad channels, (2) high pass filter of 1 Hz, (3) average reference (4) epoching EEG data (−1 to 11.5 s around email onset) and (5) removal of epochs with movement artifacts using visual inspection. Next, the EEGLAB function of runica with the extended option was used to perform independent component analysis (ICA). The ICA components generated were reviewed with those components showing eye movement, single channel localization or a residual variance (RV) above 15% were deselected. For each independent component, event related spectral perturbations (ERSPs) were computed using a Morlet wavelet and baselined from −1 to 0 s across a 3–50 Hz frequency range. The independent components from each participant were then clustered based off of similar dipole location and ERSP responses. Note that, because independent components reflect individual differences in brain neurophysiology not all subjects will have the same components. Thus, analyses of components typically have fewer subjects than the entire group of 24. Besides the benefit of measuring detailed neuropysiological activity, such sensitivity to processing in individual brains may prove to be important in understanding individual differences in susceptibility to phishing and tailoring countermeasures to the individual. Finally, the data was separated into four different categories based on email type and participant response. These categories were Phish Correct (Hit), Phish Incorrect (Miss), Spam Correct (Correct rejection) and Spam Incorrect (False Alarm). The terms used to categorize choices (hit, miss, correct rejection, false alarm) are by convention from signal detection theory [9].

3.5 Statistics

Response accuracy was categorized as hits and misses for phish emails, and correct rejections and false alarms for spam. Reaction Time and ICA components were analyzed using a repeated measure analysis of variance with factors of email type (spam, phish) and classification (correct, incorrect). Only neuronal signals that can identify when a participant incorrectly categorizes a phish email as spam will be presented.

4 Results

Both a right inferior frontal and central parietal cluster showed specificity by identifying when a participant incorrectly categorized a phish email as spam (phish incorrect). Red dotted lines represent email onset and offset.

4.1 Behavioral Data

Participants correctly classified 69 ± 4% of phish emails, with the 31% balance being misses. Participants correctly classified 63 ± 4% of spam emails, with 37% being false alarms. Reaction times were examined using a 2 (email type) × 2 (classification) repeated measure MANOVA. There were main effects of email type ($F_{(1,23)} = 27.5$, $p < 0.0001$) and classification ($F_{(1,23)} = 13.77$, $p = 0.001$). Participants were slower when responding to phish versus spam emails (phish = 4.4 ± 0.2 s, spam = 3.8 ± 0.2 s) and when incorrectly classifying email type (correct = 3.9 ± 0.2 s, incorrect = 4.3 ± 0.2 s). The email type × classification interaction was not significant ($p =$ 0.07).

4.2 ICA Components—Right Inferior Frontal Cluster

A right inferior frontal cluster from 10 subjects showed an email type × classification interaction, $F_{(1,9)} = 13.5$, $p = 0.005$. Partial Eta Squared showed an email type × classification value of 0.471 for this cluster. From 1.2 to 2.2 s after email onset upper theta/lower alpha power (6–9 Hz), power bandwidth description is based off NeuroIS guideline paper [30],was greater only when participants incorrectly categorized a phish email as spam (phish incorrect, Fig. 1). This is a candidate unconscious threat signal, which is present to phish emails even though the subject misclassified the email as spam.

4.3 ICA Components—Central Parietal Cluster

A central parietal cluster from 12 subjects had a long lasting email type × classification interaction, $F_{(1,11)} = 9.78$, $p = 0.010$. Partial Eta Squared showed an email type × classification value of 0.601 for this cluster. From 0 to 6 s after email onset, the central parietal cluster showed decreased alpha power (8–13 Hz) only when participants incorrectly classified a phish email as spam (phish incorrect). There was a similar, but smaller, decrease in alpha for correct identification of spam. Thus the alpha desynchronization correlates to spam judgments, and may show threat-related activity when a spam judgment will be made to a phish email.

5 Conclusion

Phishing attacks cause global losses to individuals, organizations and economies that exceed a trillion dollars. Phishing training seldom takes into account the individual's susceptibility to phishing emails, i.e. the reasons why people fall for phishing bait, what makes them susceptible to phishing emails, or how they process deceptive messages. While many studies have examined phishing susceptibility focusing on textual and user characteristics, fewer studies have examined it in conjunction with neural measures, which can give insight into how email information is processed by the individual.

This paper shows that neuronal responses within right frontal and posterior parietal areas can identify when a participant incorrectly categorizes a phish email as benign spam. This misidentification of threat potentially causes the victim to act in favor of the attacker. Differences in lower and upper alpha synchrony appeared shortly after email onset and a full 2 s before the average reaction time. Our results agree with previous research that shows frontal-parietal networks are critical nodes for adaptive decision-making and performance monitoring, which includes error detection. Information accumulates within these frontal-posterior networks until neuronal activity reaches a threshold and triggers a behavioral response [31, 32]. The neuronal responses observed within this study may be related to threat detection signals, which do not reach a threshold sufficient to influence behavior but can still be observed in the EEG signal. These results contribute to the literature on phishing susceptibility by developing models for investigating cognitive responses and correlated brain responses along with textual and contextual pieces in the phishing context. The study also expands our understanding of why people fall prey to phishing as well as how the neural measures are able to help defend against phishing, and expand understanding of the role of neural measures in new directions.

From a practical standpoint, our study suggests that organizations could design anti-phishing strategies by understanding individual neural processing that points to his/her phishing susceptibility [33]. It also suggests that future email countermeasures should not only be developed from a technical perspective, but also be able to consider neural processes. Novel wearable biosensor technologies can enable unobtrusive direct, as well as indirect, measures of neural activity that could be used to tap-into threat signals that do not reach a threshold needed to influence behavior. Users can be trained to set aside any suspicious or questionable emails that have triggered unconscious response for further investigation by an IT team. Such a warning system would capitalize on the massively parallel processing capabilities of the human brain, of which only a tiny fraction is manifest in conscious behavior.

Acknowledgements We would like to thank the anonymous reviewers for their comments that have significantly improved the paper. We would also like to thank Sumanpreet Kaur for her help in conducting experiments. This research has been funded in part by NSF 1724725 and 1651475. The usual disclaimer applies.

References

1. Valecha, R., Chen, R., Herath, T., Vishwanath, A., Wang, J., & Rao, R. (2015). An exploration of phishing information sharing: A heuristic-systematic approach. In *Proceedings of Workshop on Information Security and Privacy*. Fort Worth, TX.
2. Wang, J., Xiao, N., & Rao, H. R. (2015). An exploration of risk characteristics of information security threats and related public information search behavior. Information Systems Research.
3. Kaspersky. (2013). http://media.kaspersky.com/pdf/kaspersky_lab_ksn_report_the_evolution_of_phishing_attacks_2011-2013.pdf.
4. Schwatz, M. (2016). *FBI: Global business email compromise losses hit $12.5 billion*. Bank Info Security. Retrieved from https://www.bankinfosecurity.com/fbi-alert-reported-ceo-fraud-losses-hit-125-billion-a-11206.
5. Kumaraguru, P., Rhee, Y., Acquisti, A., Cranor, L. F., Hong, J., & Nunge, E. (2007). Protecting people from phishing: the design and evaluation of an embedded training email system. In *Proceedings of the SIGCHI Conference on Human Factors in Computing Systems* (pp. 905–914). ACM.
6. Wang, J., Li, Y., & Rao, H. R. (2017). Coping responses in phishing detection: an investigation of antecedents and consequences. *Information Systems Research, 28*(2), 378–396.
7. Valecha, R., Chen, R., Herath, T., Vishwanath, A., Wang, J., & Rao, H. R. (2016). Reward-based and risk-based persuasion in phishing emails. In *Dewald Roode Workshop on Information Systems Security Research IFIP*. Albuquerque, NM.
8. Valecha, R., Chen, R., Herath, T., Vishwanath, A., Wang, J., & Rao, H. R. (2017). *A Multi-level model of phishing email detection*. In *Dewald Roode Workshop on Information Systems Security Research IFIP*. Tampa, FL.
9. Green, D. M., & Sweets, J. A. (1966). *Signal detection theory and psychophysics* (Vol. 1). New York: Wiley.
10. Vishwanath, A., Harrison, B., & Ng, Y. J. (2016). Suspicion, cognition, and automaticity model of phishing susceptibility. Communication Research.
11. Wang, J., Herath, T., Chen, R., Vishwanath, A., & Rao, H. R. (2012). Research article phishing susceptibility: An investigation into the processing of a targeted spear phishing email. *IEEE Transactions on Professional Communication, 55*(4), 345–362.
12. Wang, J., Li, Y., & Rao, H. R. (2016). Overconfidence in phishing email detection. *Journal of the Association for Information Systems, 17*(11), 759.
13. Abbasi, A., Zahedi, F. M., Zeng, D., Chen, Y., Chen, H., & Nunamaker, J. F., Jr. (2015). Enhancing predictive analytics for anti-phishing by exploiting website genre information. *Journal of Management Information Systems, 31*(4), 109–157.

14. Wang, J., Chen, R., Herath, T., & Rao, H. R. (2009). An exploration of the design features of phishing attacks. *Information Assurance, Security and Privacy Services, 4,* 29.
15. Flores, W. R., Holm, H., Nohlberg, M., & Ekstedt, M. (2015). Investigating personal determinants of phishing and the effect of national culture. *Information & Computer Security, 23*(2), 178–199.
16. Kumaraguru, P., Cranshaw, J., Acquisti, A., Cranor, L., Hong, J., Blair, M. A., & Pham, T. (2009). School of phish: a real-world evaluation of anti-phishing training. In *Proceedings of the 5th Symposium on Usable Privacy and Security* (p. 3). ACM.
17. Sheng, S., Holbrook, M., Kumaraguru, P., Cranor, L. F., & Downs, J. (2010). Who falls for phish? A demographic analysis of phishing susceptibility and effectiveness of interventions. In *Proceedings of the 28th International Conference on Human factors in Computing Systems* (pp. 1–10).
18. Iuga, C., Nurse, J. R., & Erola, A. (2016). Baiting the hook: Factors impacting susceptibility to phishing attacks. *Human-centric Computing and Information Sciences, 6*(1), 8.
19. Butavicius, M., Parsons, K., Pattinson, M., McCormac, A., Calic, D., & Lillie, M. (2017). Understanding susceptibility to phishing emails: assessing the impact of individual differences and culture. In *Proceedings of the Eleventh International Symposium on Human Aspects of Information Security and Assurance* (pp. 12–23).
20. Williams, E. J., Hinds, J., & Joinson, A. N. (2018). Exploring susceptibility to phishing in the workplace. *International Journal of Human-Computer Studies, 120,* 1–13.
21. Parrish, J. L., Bailey, J. L., & Courtney, J. F. (2009). *A personality based model for determining susceptibility to phishing attacks*. Little Rock: University of Arkansas (pp. 285–296).
22. Neupane, A., Saxena, N., Maximo, J. O., & Kana, R. (2016). Neural Markers of Cybersecurity: An fMRI Study of Phishing and Malware Warnings. *IEEE Transactions on Information Forensics and Security, 11*(9), 1970–1983.
23. Neupane, A., Saxena, N., & Hirshfield, L. (2017). Neural Underpinnings of website legitimacy and familiarity detection: An fNIRS study. In *Proceedings of the 26th International Conference on World Wide Web* (pp. 1571–1580). International World Wide Web Conferences Steering Committee.
24. Downs, J. S., Holbrook, M. B., & Cranor, L. F. (2006). Decision strategies and susceptibility to phishing. In *Proceedings of the 2nd Symposium on Usable Privacy and Security* (pp. 79–90). Pittsburg, PA: ACM Press.
25. Vishwanath, A., Herath, T., Chen, R., Wang, J., & Rao, H. R. (2011). Why do people get phished? Testing individual differences in phishing vulnerability within an integrated, information processing model. *Decision Support Systems, 51*(3), 576–586.
26. Anandpara, V., Dingman, A., Jakobsson, M., & Liu, D. (2007). Phishing IQ tests measure fear, not ability. In *Proceedings of the 11th International Conference on Financial cryptography* (pp. 362–366).
27. Dhamija, R., Tygar, J. D., & Hearst, M. (2006). Why phishing works. In *Proceedings of the SIGCHI conference on Human Factors in computing systems* (pp. 581–590). ACM.
28. Furnell, S. (2007). Phishing: Can we spot the signs? *Computer Fraud and Security, 2007*(3), 10–15.
29. Adware. (2016). The Big Three Email Nuisances: Spam, Phishing and Spoofing. Adware. Retrieved from https://www.adaware.com/blog/the-big-three-email-nuisances-spam-phishing-and-spoofing.
30. Müller-Putz, G. R., Riedl, R., & Wriessnegger, S. C. (2015). Electroencephalography (EEG) as a research tool in the information systems discipline: Foundations, measurement, and applications. *Communications of the Association for Information Systems, 37*(46), 911–948.
31. Navarro-Cebrian, A., Knight, R. T., & Kayser, A. S. (2016). Frontal monitoring and parietal evidence: Mechanisms of error correction. *Journal of Cognitive Neuroscience, 28*(8), 1166–1177.
32. Gold, J. I., & Shadlen, M. N. (2007). The neural basis of decision making. *Annual Review of Neuroscience, 30,* 535–574.
33. Valecha, R., Chen, R., Herath, T., Vishwanath, A., Wang, J., & Rao, H. R. (2015). An exploration of language acts of persuasion in phishing emails. In *Dewald Roode Workshop on Information Systems Security Research IFIP*. Newark, DE.

Affective Information Processing of Fake News: Evidence from NeuroIS

Bernhard Lutz, Marc T. P. Adam, Stefan Feuerriegel, Nicolas Pröllochs and Dirk Neumann

Abstract False information such as "fake news" threatens the credibility of social media and is widely believed to affect public opinion. So far, IS literature lacks a theoretical foundation on what leads humans to classify a news item as fake. In order to shed light on this question, we performed an experiment that involved 42 subjects with both eye tracking and heart rate measurements. We find that a lower heart rate variability and a higher relative number of eye fixations per second are associated with a higher probability of fake classification. Our study contributes to IS theory by providing evidence that the decision, if a news item is real or fake, is not purely cognitive, but also involves affective information processing. Thereby, it points towards novel strategies for identifying and preventing the spread of fake news in social media.

Keywords Affective information processing · Fake news · NeuroIS

B. Lutz (✉) · D. Neumann
University of Freiburg, Freiburg, Germany
e-mail: bernhard.lutz@is.uni-freiburg.de

D. Neumann
e-mail: dirk.neumann@is.uni-freiburg.de

M. T. P. Adam
University of Newcastle, Newcastle, Australia
e-mail: marc.adam@newcastle.edu.au

S. Feuerriegel
ETH Zurich, Zurich, Switzerland
e-mail: sfeuerriegel@ethz.ch

N. Pröllochs
University of Oxford, Oxford, UK
e-mail: nicolas.prollochs@eng.ox.ac.uk

F. D. Davis et al. (eds.), *Information Systems and Neuroscience*,
Lecture Notes in Information Systems and Organisation 32,
https://doi.org/10.1007/978-3-030-28144-1_13

1 Introduction

Social media has recently witnessed a surge in misinformation, a phenomenon often referred to as "fake news" [1]. Fake news, if deliberate, threatens the reputation of individuals and can jeopardize the functioning of our political and economic system [2]. There are numerous examples of fake news, such as the *Pizzagate* conspiracy theory[1] prior to the 2016 U.S. election. In fact, it was estimated for the same election that the average U.S. citizen consumed and remembered 1.14 fake news that might have impeded an informed decision-making [1]. Today, almost 62% of adults read news on social media (primarily on Facebook), and the proportion is increasing [3]. A particular caveat of social media is that it is often challenging for users to discriminate trustworthy content from unreliable stories [4], since anyone can create and share news, and the posts spread quickly as others read and share them [5].

Earlier research has studied fake news in terms of the spreading pattern in social networks [6], presentation format [5], and repeated exposure [7]. However, it remains unclear how users arrive at a decision about whether or not to believe a news item. Specifically, research lacks a theoretical understanding regarding the involvement of affect in the decision making process. While cognitive processing yields information about the stimulus itself and its relationship to other stimuli, affective processing yields information about the relation of the stimulus to the individual [8]. In other words, cognitive processing codes *what* something is, while affective processing codes *how* something is [9]. Walla [9] refers to affective information processing as "neural activity coding for valence" (p. 147) which, once it crosses certain thresholds, leads to bodily responses. Feelings, on the other hand, are "conscious phenomena, but they are not cognitive, they are perceived bodily response" (p. 146).

In this work, we employ neurophysiological measurements to shed light on the role of affect in the human information processing of fake news [10–12]. In particular, we conducted a NeuroIS experiment involving 42 subjects that were presented 40 different news items. For each news item, subjects are first asked to state their initial belief regarding the veracity of a news item solely based on its headline. Subsequently, the news body is shown and the subjects must classify the news item as real or fake. During the experiment, we measured subjects' eye fixations and heart rate to provide insight into their affective information processing.

Our empirical evidence suggests that the initial belief towards the headline significantly explains the subsequent credibility assessment of the news body. In addition, we find that both a lower heart rate variability (HRV), as an indicator of cognitive dissonance, and a higher relative number of eye fixations per second while processing the news body, as an indicator for scrutinizing the text, are associated with a higher probability of classifying a news item as fake.

To the best of our knowledge, we present the first NeuroIS study that investigates human information processing of real and fake news. In this vein, we provide insight

[1]This story claimed that leading politicians of the Democratic Party, including Hillary Clinton, are running a child sex ring whose headquarters are located in the basement of a pizza restaurant in Washington, DC.

into the interplay of cognition and affect when users process real and fake news on social media. Complementary to existing research that highlights the importance of cognitive processes, our study contributes to IS theory by providing evidence that the decision—if a news item is real or fake—is not purely cognitive, but also involves affective information processing. Thereby, it points towards novel strategies for identifying and preventing the spread of fake news in social media.

2 Hypotheses Development

News items commonly follow a structure where a brief news headline is followed by a more comprehensive news body [13, 14]. By providing a succinct synopsis of the news item, the headline allows users to acquire a quick overview in terms of what the news item is about. However, in terms of human information processing, it is also vital to consider to what extent news headline may instigate initial beliefs about the veracity of the news item. Emphasizing the importance of this potential process, Pennycook and Rand [13] recently called for more research into "the consequences of reading news headlines for subsequent beliefs and behaviors" (p. 10). This implies that the news headline may not only serve as a brief indicator for what the news item is about but instead instigate an initial belief about its veracity. If this was to be true, the way humans process information may assign the headline a pivotal role in news classification and we should expect a positive association between the initial belief and the probability of classifying a news item as fake. Following this line of reasoning, H1 states:

Hypothesis 1 (H1). *A stronger initial belief that a news item is fake is associated with a higher probability of classifying a news item as fake.*

After processing the headline, users need to assess whether the content of the news body is in line with their own beliefs about the real world based on their prior experience. This is vital to come to a decision about whether or not to believe the content of a news item. After all, in order to make sense of an incoming news item, users need to draw connections to any relevant beliefs they hold about the categories pertinent to the news content and the relationships between them [13]. Importantly, as we elaborate in the following, there is reason to believe that this process does not only involve processes of cognitive reasoning but also affective information processing [15].

Humans experience psychological discomfort when processing materials that contradict their own beliefs, an aversive user state that is commonly referred to as cognitive dissonance [16]. Cognitive dissonance theory builds on the psychological principle that the human body continuously processes external stimuli in order to match the entire system to environmental demands. Thereby, the human body prepares a regulated response to these external stimuli through an antagonistic interaction of two autonomic systems, the excitatory sympathetic nervous system (fight or flight) and the inhibitory parasympathetic nervous system (recreation and relaxation, vagal activity) [17]. When experiencing cognitive dissonance, the balance between these two antagonistic systems, also known as sympathovagal balance, is temporarily shifted to a state of a more dominant role of the sympathetic branch.

A shift in the balance between the sympathetic and the parasympathetic nervous systems is associated with changes in HRV, that is, changes in the standard deviation of the time intervals between subsequent heart beats (or *inter-beat intervals*). In particular, states that are associated with discomfort yield *lower* HRV [17–19]. Congruently, the psychological discomfort experienced during cognitive dissonance is accompanied by lower HRV [19]. Contextualized to user classification of a news item, cognitive dissonance may arise when the content of a news item is in conflict with the beliefs of the user [20, 21]. Once this state emerges, there appear only two ways for the user to resolve the dissonance.

The user can either change their beliefs or, more likely, classify the news item as fake. Based on this reasoning, we hypothesize:

Hypothesis 2 (H2)*. A lower heart rate variability while processing the news body is associated with a higher probability of classifying a news item as fake.*

As with any written text, the processing of the news body requires visual processing of the word sequences that contain its content. It is commonly agreed that the visual and cognitive processing of this content mainly occurs during eye fixations on the respective visual elements [22]. Thereby, an eye fixation is defined as pauses over informative regions of interest [23]. In this sense, a higher number of eye fixations on the visual stimuli that carry the content is associated with a higher involvement of the user with that content [24, 25]. Conversely, a lower number of eye fixations is associated with withdrawal and avoidance of information processing.

There has been extensive research on accurately measuring eye fixations with high-definition eye tracking technology, and using these measurements to further our understanding of how humans process information [26]. Research on text comprehension suggests that a higher number of fixations and a higher number of fixations per second is linked to unexpected text segments [27–29]. In other words, a higher number of fixations are an indication that the text contains information that is in contrast to what the reader would have expected to read based on the prior beliefs that they hold. Contextualized to the classification of fake news, we can expect users to search for facts or features that are characteristic for fake news, particularly if the user forms a belief that the news item is fake. In other words, in the process of reading content that is not in line with their own worldview, we should expect a stronger scrutiny of the news body because the user has identified information elements that appear inaccurate. This scrutiny may be expressed in two different ways: a higher total number of eye fixations while processing the news body and/or a higher relative number of eye fixations per second.

Hypothesis 3a (H3a)*. A higher total number of eye fixations while processing the news body is associated with a higher probability of classifying a news item as fake.*

Hypothesis 3b (H3b)*. A higher relative number of eye fixations per second while processing the news body is associated with a higher probability of classifying a news item as fake.*

3 Method

In our experiment, each subject was shown the same set of 40 news items in random order (within-subject design). All news items were shown in a generic format using Brownie [30, 31]. The label of each news item was retrieved from the fact-checking website *politifact.com*, which assigns veracity labels to general political statements and media articles, ranging from "true" for real news items to "pants on fire" for outrageous fakes [32]. Our dataset is balanced, such that the subset of real news contains 20 news items that were labeled as true, and the subset of fake news contains three news items labeled as false and 17 news items labeled as pants on fire. Each news item is presented in two steps. In the first step, the subject only sees the headline and, based on this, rates the veracity of the new item (from (1) strongly real to (7) strongly fake). In the second step, the subject reads the news body and classifies the news item as real or fake.

Throughout the experiment, we measured electrocardiography (ECG) data using BioSignalsPlux and eye tracking gaze using Tobi Pro X3-120. Before the start of the experiment, the electrodes for the ECG measurements were attached and the eye tracking device was calibrated. Then, a trial news item was shown in order to make the subject familiar with our experimental software. Afterwards, the experiment was paused for five minutes to measure a baseline heart rate level. As such, our experiment included two neurophysiological measures.

Firstly, to determine HRV, we first performed an automatic identification of heart beats in the raw ECG signal which we later corrected with manual inspection. We then computed a sequence of inter-beat intervals based on the timestamps of heart-beats and calculated HRV (standard deviation of interbeat-intervals) using *cmetx* [33]. Secondly, to determine eye fixations, we create an index that maps relative coordinates on the screen to a word identifier. Subsequently, we use this mapping and the raw eye gaze data in order to calculate the sequence of fixated words. We define an eye fixation as focusing the same rectangle for at least 100 ms [23].

During the whole experiment, the subject is not bound to any time constraints and does not receive feedback whether a news item is actually fake or real. In addition, our subjects are not aware of the distribution between real and fake news. Our experiment is fully incentivized, i.e., subjects earn €0.50 (approx. USD 0.60) for each correct classification. A total of 42 persons participated in our study; however, we discarded two subjects and two observations, where the sensor measurements failed. The final dataset thus consists of 1598 observations from 40 subjects (all college students, 26 male, 14 female, mean age of 26.0 years).

4 Results and Discussion

We employed a mixed effects logistic regression model with subjects' news classification as the dependent variable (real = 0, fake = 1), controlling for the treatment (0 = real, 1 = fake), the news sequence number (0–39), news readability, and a subject-specific random intercept. The key variables in our model are the initial belief, HRV, and the number of total and relative eye fixations (per second).

Supporting H1, our analysis shows a significant effect of initial belief on news classification ($\beta = 0.562$, $p < 0.001$). A one unit increase in the initial belief (that a news item is fake) increases the odds of a fake classification by 75.4% ($e^{0.562} \approx 1.754$). Further, and in support for H2, we find a significant negative effect of HRV on fake classification ($\beta = -0.006, p < 0.05$). This suggests that a one unit decrease in HRV (i.e., 1 ms) while processing the news body increases the odds of a fake classification by 0.6%. Finally, we examine eye fixations. While the coefficient of the total number of eye fixations is not significant (H3a rejected), the coefficient of the relative number of eye fixations is positive and significant ($\beta = 0.393, p < 0.05$). This suggests that a one unit increase in the relative number of eye fixations per second increases the odds of a fake classification by 48.1% (H3b supported).

Despite the fact that we chose a rather general selection of news stories that have appeared on social media, our task provided considerable challenges for our subjects in determining the veracity of each item. Overall, the mean accuracy in correctly classifying veracity amounts to a mere 73.46%. This reveals the difficulty—even for college students in our study—in distinguishing real news from fake news and thus confirms the relevance of our research.

Importantly, our study provides first neurophysiological evidence for a pronounced role of affective information processing in determining whether a news item is real or fake. In particular, we find that the process of classifying news items as real or fake is not purely cognitive, but also involves affective information processing. The findings remain robust against various checks and when performing sensitivity analysis. As direct implications, our work suggests novel directions for recognizing and preventing the spread of fake news that recognize the role of affective processing. Recent advances in affective computing (e.g., deep learning for human affect recognition [34]) may contribute to furthering our understanding of the role of affective information processing in human news classification, and the development of systems that recognize and prevent the spread of fake news.

References

1. Allcott, H., & Gentzkow, M. (2017). Social media and fake news in the 2016 election. *Journal of Economic Perspectives, 31*(2), 211–236.
2. Wattal, S., Schuff, D., Mandviwalla, M., & Williams, C. B. (2010). Web 2.0 and politics: The 2008 US presidential election and an e-politics research agenda. *MIS Quarterly, 34*(4), 669–688.
3. Gottfried, J., & Shearer, E. (2016). News Use Across Social Medial Platforms 2016. Pew Research Center.
4. Luca, M., & Zervas, G. (2016). Fake it till you make it: Reputation, competition, and yelp review fraud. *Management Science, 62*(12), 3412–3427.
5. Kim, A., & Dennis, A. (2018). Says who? How news presentation format influences perceived believability and the engagement level of social media users. In *Proceedings of the 51st Hawaii International Conference on System Sciences* (pp. 3955–3965).

6. Vosoughi, S., Roy, D., & Aral, S. (2018). The spread of true and false news online. *Science, 359*(6380), 1146–1151.
7. Pennycook, G., Cannon, T. D., & Rand, D. G. (2018). Prior exposure increases perceived accuracy of fake news. *Journal of Experimental Psychology: General, 147*(12), 1865–1880.
8. LeDoux, J. E. (1989). Cognitive-emotional interactions in the brain. *Cognition and Emotion, 3*(4), 267–289.
9. Walla, P. (2018). Affective processing guides behavior and emotions communicate feelings: Towards a guideline for the NeuroIS community. In *Information Systems and Neuroscience* (pp. 141–150). Berlin: Springer.
10. Dimoka, A., Davis, F. D., Gupta, A., Pavlou, P. A., Banker, R. D., Dennis, A. R., et al. (2012). On the use of neurophysiological tools in IS research: Developing a research agenda for NeuroIS. *MIS Quarterly, 36*(3), 679–702.
11. Riedl, R., Davis, F. D., & Hevner, A. R. (2014). Towards a neurois research methodology: Intensifying the discussion on methods, tools, and measurement. *Journal of the Association for Information Systems, 15*(10), 1–35.
12. vom Brocke, J., & Liang, T. P. (2014). Guidelines for neuroscience studies in information systems research. *Journal of Management Information Systems, 30*(4), 211–234.
13. Pennycook, G., & Rand, D. G. (2018). Lazy, not biased: Susceptibility to partisan fake news is better explained by lack of reasoning than by motivated reasoning. *Cognition* (in press).
14. Pennycook, G., & Rand, D. G. (2019). Who falls for fake news? The roles of analytic thinking, motivated reasoning, political ideology, and bullshit receptivity. *Journal of Personality* (forthcoming).
15. Walla, P., & Panksepp, J. (2013). Neuroimaging helps to clarify brain affective processing without necessarily clarifying emotions. In K. N. Fountas (Ed.), *Novel Frontiers of Advanced Neuroimaging* (pp. 93–118). IntechOpen.
16. Festinger, L. (1962). *A theory of cognitive dissonance* (Vol. 2). Stanford University Press.
17. Appelhans, B. M., & Luecken, L. J. (2006). Heart rate variability as an index of regulated emotional responding. *Review of General Psychology, 10*(3), 229–240.
18. Eckberg, D. (1997). Sympathovagal balance: A critical appraisal. *Circulation, 96*(9), 3224–3232.
19. Pumprla, J., Howorka, K., Groves, D., Chester, M., & Nolan, J. (2002). Functional assessment of heart rate variability: Physiological basis and practical applications. *International Journal of Cardiology, 84*(1), 1–14.
20. Berghel, H. (2017). Lies, damn lies, and fake news. *Computer, 50*(2), 80–85.
21. Lazer, D. M. J., Baum, M. A., Benkler, Y., Berinsky, A. J., Greenhill, K. M., Menczer, F., et al. (2018). The science of fake news. *Science, 359*(6380), 1094–1096.
22. Just, M. A., & Carpenter, P. A. (1984). Using eye fixations to study reading comprehension. *New Methods in Reading Comprehension Research,* 151–182.
23. Salvucci, D. D., Goldberg, J. H. (2000). Identifying fixations and saccades in eye-tracking protocols. In *Proceedings of the 2000 Symposium on Eye Tracking Research and Applications* (pp. 71–78).
24. Loftus, G. R. (1981). Tachistoscopic simulations of eye fixations on pictures. *Journal of Experimental Psychology: Human Learning and Memory, 7*(5), 369–376.
25. Poole, A., & Ball, L. J. (2006). Eye tracking in HCI and usability research. In *Encyclopedia of Human Computer Interaction* (pp. 211–219).
26. Léger, P. M., Sénecal, S., Courtemanche, F., de Guinea, A. O., Titah, R., Fredette, M., & Labonte-LeMoyne, É. (2014). Precision is in the eye of the beholder: Application of eye fixation-related potentials to information systems research. *Journal of the Association for Information Systems, 15*(10), 651–678.
27. Carpenter, P. A., & Just, M. A. (1977). Reading comprehension as eyes see it. In *Cognitive processes in comprehension* (pp. 109–139). New York, NY: Psychology Press.
28. Ooms, K., de Maeyer, P., & Fack, V. (2014). Study of the attentive behavior of novice and expert map users using eye tracking. *Cartography and Geographic Information Science, 41*(1), 37–54.

29. Al-Moteri, M. O., Symmons, M., Plummer, V., & Cooper, S. (2017). Eye tracking to investigate cue processing in medical decision-making: A scoping review. *Computers in Human Behavior, 66,* 52–66.
30. Hariharan, A., Adam, M. T. P., Lux, E., Pfeiffer, J., Dorner, V., Müller, M. B., & Weinhardt, C. (2017). Brownie: A platform for conducting NeuroIS experiments. *Journal of the Association for Information Systems, 18*(4), 264–296.
31. Jung, D., Adam, M. T. P., Dorner, V., & Hariharan, A. (2017). A practical guide for human lab experiments in information systems research: A tutorial with brownie. *Journal of Systems and Information Technology, 19*(3/4), 228–256.
32. Graves, L. (2016). Boundaries not drawn: Mapping the institutional roots of the global fact-checking movement. *Journalism Studies, 19*(5), 613–631.
33. Allen, J. J. B., Chambers, A. S., & Towers, D. N. (2007). The many metrics of cardiac chronotropy: A pragmatic primer and a brief comparison of metrics. *Biological Psychology, 74*(2), 243–262.
34. Rouast, P. V., Adam, M. T. P., Chiong, R. (2019). Deep learning for human affect recognition: Insights and new developments. *IEEE Transactions on Affective Computing,* 1–20.

What Can NeuroIS Learn from the Replication Crisis in Psychological Science?

Colin Conrad and Lyam Bailey

Abstract The Reproducibility Crisis is a phenomenon that has gained considerable attention in the psychological sciences. Scholars in these fields have found that many high profile findings are either difficult to reproduce or could not be replicated. These findings have ultimately encouraged researchers to adopt pre-registered results, replication in study design and open data. As an emerging field, NeuroIS has an opportunity to learn from this crisis and adopt new practices based on the lessons learned in the psychological sciences. We explored the current state of NeuroIS research from the perspective of reproducibility by conducting a survey of the extant NeuroIS literature. We conclude by suggesting two practices that the NeuroIS community can undertake to help address the replication problem.

Keywords NeuroIS · Replication · Research methods · Construct validity

1 Introduction

Reproducibility is often regarded to be one of the defining characteristics of hypothesis-driven science. When empirical observations are reproduced through experimentation, observers gain evidence for the relationships between observed phenomena. Additional empirical observations decrease the likelihood of a false positive or false negative result, allowing scientists to be more confident about the relationship being observed. It is through this process that scientists often induce causal relationships [1]. The managerial sciences are no exception; the logical validity of our work as Information Systems scholars is improved by its reproducibility. Unfortunately, the benefits of replication are often overshadowed by motivations to publish novel, high-impact work. Many scholars will therefore forego replication of a previously observed phenomenon in favor of new, more exciting research. We posit

C. Conrad (✉) · L. Bailey
Dalhousie University, Halifax, Canada
e-mail: colin.conrad@dal.ca

L. Bailey
e-mail: lyam.bailey@dal.ca

F. D. Davis et al. (eds.), *Information Systems and Neuroscience*,
Lecture Notes in Information Systems and Organisation 32,
https://doi.org/10.1007/978-3-030-28144-1_14

that this may become a non-trivial concern for research in NeuroIS. As an emergent field, NeuroIS holds the potential for large or surprising gains in understanding of information systems and information technology phenomena. Scholars in this field, therefore, would be well-served to ensure that any such gains in understanding are replicable from the outset.

To tangibly demonstrate the importance of this issue, and why it should not be left unchecked, we point to the ongoing Replication Crisis in psychological science. In August 2015, The Open Science Collaboration published a research article demonstrating that out of a selection of 100 experiments published in prominent psychological journals, 97 of which originally yielded significant effects, only 35 of the original observations (i.e. less than half) were replicable [2]. This paper, among others, sparked a wider conversation about reproducibility and openness in science broadly [3]. Fortunately, scholarly responses to these concerns have been largely positive. For example, one of the largest barriers to replication-oriented research is publication biases favoring novel and significant results. In simple terms, a replication experiment is considerably less likely to be published if it fails to reproduce a previously observed finding (i.e. obtains a null result) [4, 5]. Many psychological journals have taken steps to combat this issue by endorsing so called "pre-registered reports"—dedicated article types for which the decision to accept or reject is made prior to data collection and is based entirely on the merits of the proposed methods [6]. In this vein, some journals even endorse pre-registered *replication* articles, which carry the benefits of 'regular' pre-registered reports, but which specifically foster replication-oriented research. Moreover, recent years have seen movements towards greater transparency in science. Most notably, the recently established Open Science Framework provides an online platform for researchers to pre-register experimental designs and hypotheses, and to store and share their data [7].

The field of neuroimaging faces its own unique concerns surrounding replicability [8, 9]. Neuroimaging experiments require large financial investment (owing to the costs of equipment maintenance and technical staff), meaning that the feasibility of replication is constrained by available funds. Moreover, neuroimaging studies often make large numbers of comparisons. Despite these issues, replication in neuroimaging is by no means a lost cause. The recent emergence of publicly accessible neuroimaging datasets such as the Human Connectome Project [10] the Cambridge Centre for Ageing and Neuroscience database [11], and the Alzheimer's Disease Neuroimaging Initiative [12] has allowed researchers to explore new lines of research *and* explore the replicability of old ones without incurring prohibitive financial costs.

We posit that the steps taken in psychological science to counter the replicability crisis may also be applied to the emergent field of NeuroIS. NeuroIS researchers have recognized the need to adopt best practices and use multiple methods to understand the relationship between neurophysiological observations and IS constructs [13, 14]. However, to the best of our knowledge there has been limited discussion about reproducibility in NeuroIS specifically. Though there has been some recognition of the importance of replication in the broader Information Systems community [15], we believe that a conversation about the value of replication in NeuroIS research is warranted. In this paper, we briefly explore the extent that replication has been incorporated in past NeuroIS research by observing the methods employed by published NeuroIS papers originally described by Riedl et al. [16]. After assessing the findings of the review, we conclude by outlining two potential practices that can be adopted in future NeuroIS research.

2 Replication and NeuroIS

In order to assess how NeuroIS researchers have incorporated replication in the past, we manually reviewed the 164 papers specified by Riedl et al. [16]. We opted to observe the previously reported works because, to the best of our knowledge, this work represents the only comprehensive literature review on the NeuroIS discipline and extending a comprehensive literature review is outside of the scope of this short paper. Given that new NeuroIS research involves either novel measures or novel applications, we opted to investigate studies that incorporated elements of replication in their research design. We identified papers based on whether they conducted empirical studies, which methods were used, whether they incorporated multiple experiments in their research, and whether the authors reported replicating their neurophysiological findings. Our analysis similarly found 103 empirical NeuroIS papers, summarized in Table 1.[1]

Though we identified 20 studies which incorporated multiple experiments, 9 of the studies described multiple experiments using different methods (e.g. neurophysiological, self-report) in an effort to triangulate findings with prior IS constructs [17–25]. We identified 10 studies which employed multiple experiments and reported successfully replicating observed results through a later experiment [26–36]. We also observed 3 papers which employed a single study but directly replicated an experiment previously reported in another paper [37–39]. Furthermore, we observed a trend that research published in the AIS Senior Scholars Basket [40] or highly recognized conferences were more likely to report multiple experiments or replicate NeuroIS phenomena. Table 2 summarizes findings from the AIS Senior Scholar's Basket, a widely recognized basket of quality journals in the IS discipline. We observe that studies published in the basket are more likely to report multiple experiments (though not necessarily multiple neurophysiological experiments).

Does NeuroIS have a reproducibility problem?

The results of our investigation suggest that some NeuroIS researchers have taken steps to include multiple experiments or replication in their designs when appropriate, especially when publishing in highly recognized venues. Notably, we did not observe any fMRI studies which replicate their neurophysiological findings, as the

[1] Discrepancies between the findings of this paper and Riedl et al. [16] can be attributed to differences in how authors interpreted studies as complete or empirical, and the subjective judgement employed by us when identifying a primary research method for each study.

Table 1 NeuroIS studies identified which reported multiple experiments or replication

Primary tool	Number	No. reporting multiple experiments	No. reporting measure replication
Eye tracking	46	7	5
EEG	22	5	5
fMRI	9	2	1
Other than above	26	5	2
Total	103	20	13

Table 2 Summary of empirical studies published in the AIS Senior Scholar's Basket reporting multiple experiments or replication

Primary tool	Number	No. reporting multiple experiments	No. reporting measure replication
Eye tracking	4	1	1
EEG	7	2	1
fMRI	6	2	0
Other than above	5	3	2
Total	22	8	4

one study identified replicated behavioural results [38]. This observation is likely attributable to high cost of running fMRI experiments—often around $500 USD per hour—alluded to earlier. Given that our investigation concerned publications published prior to 2017 however, we are led to conclude that NeuroIS has largely conformed to the standards in Neuroimaging, which similarly did not concern itself with replication until recently. NeuroIS would nonetheless benefit from similarly addressing the problem.

3 Recommendations for Future NeuroIS Research

Consider incorporating replication in study designs

We encourage NeuroIS researchers to test the replicability of their observed findings wherever possible. Such efforts might not necessarily take the form of a so-called "straight replication"—that is, repeating a study using precisely the same methods, on a new sample of participants. Replication may instead be achieved by further exploring an observed finding. For example, a research team might observe that a particular stimulus elicits a particular neural response, as measured with non-invasive neuroimaging. The team could then explore whether this neural response is modulated by certain experimental parameters, by the demographic/cognitive characteristics of participants, or is corroborated by psychometric measures (when possible).

Such a follow-up experiment would not only serve to replicate the first, but would provide a novel advancement in understanding, with respect to the neural response in question.

Such efforts might also be incentivized by prominent IS journals and conferences, which could consider special treatment for research articles aiming to test the replicability of previously observed findings. For example, having a dedicated article type for replication or for the acceptance of pre-registered reports, as is becoming more common practice in psychological science. Furthermore, NeuroIS researchers could consider disseminating replication experiments in the *AIS Transactions on Replication Research* [15] or similar venues dedicated to supporting replication.

Towards an open NeuroIS data repository

We also encourage NeuroIS researchers to explore collaborative data-sharing initiatives. In the field of Neuroimaging, publicly available repositories such as the Human Connectome Project [10], amongst others [11, 12] enable researchers to test the replicability of previously reported effects, often with considerably larger samples. NeuroIS may similarly benefit from the creation of a repository of publicly accessible experiment data or a publicly available dataset. Such a resource might be achieved by means of a dedicated project (as is the case for the databases cited previously), from multiple collaborators pooling their anonymized data collected for the purposes of ongoing research studies, or even simply by making past data accessible to the NeuroIS community through the web. Calls for open software and data have been made in the broader IS community [41]; in the NeuroIS context, such an undertaking has the further potential to foster sharing of best practices and provide a resource for training research students. The final result of such an undertaking would not merely be increased transparency of NeuroIS research, but also the wider dissemination of NeuroIS research to other academic communities.

References

1. Hempel, C. G. (1968). Maximal specificity and lawlikeness in probabilistic explanation. *Philosophy of Science, 35*(2), 116–133.
2. Open Science Collaboration. (2015). Estimating the reproducibility of psychological science. *Science, 349*(6251), aac4716.
3. Baker, M. (2016). Is There a Reproducibility Crisis? *Nature, 533.*
4. Francis, G. (2012). Too good to be true: Publication bias in two prominent studies from experimental psychology. *Psychonomic Bulletin & Review, 19*(2), 151–156.
5. Francis, G. (2012). Publication bias and the failure of replication in experimental psychology. *Psychonomic Bulletin & Review, 19*(6), 975–991.
6. Gonzales, J. E., & Cunningham, C. A. (2015). *The promise of pre-registration in psychological research.* https://www.apa.org/science/about/psa/2015/08/pre-registration.
7. Foster, E. D., & Deardorff, A. (2017). Open science framework (OSF). *Journal of the Medical Library Association, 105*(2), 203.
8. Luck, S. J., & Gaspelin, N. (2017). How to get statistically significant effects in any ERP experiment (and why you shouldn't). *Psychophysiology, 54,* 146–157.

9. Poldrack, R. A., Baker, C. I., Durnez, J., Gorgolewski, K. J., Matthews, P. M., Munafo, M. R., et al. (2017). Scanning the horizon: Towards transparent and reproducible neuroimaging research. *Nature Reviews, 18.*
10. Van Essen, D. C., Ugurbil, K., Auerbach, E., Barch, D., Behrens, T. E. J., Bucholz, R., et al. (2012). The Human Connectome Project: A data acquisition perspective. *Neuroimage, 62*(4), 2222–2231.
11. Taylor, J. R., Williams, N., Cusack, R., Auer, T., Shafto, M. A., Dixon, M., et al. (2017). The Cambridge Centre for Ageing and Neuroscience (CamCAN) data repository: Structural and functional MRI, MEG, and cognitive data from a cross-sectional adult lifespan sample. *NeuroImage, 144,* 262–269.
12. Mueller, S. G., Weiner, M. W., Thal, L. J., Petersen, R. C., Jack, C. R., Jagust, W., et al. (2005). Ways toward an early diagnosis in Alzheimer's disease: The Alzheimer's Disease Neuroimaging Initiative (ADNI). *Alzheimer's and Dementia, 1*(1), 55–66.
13. Dimoka, A., Davis, F. D., Gupta, A., Pavlou, P. A., Banker, R. D., Dennis, A. R., et al. (2012). On the use of neurophysiological tools in IS research: Developing a research agenda for NeuroIS. *MIS Quarterly, 36*(3), 679–702.
14. vom Brocke, J., & Liang, T.-P. (2014). Guidelines for neuroscience studies in information systems research. *Journal of Management Information Systems.*
15. Dennis, A. R., & Valacich, J. S. (2014). A replication manifesto. *AIS Transactions on Replication Research, 1,* 1–4.
16. Riedl, R., Fischer, T., & Léger, P.-M. (2017). A decade of NeuroIS research: Status quo, challenges, and future directions. In *Proceedings of the 38th International Conference on Information Systems*, Seoul.
17. Dimoka, A. (2010). What does the brain tell us about trust and distrust? Evidence from a functional neuroimaging study. *MIS Quarterly, 34*(2), 373–396.
18. De Guinea, A. O., & Webster, J. (2013). An investigation of information systems use patterns: Technological events as triggers, the effect of time, and consequences for performance. *MIS Quarterly, 38*(4), 1165–1188.
19. Goggins, S. P., Schmidt, M., Guajardo, J., & Moore, J. (2010). Assessing multiple perspectives in three dimensional virtual worlds: Eye tracking and all views qualitative analysis (AVQA). In *2010 43rd Hawaii International Conference on System Sciences* (pp. 1–10). IEEE.
20. Gregor, S., Lin, A. C., Gedeon, T., Riaz, A., & Zhu, D. (2014). Neuroscience and a nomological network for the understanding and assessment of emotions in information systems research. *Journal of Management Information Systems, 30*(4), 13–48.
21. Perrin, J. L., Paillé, D., & Baccino, T. (2014). Reading tilted: Does the use of tablets impact performance? An oculometric study. *Computers in Human Behavior, 39,* 339–345.
22. Clayton, R. B., Leshner, G., & Almond, A. (2015). The extended iSelf: The impact of iPhone separation on cognition, emotion, and physiology. *Journal of Computer-Mediated Communication, 20*(2), 119–135.
23. Cole, M. J., Hendahewa, C., Belkin, N. J., & Shah, C. (2015). User activity patterns during information search. *ACM Transactions on Information Systems (TOIS), 33*(1), 1–39.
24. Jenkins, J. L., Anderson, B. B., Vance, A., Kirwan, C. B., & Eargle, D. (2016). More harm than good? How messages that interrupt can make us vulnerable. *Information Systems Research, 27*(4), 880–896.
25. Luan, J., Yao, Z., Zhao, F., & Liu, H. (2016). Search product and experience product online reviews: An eye-tracking study on consumers' review search behavior. *Computers in Human Behavior, 65,* 420–430.
26. Bahr, G. S., & Ford, R. A. (2011). How and why pop-ups don't work: Pop-up prompted eye movements, user affect and decision making. *Computers in Human Behavior, 27*(2), 776–783.
27. Nunamaker, J. F., Derrick, D. C., Elkins, A. C., Burgoon, J. K., & Patton, M. W. (2011). Embodied conversational agent-based kiosk for automated interviewing. *Journal of Management Information Systems, 28*(1), 17–48.
28. Astor, P. J., Adam, M. T., Jerčić, P., Schaaff, K., & Weinhardt, C. (2013). Integrating biosignals into information systems: A NeuroIS tool for improving emotion regulation. *Journal of Management Information Systems, 30*(3), 247–278.

29. De Guinea, A. O., Titah, R., & Léger, P. M. (2013). Measure for measure: A two study multi-trait multi-method investigation of construct validity in IS research. *Computers in Human Behavior, 29*(3), 833–844.
30. Jay, C., Brown, A., & Harper, S. (2013). Predicting whether users view dynamic content on the world wide web. *ACM Transactions on Computer-Human Interaction (TOCHI), 20*(2), 1–33.
31. Dogusoy-Taylan, B., & Cagiltay, K. (2014). Cognitive analysis of experts' and novices' concept mapping processes: An eye tracking study. *Computers in Human Behavior, 36,* 82–93.
32. Molina, A. I., Redondo, M. A., Lacave, C., & Ortega, M. (2014). Assessing the effectiveness of new devices for accessing learning materials: An empirical analysis based on eye tracking and learner subjective perception. *Computers in Human Behavior, 31,* 475–490.
33. Wang, C. C., & Hsu, M. C. (2014). An exploratory study using inexpensive electroencephalography (EEG) to understand flow experience in computer-based instruction. *Information & Management, 51*(7), 912–923.
34. Galluch, P. S., Grover, V., & Thatcher, J. B. (2015). Interrupting the workplace: Examining stressors in an information technology context. *Journal of the Association for Information Systems, 16*(1), 1–47.
35. Huang, Y. F., Kuo, F. Y., Luu, P., Tucker, D., & Hsieh, P. J. (2015). Hedonic evaluation can be automatically performed: An electroencephalography study of website impression across two cultures. *Computers in Human Behavior, 49,* 138–146.
36. Hu, Q., West, R., & Smarandescu, L. (2015). The role of self-control in information security violations: Insights from a cognitive neuroscience perspective. *Journal of Management Information Systems, 31*(4), 6–48.
37. Adam, M. T., Krämer, J., & Weinhardt, C. (2012). Excitement up! Price down! Measuring emotions in Dutch auctions. *International Journal of Electronic Commerce, 17*(2), 7–40.
38. Riedl, R., Mohr, P. N., Kenning, P. H., Davis, F. D., & Heekeren, H. R. (2014). Trusting humans and avatars: A brain imaging study based on evolution theory. *Journal of Management Information Systems, 30*(4), 83–114.
39. Labonté-LeMoyne, É., Santhanam, R., Léger, P. M., Courtemanche, F., Fredette, M., & Sénécal, S. (2015). The delayed effect of treadmill desk usage on recall and attention. *Computers in Human Behavior, 46,* 1–5.
40. Association for Information Systems: Senior Scholars' Basket of Journals. https://aisnet.org/page/SeniorScholarBasket.
41. van der Aalst, W., Bichler, M., & Heinzl, A. (2016). Open research in business and information systems engineering. *Business & Information Systems Engineering, 58*(6), 375–379.

Techno-Unreliability: A Pilot Study in the Field

Thomas Kalischko, Thomas Fischer and René Riedl

Abstract We report on a pilot study, with the aim of investigating techno-unreliability in the field. Over the course of three days, we collected physiological data (heart rate) from four participants, which experienced manipulations that rendered a variety of work-related systems unusable. Although we found that it is feasible to also investigate technostress in this way in the field, more data is needed to interpret the captured physiological reactions. In particular, we need data that allows us to investigate stress appraisal, ideally at a sampling rate that is comparable to those that we get from capturing physiological data.

Keywords Unreliability · Technostress · Heart rate · Field experiment

1 Introduction and Study Background

Technostress has become an important research topic in the Information Systems (IS) discipline (for a recent review, see [1]), and many NeuroIS papers have been published with a focus on the physiological mechanisms underlying technostress [2–10]. Yet, in previous technostress research, self-report instruments have been the main means of data collection [1, 11] and particularly the technostress creators scale was used [1, 12, 13], which includes five main stressor categories (i.e., techno-overload, techno-invasion, techno-complexity, techno-uncertainty, techno-insecurity), but is missing one important stressor, namely techno-unreliability (e.g., [14, 15]).

T. Kalischko
Smarter Ecommerce GmbH, Linz, Austria
e-mail: thomas.kalischko@smarter-ecommerce.com

T. Fischer (✉) · R. Riedl
University of Applied Sciences Upper Austria, Steyr, Austria
e-mail: thomas.fischer@fh-steyr.at

R. Riedl
e-mail: rene.riedl@fh-steyr.at

R. Riedl
Johannes Kepler University, Linz, Austria

F. D. Davis et al. (eds.), *Information Systems and Neuroscience*,
Lecture Notes in Information Systems and Organisation 32,
https://doi.org/10.1007/978-3-030-28144-1_15

Unreliability of information and communication technologies (ICT) is a highly prevalent stressor in our modern society (e.g., [16, 17]) and it can take many forms. For example, as the stress-inducing effects of long and variable ICT response times [18, 19], or breakdowns of systems [7, 8], have been investigated. It can be argued that, to some extent, we have learned to live with and accept that some systems or even ICT in general may never be fully reliable [20]. Yet, individuals differ in this regard and particularly the appraisal of one and the same stressor can change (e.g., due to experiences with the stressor and how to handle it in the context of smartphone failure [21]). Against this background, we are therefore interested in exploring whether this is also true for the physiological effects of ICT unreliability.

We report on a pilot study that had the goal to derive avenues for future studies on the physiological effects of ICT unreliability. What distinguishes our approach from previous studies (e.g., [7, 8, 18, 19]) though, is that our pilot study was designed as a field experiment. Due to the significant lack of field studies in the extant NeuroIS literature [22], we also seek to make an initial methodological contribution, by reporting on experiences with our field study design.

2 Methodology

Setting. As we wanted to gain practical insights into the physiological effects of ICT unreliability, we conducted a study in the field over a period of three days. The context for our study was a company in Linz, Austria (smec—Smarter Ecommerce GmbH[1]), which offers services for the optimization of online advertisements (i.e., pay-per-click automation) on various platforms (e.g., Google or Facebook). Around 100 individuals are employed at the headquarters in Linz, with almost every staff member being a knowledge worker. For our experiment, we recruited eight individuals, four of which actually received techno-unreliability manipulations and the remaining four served as controls.

Stressor Selection. Initially, the first author, who is also an employee of the company, conducted interviews with eight staff members (3f, 5m; from 11/02/2017 to 12/20/2017) to evaluate whether ICT unreliability constitutes a prevalent stressor in the organization. Using eight stressor categories as basis for the interviews (i.e., overload, invasion, complexity, uncertainty, insecurity, unreliability, monitoring, cyberbullying [16]), we found that unreliability and overload were previously experienced as stressor at work by all interviewed staff members.

We then moved on to select specific stressors as stimuli for our study, as unreliability can include a wide area of hassles that are relevant candidates for experimental manipulation (e.g., slow program speed, slow computer speed, or keyboard typing errors [23]). Through the previously conducted interviews, we learned that complete system breakdowns are not common in this organization, but the temporary

[1]https://smarter-ecommerce.com/en [03/03/2019].

Fig. 1 Example for an error message

unavailability of certain online services is rather common. In addition, as most used systems are not installed locally, but mostly used as cloud-based services, their availability could be manipulated relatively easily by the company's IT administrator.

In total, six types of systems were manipulated during the experiment. If one of the participants tried to access Facebook or YouTube (both used to post social media ads), Gmail (the mail program used by all employees), Google AdWords (used to manage Google ads), Sage (used for internal time management, e.g., requesting vacation) or Salesforce (used to keep track of time worked on a project that is then charged to a client) a common error message was displayed by the browser. To compare the effect of these manipulations to an arguably more extreme case of unreliability, we also locked access to the laptop for two of our participants at one point. An example for the error message that participants received when they tried to reach one of the manipulated services through their online browser is displayed in Fig. 1.

Measures. We decided to collect heart rate, as other physiological measures previously used in technostress research [24] such as cortisol from saliva [8], blood pressure [25] or skin conductance [7] would have been too obtrusive to measure in this setting. Participants had to wear a chest belt (Polar H7) that was linked to a smartphone app[2] that collected the captured data. In previous studies it was shown that this setting can deliver data that is comparable in quality to an ECG [26]. At the end of the manipulation period, participants also filled out a questionnaire, which included questions related to the individual's personality ([27]; five-point Likert scale ranging from "strongly agree" to "strongly disagree") (note that personality has previously been linked to technostress, e.g., [28, 29]), and a self-assessment of technology self-efficacy ([30]; 10-point scale ranging from "1—no confidence" to "10—totally confident"; 12 out of 15 items have been used to improve content valid-

[2]https://itunes.apple.com/at/app/heart-rate-variability-logger/id683984776?mt=8 [03/03/2019].

ity), which has been applied in technostress research before, albeit using a previous version of the instrument (e.g., [31, 32]).

Participants. Although physiological and self-report measures were applied on eight individuals, only the systems of four employees were manipulated during the experiment. Of the four participants three were male (ID 1, 3, 4) and one was female (ID 2). One of the participants was a senior employee (ID 1, 36 years old; had been in the company for five years) who was mainly responsible for larger customers, while the other three participants (ID 2, 24 years; ID 3, 30 years; ID 4, 26 years) were mainly responsible for smaller customers and worked for the company for about one year.

Procedure. Data was collected from 03/20/2018 to 03/22/2018 with five-minute baselines of heart rate being taken in the morning of March 20 (average: ID 1: 55.89 bpm; ID 2: 68.81 bpm; ID 3: 61.50 bpm; ID 4: 66.62 bpm; bpm refers to beats-per-minute, which is the number of heart beats in 60 s). We informed our participants beforehand that the study would focus on stress caused by ICT, but did not indicate that there would be any controlled manipulations. The four participants then received the manipulations (controlled by the IT administrator and approved by the company's top management). During times when participants had to interact with uninvolved outsiders (e.g., customers) and/or during meetings, the manipulations were inactive. In some cases, participants reached out to the system administrator or even the company's CTO to receive support regarding the ICT unreliability, though even in these cases the manipulations were still continued.

At the end of the last day, participants filled out the questionnaires and were debriefed about the manipulations. In the questionnaire we also included a manipulation check and asked participants (including the four individuals who had not experienced any manipulations) whether they had perceived more technical difficulties than usual. Only the four manipulated individuals answered with "Yes, significantly more" as opposed to "Not more than usual" or "None".

Data Analysis. The mobile app provided us with an estimate of the beats-per-minute (bpm) once every second. Metrics related to heart-rate variability (HRV) have not been analyzed thus far, though increases in bpm can also indicate individual stress (e.g., [33, 34]). Using the baseline bpm values, we calculated relative values from our data (i.e., relative percentage as compared to the baseline bpm). Through the web server (browser data) we were the able to track interactions with manipulated systems down to the second level. We then calculated the area-under-the-curve (AUC) for 5-min periods after each manipulation based on relative bpm values (for a comparable approach in the context of cortisol levels, see [35]) to capture the employees' stress reaction and subsequent relaxation in one metric.

3 Initial Results and Potential for Future Research

In Table 1, we summarize the number of stressors per technology that each participant experienced and the average AUC as well as standard deviation of AUCs. It has to be noted that we did not control the number of stressors per participant for the sake of external validity. Instead, it was coincidental if and when a participant would encounter a technological stressor and therefore not all participants encountered all stressors.

Initially, we expected that technologies that are highly critical for the work processes of participants would lead to more pronounced stress reactions (i.e., greater AUC). In particular, the availability of Salesforce was expected to be highly relevant, as it has direct monetary implications if working hours are not tracked correctly. This expectation was fulfilled for the most senior employee (ID 1) and based on this observation we formulate the hypotheses that:

H1: *Importance of the unreliable technology for work tasks will positively moderate the relationship between unreliability and physiological stress reactions.*

This hypothesis is also in line with cybernetic stress theories, in which the importance of a discrepancy moderates the stressor-strain relationship (e.g., [36]). As can be seen for other participants though, there seem to be additional influencing mechanisms. This is particularly captured in one instance, where the person's laptop was completely locked (i.e., user credentials did not work any longer until the system administrator was contacted) for two participants (ID 1, 4). While this issue led to the highest stress levels in participant 1, participant 4 was not substantially stressed by it. Hence, further situational and/or individual characteristics seem to be at work.

Regarding individual characteristics, we observed that in some cases (e.g., for Salesforce related unreliabilities), participant ID 1 showed higher stress levels if compared to the other participants. As Salesforce is not directly crucial for work tasks, but has monetary implications for the organization (working hours are tracked

Table 1 Overview of AUCs per participant and technology type

Type of stressor	ID 1	ID 2	ID 3	ID 4
AdWords	1 (399.02; N/A)	2 (368.05; 13.74)	–	–
Facebook	4 (366.97; 26.04)	–	9 (364.99; 32.37)	–
Gmail	2 (359.82; 0.89)	2 (358.05; 11.16)	3 (346.95; 11.16)	–
Sage	1 (349.45; N/A)	1 (382.95; N/A)	3 (366.01; 2.49)	–
Salesforce	2 (420.02; 11.70)	5 (319.95; 18.01)	3 (348.53; 39.77)	6 (350.31; 33.97)
YouTube	6 (376.66; 42.71)	–	–	2 (366.59; 15.83)
Locked Laptop	1 (484.33; N/A)	–	–	1 (355.31; N/A)
Overall	383.55 (42.47)	343.28 (28.82)	359.41 (29.63)	354.38 (31.27)

Table 2 Individual characteristics of participants

ID/construct	EX	AG	NE	OP	CO	TSE
1	3.5	3.0	2.5	4.0	3.0	7.9
2	4.0	3.0	3.5	4.0	3.5	3.0
3	3.0	3.0	2.0	1.5	2.5	5.3
4	4.0	3.5	3.0	3.0	3.0	9.1

Big five: *EX* extraversion; *AG* agreeableness; *NE* neuroticism; *OP* openness; *CO* conscientiousness; *TSE* technology self-efficacy; big five: 1–5; *TSE* 1–10

in the system), we suspect that the commitment to the organization is particularly high in case of the senior employee. Hence, we hypothesize that:

H2: *(a) Organizational commitment and/or (b) organizational tenure will positively moderate the relationship between unreliability and physiological stress reactions.*

For personality characteristics (i.e., Big Five) and technology self-efficacy there are some differences between the participants (see Table 2), though they do not allow for conclusive statements. For example, while participants ID 2 and ID 3 show lower technology self-efficacy as the other participants, their stress levels were not substantially higher in any instance as we would expect based on previous research (e.g., [32]).

Regarding situational characteristics, we looked at the temporal order of stressors and particularly at the frequency of the same stressor and whether it had an influence on the physiological reaction. In some instances, we found patterns that could be indicative of possible relationships. For example, for participant ID 1, physiological reactions to YouTube not working decreased over time, while they increased steadily for Salesforce related problems. Returning to the same technology several times initially indicates that it is important for the current task, yet while in one case this might simply be a type of ancillary fact-checking (e.g., Does YouTube work again, so I can quickly perform a related task?), in the other case the increase in physiological stress reactions might be indicative of the preparation for an escalation in coping mechanisms (e.g., to see if the technology works again and whether the system administrator has to be contacted). Again though, this observation is not consistent between participants, as participants ID 2 and ID 3 showed a steady decline in physiological reactions towards Salesforce not working (with the exception of the first check for participant ID 2 the next morning, which led to a substantially stronger physiological reaction if compared to the problems of the day before). Based on these patterns, we can formulate an initial hypothesis, at least for technologies that seem to be pivotal for work at a specific moment. Yet, we cannot indicate a specific direction for the relationship and instead formulate that:

H3: *Repeated encounters of a specific unreliability stressor influence the level of physiological reactions to it.*

4 Limitations and Concluding Statement

Several limitations have to be noted. First, our results are based on a small sample size and hence they are preliminary. Second, heart-rate variability (HRV) metrics should be used as further stress indicators, in addition to pure heart rate (HR) (e.g., RMSEA or LF/HF ratio [37–39]). Third, in previous research it was also highlighted that the measurement setting could become uncomfortable and tedious (e.g., due to the chest strap and extra phone, [40]), although the participants of the present study did not indicate any problems related to the measurement setting. Yet, in some cases the export of N-N intervals (which are saved in a separate file from HR values and HRV values) did not work properly and therefore missing data would have been present for HRV calculations (N-N intervals, which are also referred to as normal-to-normal-intervals or interbeat intervals, are the millisecond periods between successive heart beats). Fourth, due to our study design, the exact controlling of the timing and types of stressors was not possible.

What we learned though is that it is feasible to study technostress encounters in the field, here related to ICT unreliability. In addition, we observed a wide variety of reactions to largely similar stressors. Hence, findings from previous research which have shown that particularly contextual differences (e.g., differences in tasks [41], decision latitude to handle the tasks [6], or time pressure to execute the tasks [42]) are important for technostress research have to be emphasized based on our findings. Moreover, future research should capture stress appraisal processes too (i.e., Is an encounter perceived as a source of distress or not?). Hence, we call for further field research to better understand how technostress unfolds outside of the laboratory.

References

1. Fischer, T., & Riedl, R. (2017). Technostress research: A nurturing ground for measurement pluralism? *Communications of the Association for Information Systems, 40,* 375–401.
2. Riedl, R., Kindermann, H., Auinger, A., & Javor, A. (2012). Technostress from a neurobiological perspective—System breakdown increases the stress hormone cortisol in computer users. *Business & Information Systems Engineering, 4,* 61–69.
3. Riedl, R., Kindermann, H., Auinger, A., & Javor, A. (2013). Computer breakdown as a stress factor during task completion under time pressure: Identifying gender differences based on skin conductance. *Advances in Human-Computer Interaction,* 1–8.
4. Tams, S., Hill, K., de Guinea, A. O., Thatcher, J., & Grover, V. (2014). NeuroIS—Alternative or complement to existing methods? Illustrating the holistic effects of neuroscience and self-reported data in the context of technostress research. *Journal of the Association for Information Systems, 15,* 723–753.
5. Clayton, R. B., Leshner, G., & Almond, A. (2015). The extended iSelf: The impact of iPhone separation on cognition, emotion, and physiology. *Journal of Computer-Mediated Communication, 20,* 119–135.
6. Galluch, P., Grover, V., & Thatcher, J. B. (2015). Interrupting the workplace: Examining stressors in an information technology context. *Journal of the Association for Information Systems, 16,* 1–47.

7. Anderson, B., Vance, A., Kirwan, C. B., Eargle, D., & Jenkins, J. L. (2016). How users perceive and respond to security messages: A NeuroIS research agenda and empirical study. *European Journal of Information Systems, 25,* 364–390.
8. Kothgassner, O. D., Felnhofer, A., Hlavacs, H., Beutl, L., Palme, R., Kryspin-Exner, I., & Glenk, L. M. (2016). Salivary cortisol and cardiovascular reactivity to a public speaking task in a virtual and real-life environment. *Computers in Human Behavior, 62,* 124–135.
9. González-Cabrera, J., Calvete, E., León-Mejía, A., Pérez-Sancho, C., & Peinado, J. M. (2017). Relationship between cyberbullying roles, cortisol secretion and psychological stress. *Computers in Human Behavior, 70,* 153–160.
10. Konok, V., Pogány, Á., & Miklósi, Á. (2017). Mobile attachment. Separation from the mobile phone induces physiological and behavioural stress and attentional bias to separation-related stimuli. *Computers in Human Behavior, 71,* 228–239.
11. Riedl, R. (2013). On the biology of technostress: literature review and research agenda. *DATA BASE for Advances in Information Systems, 44,* 18–55.
12. Agogo, D., Hess, T. J., Te'eni, D., & McCoy, S. (2018). "How does tech make you feel?". A review and examination of negative affective responses to technology use. *European Journal of Information Systems, 27,* 570–599.
13. Sarabadani, J., Carter, M., & Compeau, D. R. (2018). 10 years of research on technostress creators and inhibitors: Synthesis and critique. In *Proceedings of AMCIS 2018*. AIS.
14. Adam, M. T. P., Gimpel, H., Maedche, A., & Riedl, R. (2017). Design blueprint for stress-sensitive adaptive enterprise systems. *Business & Information Systems Engineering, 59,* 277–291.
15. Fischer, T., & Riedl, R. (2015). Theorizing technostress in organizations: A cybernetic approach. In O. Thomas & F. Teuteberg (Eds.), *Proceedings of the 12th International Conference on Wirtschaftsinformatik* (pp. 1453–1467).
16. Gimpel, H., Lanzl, J., Manner-Romberg, T., & Nüske, N. (2018). Digitaler Stress in Deutschland. Eine Befragung von Erwerbstätigen zu Belastung und Beanspruchung durch Arbeit mit digitalen Technologien, Düsseldorf, Germany.
17. Fischer, T., Pehböck, A., & Riedl, R. (2019). Is the technostress creators inventory still an up-to-date measurement instrument? Results of a large-scale interview study. In T. Ludwig & V. Pipek (Eds.), *14. Internationale Tagung Wirtschaftsinformatik* (pp. 1834–1845).
18. Emurian, H. H. (1991). Physiological responses during data retrieval: Comparison of constant and variable system response times. *Computers in Human Behavior, 7,* 291–310.
19. Emurian, H. H. (1993). Cardiovascular and electromyograph effects of low and high density work on an interactive information system. *Computers in Human Behavior, 9,* 353–370.
20. Butler, B. S., & Gray, P. H. (2006). Reliability, mindfulness, and information systems. *MIS Quarterly, 30,* 211.
21. Salo, M., Pirkkalainen, H., Makkonen, M., & Hekkala, R. (2018). Distress, eustress, or no stress? Explaining smartphone users different technostress responses. In *Proceedings of ICIS 2018*. AIS.
22. Riedl, R., Fischer, T., & Léger, P.-M. (2017). A decade of NeuroIS research: Status quo, challenges, and future directions. In *Proceedings of ICIS 2017*. AIS.
23. Hudiburg, R. A. (1995). Psychology of computer use. XXXIV. The computer hassles scale: Subscales, norms, and reliability. *Psychological Reports, 77,* 779–782.
24. Fischer, T., & Riedl, R. (2015). The status quo of neurophysiology in organizational technostress research: A review of studies published from 1978 to 2015. In F. D. Davis, R. Riedl, J. Vom Brocke, P.-M. Léger, & A. B. Randolph (Eds.), *Information systems and neuroscience* (Vol. 10, pp. 9–17). Cham: Springer International Publishing.
25. Fischer, T., Halmerbauer, G., Meyr, E., & Riedl, R. (2017). Blood pressure measurement: A classic of stress measurement and its role in technostress research. In F. D. Davis, R. Riedl, J. Vom Brocke, P.-M. Léger, & A. B. Randolph (Eds.), *Information Systems and Neuroscience. Gmunden Retreat on NeuroIS 2017* (pp. 25–35). Cham: Springer.
26. Plews, D. J., Scott, B., Altini, M., Wood, M., Kilding, A. E., & Laursen, P. B. (2017). Comparison of heart-rate-variability recording with smartphone photoplethysmography, Polar H7 chest

strap, and electrocardiography. *International Journal of Sports Physiology and Performance, 12,* 1324–1328.
27. Rammstedt, B., & John, O. P. (2007). Measuring personality in one minute or less: A 10-item short version of the Big Five Inventory in English and German. *Journal of Research in Personality, 41,* 203–212.
28. Srivastava, S. C., Chandra, S., & Shirish, A. (2015). Technostress creators and job outcomes: Theorising the moderating influence of personality traits. *Information Systems Journal, 25,* 355–401.
29. Pflügner, K., Mattke, J., & Maier, C. (2019). Who is stressed by using ICTs? A qualitative comparison analysis with the big five personality traits to understand technostress. In T. Ludwig & V. Pipek (Eds.), *14. Internationale Tagung Wirtschaftsinformatik* (Vol. 14, pp. 1175–1189).
30. Compeau, D. R., Correia, J., & Thatcher, J. B. (2017). Implications of technological progress for the measurement of technology acceptance variables: The case of self-efficacy. In *Proceedings of ICIS 2017*. AIS.
31. Shu, Q., Tu, Q., & Wang, K. (2011). The impact of computer self-efficacy and technology dependence on computer-related technostress: A social cognitive theory perspective. *International Journal of Human-Computer Interaction, 27,* 923–939.
32. Tarafdar, M., Pullins, E. B., & Ragu-Nathan, T. S. (2015). Technostress: Negative effect on performance and possible mitigations. *Information Systems Journal, 25,* 103–132.
33. Kudielka, B. M., Schommer, N. C., Hellhammer, D. H., & Kirschbaum, C. (2004). Acute HPA axis responses, heart rate, and mood changes to psychosocial stress (TSST) in humans at different times of day. *Psychoneuroendocrinology, 29,* 983–992.
34. Knight, W. E. J., & Rickard, N. S. (2001). Relaxing music prevents stress-induced increases in subjective anxiety, systolic blood pressure, and heart rate in healthy males and females. *Journal of Music Therapy, 38,* 254–272.
35. Pruessner, J. C., Kirschbaum, C., Meinlschmid, G., & Hellhammer, D. H. (2003). Two formulas for computation of the area under the curve represent measures of total hormone concentration versus time-dependent change. *Psychoneuroendocrinology, 28,* 916–931.
36. Edwards, J. R. (1992). A cybernetic theory of stress, coping and well-being in organizations. *The Academy of Management Review, 17,* 238–274.
37. Hjortskov, N., Rissén, D., Blangsted, A. K., Fallentin, N., Lundberg, U., & Søgaard, K. (2004). The effect of mental stress on heart rate variability and blood pressure during computer work. *European Journal of Applied Physiology, 92,* 84–89.
38. Vrijkotte, T. G. M., van Doornen, L. J. P., & de Geus, E. J. C. (2000). Effects of work stress on ambulatory blood pressure, heart rate, and heart rate variability. *Hypertension, 35,* 880–886.
39. Baumgartner, D., Fischer, T., Riedl, R., & Dreiseitl, S. (2018). Analysis of heart rate variability (HRV) feature robustness for measuring technostress. In F. D. Davis, R. Riedl, J. Vom Brocke, P.-M. Léger, & A. Randolph (Eds.), *Information systems and neuroscience: NeuroIS retreat 2018* (pp. 221–228). Berlin Heidelberg: Springer.
40. Calikli, G., Price, B., Andersen, M. S., Nuseibeh, B., & Bandara, A. (2014). Personal informatics for non-geeks. Lessons learned from ordinary people. In A. J. Brush, A. Friday, J. Kientz, J. Scott, & J. Song (Eds.), *Proceedings of 2014 ACM International Joint Conference on Pervasive and Ubiquitous Computing Adjunct Publication (UbiComp 2014 Adjunct)* (pp. 683–686).
41. Bailey, B. P., & Konstan, J. A. (2006). On the need for attention-aware systems: Measuring effects of interruption on task performance, error rate, and affective state. *Computers in Human Behavior, 22,* 685–708.
42. Boucsein, W. (2009). Forty years of research on system response times—What did we learn from it? In C. M. Schlick (Ed.), *Industrial engineering and ergonomics* (pp. 575–593). Berlin: Springer.

Wavelet Transform Coherence: An Innovative Method to Investigate Social Interaction in NeuroIS

Paul Léné, Alexander J. Karran, Elise Labonté-Lemoyne, Sylvain Sénécal, Marc Fredette, Kevin J. Johnson and Pierre-Majorique Léger

Abstract Neuro-information system researchers called for the development of tools and methods of analysis that allows for the assessment of social interactions that considers the full range of dynamics and complexities. Recent work and progress in hyperscanning—the simultaneous recording of multiple subjects—offers the possibility to develop such methods and new research paradigms to respond to that need. Among these methods, Wavelet Transform Coherence (WTC) analysis is gaining in popularity and accessibility. However, hyperscanning methods—including WTC—remain a challenging experimental paradigm and analysis method, requiring careful preparation of data and consideration of the constraints of each neuroimaging technique. This manuscript aims to introduce the Wavelet Transform Coherence method of analysis as an innovative approach to Neuro-Information Systems research. We present practical examples and results in order to highlight the potential and complexity of this approach.

P. Léné (✉) · K. J. Johnson
Département de Management, HEC Montréal, Montréal, Canada
e-mail: paul.lene@hec.ca

K. J. Johnson
e-mail: kevin.johnson@hec.ca

A. J. Karran · E. Labonté-Lemoyne · S. Sénécal · M. Fredette · P.-M. Léger
Département des Technologies de l'information, HEC Montréal, Montréal, Canada
e-mail: alexander-john.karran@hec.ca

E. Labonté-Lemoyne
e-mail: elise.labonte-lemoyne@hec.ca

S. Sénécal
e-mail: sylvain.senecal@hec.ca

M. Fredette
e-mail: marc.fredette@hec.ca

P.-M. Léger
e-mail: pierre-majorique.leger@hec.ca

F. D. Davis et al. (eds.), *Information Systems and Neuroscience*,
Lecture Notes in Information Systems and Organisation 32,
https://doi.org/10.1007/978-3-030-28144-1_16

Keywords EEG · Neuro-information system · Social neuroscience · Hyperscanning · Wavelet transform coherence · Wavelet analysis

1 Introduction

In the past decade, researchers from social and cognitive neuroscience displayed a growing interest in the exploration and understanding of social interactions and their associated brain networks [1]. A recent review [2] outlined the breadth and depth of this type of research showing that a growing number of more recent studies utilised hyperscanning as a breakthrough experimental paradigm. Hyperscanning is a neuroimaging technique that consists of the simultaneous recording of multiple subjects in parallel and ensures the precise synchronisation of data so that recorded signals can be compared and contrasted between individuals [3]. The development, improvement and utilisation of hyperscanning has allowed a number of significant contributions in various fields of research which address the neurophysiological basis and dynamics of social interactions, such as in a context of neuroeconomy, sociology and team coordination [1, 3–7].

In parallel with the development of hyperscanning, a growing interest in neuroimaging tools and methods from the domain of information technology and more specifically neuro-information system (NeuroIS) was observed [8, 9]. Calls for research and early work were made concerning topic such as collaboration, flow, trust and digital interaction [8, 10–13]. Since hyperscanning allows the recording of multiple subjects with precise synchronization of their signals while performing tasks involving information systems, it appears timely to combine the progress in hyperscanning with the growth of NeuroIS. The goal of this paper is to introduce the Wavelet Transform Coherence method of analysis and to demonstrate the exciting potential that it represents for NeuroIS research.

2 Hyperscanning Methods

The hyperscanning methodology can be applied using any neuroimaging tool. From the standpoint of an experimental paradigm, hyperscanning "only" consists of the simultaneous recording of multiple subjects which considers two brains as a "two-in-one" system such as in the case of teamwork using information systems to complete a task. Critically this paradigm creates several factors that require consideration, the first of which is how to calculate the correlation between neuroimaging data acquired from two or more individuals. Secondly, is equipment availability, having access to multiple recording devices at the same time (e.g., 2 MRIs in one place is rare) can prove problematic. The final factors to be considered are risk and cost, there are costs associated with imaging equipment and participant recruitment and then there is the risk associated with data loss, to lose one or more participants data is to lose either fidelity of signal (in the case of multiple participants) or complete loss during analysis (in the case of 2 participants). These factors are discussed in detail in the literature through extensive reviews (see Babiloni et al. [1] and Mu et al. [14]). However, as previously discussed, the primary critical factor associated with hyperscanning and one that is most frequently underestimated in hyperscanning lies in the analysis process.

Within the last decade, different types of hyperscanning methods of analysis have been developed, such as the Phase Lag Index [15] which can be applied to EEG or MEG data. Most approaches are specific to a given neuroimaging tool due to the data produced and nature of effect under observation (e.g., NIRS—blood flow). This variety of methods offers various approaches to analysing a unique phenomenon. However, the interpretation of results will differ significantly depending on the neuroimaging tool and method of analysis that is used. Therefore, a new analysis method for hyperscanning data that is not specific to any tool may benefit researchers in the field of neuroIS.

Recently, wavelet transform coherence (WTC) [4, 16]. has been adapted for use in the neurosciences. The WTC method was used initially in the field of meteorology in the early 1990s [17] to study the EL-Nino-Southern Oscillation. It has since been successfully adapted for neuroscience as a method of study the time-frequency dynamics of resting-state brain activity using fMRI [18] and has since gained popularity in hyperscanning research [4, 19]. The following section will present a broad outline of the WTC analysis method.

3 Wavelet Transform Coherence

WTC is a method of analysis that measures the cross-correlation between two signals as a function of frequency and time [17]. It calculates the localized coherence coefficient as a value between 0 and 1. Mapping this coherence value uncovers inter-signal phenomenon that might not be discoverable through traditional time-series analysis [4, 16]. Figure 1 shows a practical example which highlights the difference between individual and WTC data visualization. Shown in blue and red (Fig. 1a) is a NIRS trace of two participants performing an information systems task. From this, no discernable synchronous pattern between the two signals can be seen. However, when WTC is applied to the same data, it generates a coherence heat map (Fig. 1b) showing the coherence coefficient. In this plot, yellow areas represent a high degree of correlation between the two signals (coherence coefficient ≥ 1), and blue areas represent the absence of synchronicity of the two signals (coherence coefficient $=$ 0). The WTC output suggests that the highest coherence between the brain signals occurring within the 60 s period, which indicates that participants are performing very similar actions at similar points in time and utilizing similar amounts of cognitive effort. The thick yellow bar visible at 1 s period is an artefact associated with heart

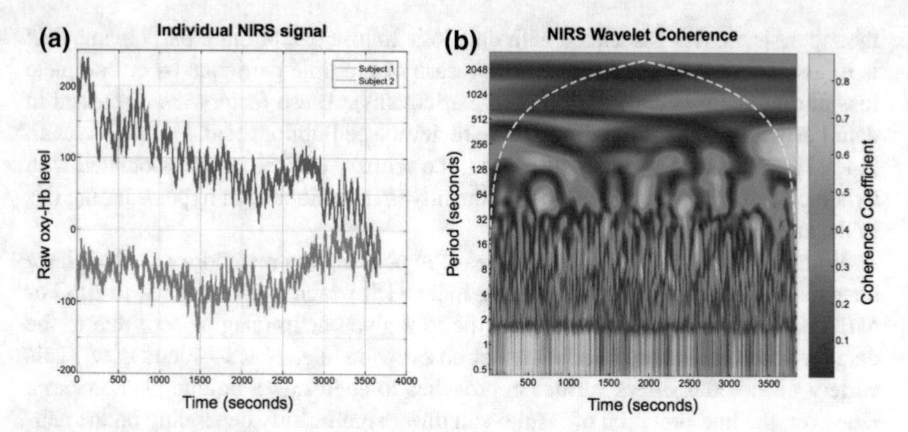

Fig. 1 **a** Individual NIRS Signal. The horizontal axis represents the time of the experimental task and the vertical axis represents the oxygenated hemoglobin level for two subjects. **b** Wavelet Transform Coherence. The horizontal axis represents the time of the experimental task and the vertical axis represents the period(s) of wavelets—the intervals at which synchronicity is potentially observed

rate. This example highlights the difference in analysis and interpretation presented by WTC when compared to traditional trace plots.

The fact that WTC is a neuroimaging-tool agnostic analysis method makes it highly appealing as a means for replicating or adding value to studies that utilised a variety of neuroimaging tools, potentially providing additional rigour and insight to already completed work that involved one or more participants working in parallel. WTC as an analytical approach is gaining in popularity through increased accessibility and standardisation, with many research teams developing open access code and toolboxes for use. For the examples shown in this manuscript, we utilised the Matlab® Wavelet Toolbox (4.1.9, MathWorks Inc., Natick MA).

4 Illustrative Research

To further illustrate the utility of WTC, we present comparative research that replicates the work of Cui et al. [4] who utilised NIRS to simultaneously measure brain activity in two people while they played a computer-based cooperation game side by side and investigate inter-brain activity coherence. In our study, we investigate the neurophysiological basis of cooperative task completion using EEG and apply the WTC method to interpret the hyperscanning dynamics (under review). The following section presents a summary of our results presented as an example of the holistic representation that single subject results associated with the insight from WTC offers. The experiment consisted of two subjects, sited side by side, performing an IT task (fast-button response) with a period of approximately 7 s. In one condition subjects were instructed to be as fast as possible (competition) and in another condition instructed to synchronise clicks as much as possible with the second participant (collaboration); a third control condition was introduced—in which one subject performs the task while the other passively obverses and inversely. Each condition was composed of two blocks with a resting period of 30 s. EEG data were treated using Matlab® 2018A environment (Mathworks, Hyderabad, Indiana, United-States) and the EEGLab Toolbox [20]. The EEG signal was filtered to remove noisy channels or data segments using the EEGLab artefact subspace reconstruction. A band-pass filter (1–100 Hz) was applied along with a 60 Hz notch filter. Blinks and muscular artefacts were removed through an ICA [21]. EEG signal was re-referenced to the common average reference. The EEG signal was then divided into frequency bands: theta (4–7 Hz), alpha (8–12 Hz) and beta (13–30 Hz). The recordings were segmented into 3 s epochs starting one second before the stimulus presentation. We performed both single subject and WTC analysis.

Through analysis of single-subject data, we observed significant modulation of brain wave activity for each condition. From this, we concluded that the social conditions (competition and collaboration) could be associated with specific brain activity located primarily in the frontal cortex with higher frequency band power when compared to single conditions. More precisely, competition and collaboration showed significant differences in frontal and prefrontal regions within all frequency bands. This result is in line with results reported by Cui et al. in their earlier study.

However, when WTC was applied to those same data the results showed that coherence between dyad member's alpha band cortical activity increased in their prefrontal cortex only during the collaboration condition (As visible in Fig. 2a with the coherence appearing around 7 s corresponding to the task period). We also observed

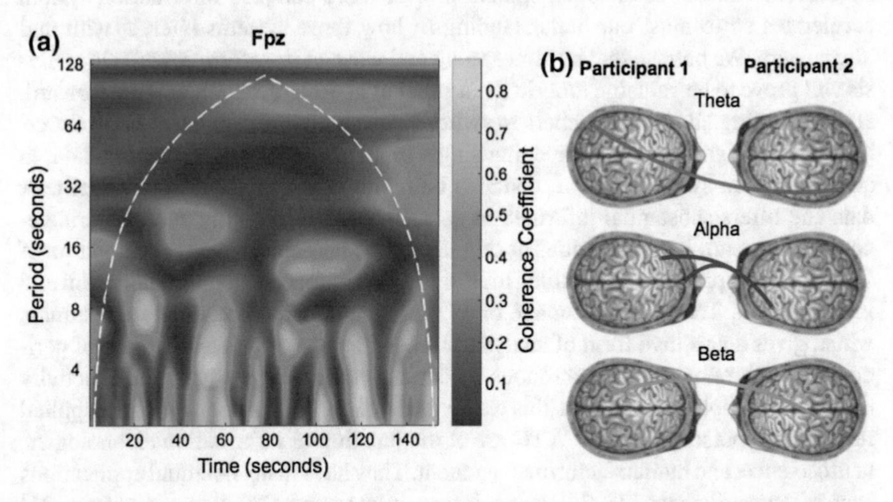

Fig. 2 **a** Wavelet transform coherence in Fpz between two subjects. The horizontal axis represents the time of the experimental task and the vertical axis represents the period(s) of wavelets. **b** Topographic representation of coherence results. Connections represent significant level of coherence between the two participants ($p < 0.05$). Each color is associated with a frequency band: purple—theta, blue—alpha, green—beta

that social conditions appeared to highlight a synchronicity of cortical activity in the theta frequency band. Results showing increased coherence in the alpha band over the frontal cortex during collaboration can potentially be attributed to either higher cognitive load or inhibition, and this coupled with activity in the theta band over the parietal regions, is compatible with the hypothesis of an increased cognitive resource demand required for information processing and adjustments to the partner [22]. Brain activity in the other experimental conditions demonstrated significant synchronicity over occipital and parietal regions that are possibly task-related. However, this inference is under investigation.

The application of WTC as an additional analysis tool in this instance provided far deeper insight than single subject analysis alone, without WTC our interpretation of the results would not have allowed us to differentiate brain activity between competition and collaboration. While the single subject analysis provided essential information about the cortical areas involved in social interaction, WTC offered a unique snapshot of the interaction between subjects that proved fundamental to our understanding and interpretation of social interactions during task completion. Thus, without both types of analysis, our interpretation of the results would have lacked depth and the extra dimensionality, that is the bedrock of neurophysiological inference.

5 Conclusion

As societies needs for the development of ever more complex information system accelerates so to must our understanding of how those systems interact with and affect users. We believe that hyperscanning and wavelet transform coherence analysis will prove to be valuable tools in the arsenal of neuroIS research moving forward. Hyperscanning allows researchers to switch from single-brain to multi-brain experimental paradigms, without impeding single brain analysis. Hyperscanning data, in an information system context, contains both single and multiple concurrent user data and offers substantial information gain, in terms of usability and cooperative-competitive team-based activities during IS use and task completion. The addition of WTC analysis provides a versatile "tool" agnostic approach to data visualisation and interpretation. The main advantage of WTC is the coherence coefficient heatmap, which gives a narrative form of analysis allowing for the easy identification of periods of synchronous or asynchronous behaviour and the intensity of brain activity during the completion of a task, this we believe makes WTC a strong tool for applied research. Hyperscanning and WTC are at the junction of information technologies, neuroscience, and human factor/management. They have many potential applications such as team efficiency [23, 24], neuro-information system [25], user experience [26] and management [27, 28].

References

1. Babiloni, F., & Astolfi, L. (2014). Social neuroscience and hyperscanning techniques: Past, present and future. *Neuroscience and Biobehavioral Reviews, 44,* 76–93.
2. Lieberman, M. D. (2007). Social cognitive neuroscience: A review of core processes. *Annual Review of Psychology, 58,* 259–289.
3. Montague, P. R., Berns, G. S., Cohen, J. D., McClure, S. M., Pagnoni, G., Dhamala, M. … Fisher, R. E. (2002). Hyperscanning: Simultaneous fMRI during linked social interactions. *NeuroImage, 16*(4), 1159–1164.
4. Cui, X., Bryant, D. M., & Reiss, A. L. (2012). NIRS-based hyperscanning reveals increased interpersonal coherence in superior frontal cortex during cooperation. *NeuroImage, 59*(3), 2430–2437.
5. Dumas, G., Nadel, J., Soussignan, R., Martinerie, J., & Garnero, L. (2010). Inter-brain synchronization during social interaction. *PLoS ONE, 5*(8), e12166.
6. Horat, S. K., Prévot, A., Richiardi, J., Herrmann, F. R., Favre, G., Merlo, M. C. G., & Missonnier, P. (2017). Differences in social decision-making between proposers and responders during the ultimatum game: An EEG study. *Frontiers in Integrative Neuroscience, 11*(July).
7. Toppi, J., Borghini, G., Petti, M., He, E. J., De Giusti, V., He, B. … Babiloni, F. (2016). Investigating cooperative behavior in ecological settings: An EEG hyperscanning study. *PLOS One, 11*(4), 1–26.
8. Loos, P., Riedl, R., Müller-Putz, G. R., Vom Brocke, J., Davis, F. D., Banker, R. D & Léger, P.-M. (2010). NeuroIS: Neuroscientific approaches in the investigation and development of information systems. *Business & Information Systems Engineering, 2*(6), 395–401.
9. Riedl, R., & Léger, P.-M. (2016). Fundamentals of NeuroIS. In *Studies in neuroscience, psychology and behavioral economics*. Berlin, Heidelberg: Springer.
10. Bastarache-Roberge, M.-C., Léger, P.-M., Courtemanche, F., Sénécal, S., & Fredette, M. (2015). Measuring flow using psychophysiological data in a multiplayer gaming context. In *Information systems and neuroscience* (pp. 187–191). Berlin: Springer.
11. Dimoka, A., Benbasat, I., Davis, F. D., Dennis, A. R., Gefen, D., & Weber, B. (2012). On the use of neurophysical tools in IS research: Developing a research agenda for NeuroIS. *MIS Qarterly, 36*(3), 679–702.
12. Labonté-LeMoyne, É., Léger, P. M., Resseguier, B., Bastarache-Roberge, M. C., Fredette, M., Sénécal, S., & Courtemanche, F. (2016, May). Are we in flow neurophysiological correlates of flow states in a collaborative game. In *Proceedings of the 2016 CHI Conference Extended Abstracts on Human Factors in Computing Systems* (pp. 1980–1988). ACM.
13. Léger, P.-M., Sénécal, S., Aubé, C., Cameron, A.-F., de Guinea, A. O., Brunelle, E., et al. (2013). The influence of group flow on group performance: A research program. *Proceedings of the Gmunden Retreat on NeuroIS, 13.*
14. Mu, Y., Cerritos, C., & Khan, F. (2018). Neural mechanisms underlying interpersonal coordination: A review of hyperscanning research. *Social and Personality Psychology Compass, 12*(11).
15. Stam, C. J., Nolte, G., & Daffertshofer, A. (2007). Phase lag index: assessment of functional connectivity from multi channel EEG and MEG with diminished bias from common sources. *Human Brain Mapping, 28*(11), 1178–1193.
16. Grinsted, A., Moore, J. C., & Jevrejeva, S. (2004). Application of the cross wavelet transform and wavelet coherence to geophysical time series. *Nonlinear Processes in Geophysics, 11*(5/6), 561–566.
17. Torrence, C., & Compo, G. P. (1998). A practical guide to wavelet analysis. *Bulletin of the American Meteorological Society, 79*(1), 61–78.
18. Chang, C., & Glover, G. H. (2010). Time–frequency dynamics of resting-state brain connectivity measured with fMRI. *Neuroimage, 50*(1), 81–98.
19. Addison, P. S. (2017). *The illustrated wavelet transform handbook: Introductory theory and applications in science, engineering, medicine and finance*. CRC Press.

20. Delorme, A., & Makeig, S. (2004). EEGLAB: An open source toolbox for analysis of single-trial EEG dynamics including independent component analysis. *Journal of Neuroscience Methods, 134*(1), 9–21.
21. Jung, T. P., Makeig, S., Westerfield, M., Townsend, J., Courchesne, E., & Sejnowski, T. J. (2000). Removal of eye activity artifacts from visual event-related potentials in normal and clinical subjects. *Clinical Neurophysiology, 111*(10), 1745–1758.
22. Klimesch, W. (1999). EEG alpha and theta oscillations reflect cognitive and memory performance: A review and analysis. *Brain Research Reviews, 29*(2–3), 169–195.
23. Dodel, S., Cohn, J., Mersmann, J., Luu, P., Forsythe, C., & Jirsa, V. (2011). Brain signatures of team performance. In D. D. Schmorrow & C. M. Fidopiastis (Eds.), *Foundations of augmented cognition. Directing the future of adaptive systems* (pp. 288–297). Berlin, Heidelberg: Springer.
24. Stevens, R., Galloway, T., Wang, P., Berka, C., Tan, V., Wohlgemuth, T., … Buckles, R. (2013). Modeling the neurodynamic complexity of submarine navigation teams. *Computational and Mathematical Organization Theory*, *19*(3), 346–369.
25. Dimoka, A., Pavlou, P. A., & Davis, F. D. (2011). Research commentary—NeuroIS: The potential of cognitive neuroscience for information systems research. *Information Systems Research, 22*(4), 687–702.
26. Pan, Y., Cheng, X., Zhang, Z., Li, X., & Hu, Y. (2017). Cooperation in lovers: An fNIRS-based hyperscanning study. *Human Brain Mapping, 38*(2), 831–841.
27. Astolfi, L., Toppi, J., Borghini, G., Vecchiato, G., He, E. J., Roy, A. … Babiloni, F. (2012). Cortical activity and functional hyperconnectivity by simultaneous EEG recordings from interacting couples of professional pilots. In *Proceedings of the Annual International Conference of the IEEE Engineering in Medicine and Biology Society, EMBS* (pp. 4752–4755).
28. Bezerianos, A., Sun, Y., Chen, Y., Woong, K. F., Taya, F., Arico, P. … Thakor, N. (2015). Cooperation driven coherence: Brains working hard together. In *Proceedings of the Annual International Conference of the IEEE Engineering in Medicine and Biology Society, EMBS, 2015–Novem* (pp. 4696–4699).

Towards a Software Architecture for Neurophysiological Experiments

Constantina Ioannou, Ekkart Kindler, Per Bækgaard, Shazia Sadiq and Barbara Weber

Abstract Despite their wide adoption for conducting experiments in numerous domains, neurophysiological measurements often are time consuming and challenging to interpret because of the inherent complexity of deriving measures from raw signal data and mapping measures to theoretical constructs. While significant efforts have been undertaken to support neurophysiological experiments, the existing software solutions are non-trivial to use because often these solutions are domain specific or their analysis processes are opaque to the researcher. This paper proposes an architecture for a software platform that supports experiments with multi-modal neurophysiological tools through extensible, transparent and repeatable data analysis and enables the comparison between data analysis processes to develop more robust measures. The identified requirements and the proposed architecture are intended to form a basis of a software platform capable of conducting experiments using neurophysiological tools applicable to various domains.

Keywords Neurophysiological tools · Software architecture · Neurophysiological experiments

C. Ioannou (✉) · E. Kindler · P. Bækgaard · B. Weber
Technical University of Denmark, Copenhagen, Denmark
e-mail: coio@dtu.dk

E. Kindler
e-mail: ekki@dtu.dk

P. Bækgaard
e-mail: pgba@dtu.dk

B. Weber
e-mail: barbara.weber@unisg.ch

S. Sadiq
The University of Queensland, Brisbane, Australia
e-mail: shazia@itee.uq.edu.au

B. Weber
University of St. Gallen, St. Gallen, Switzerland

F. D. Davis et al. (eds.), *Information Systems and Neuroscience*,
Lecture Notes in Information Systems and Organisation 32,
https://doi.org/10.1007/978-3-030-28144-1_17

1 Introduction

Neurophysiological tools are used by researchers in numerous domains such as information systems (IS) and software engineering (SE) to measure human responses when people engage with information technology (IT) artifacts to perform a task (e.g. code comprehension, web browsing). The advantages of using these tools to measure human responses in the context of studies are the following: they provide objective measures, complementing subjective, perception-based measures and they enable continuous real-time data collection [6].

In addition, the availability of neurophysiological tools at a better quality, lower cost and reduced intrusiveness motivated researchers to apply such measurements (e.g., to assess cognitive load). Cognitive load provides insights into a person's processing capabilities while performing a task which can support personalized IS design and adaptation to the users' cognitive needs (e.g., in a digital learning context the task difficulty could be decreased when high cognitive load is observed to avoid stress, frustration, and errors) [6]. Therefore, research has focused on creating reliable and objective measures of cognitive load [4]. For instance, a research focused on pupillary response data for identifying an indicator of cognitive load [7], while another study used multi-modal measurements (e.g., the combination pupillary responses with EDA signals) to derive cognitive load and to assess task difficulty [8].

In response to these trends, software solutions have evolved to support researchers during the execution of experiments including the collection and synchronisation of neurophysiological measurements as well as the (online) analysis of data. Software solutions like Brownie, iMotions, CubeHX, Noldus Observer, and OpenVibe support the design of experiments and the collection and synchronisation of neurophysiological measurements including their visualization [9, 10, 12–14]. Moreover, software solutions like EEGLAB and OpenVibe provide extensible and transparent ways to analyse data and partially support repeatable data analysis pipelines [5, 14].

However, current solutions do not provide any automatic means to compare different data analysis pipelines to obtain robust measures (e.g., of cognitive load). In order to close this gap, we aim to identify a set of requirements and develop a software architecture that not only supports comparing different analysis pipelines, but additionally also supports multi-modal neurophysiological measurements in an extensible, transparent, and repeatable way.

2 Requirements

In recent years, an increasing body of literature applies neurophysiological measurements to investigate cognitive and emotional states of a developer during software engineering tasks [7, 8, 11]. By using this domain as a basis, we derived a group of selected requirements for a software architecture that aims to provide support for researchers in defining, deploying, executing, and analysing experiments using neurophysiological measurements. In this section, first we explain the requirements and briefly discuss related work.

Multi-modal measurements collection. Various studies aimed at better understanding the cognitive load of developers while engaging with different software artifacts using neurophysiological measurements. For example, one study used pupillary response data, electroencephalography (EEG) and galvanic skin response (GSR) to assess task difficulty during change tasks [8]. Similarly, another study combined EEG and pupillary response data to predict task difficulty during program comprehension tasks [11]. In all these studies multiple neurophysiological measurements were collected and had to be associated with the task context. Therefore, we formulate our first requirement as follows:

R1: A software architecture for neurophysiological experiments needs the ability to support the simultaneous multi-modal measurement collection, including their synchronisation and their association with the experiment's task context.

Considering the need for synchronising different neurophysiological measurements, several solutions have evolved (e.g., Brownie, iMotions, Noldus Observer, CubeHX, and OpenVibe) [9, 10, 12–14]. These solutions ease the complexity and the time required to collect multi-modal data and associate it with the task context.

Extensible data analysis. Neurophysiological tools enable to obtain objective, continuous real-time measurements as a basis to determine the cognitive load of a subject. However, analysing such data is often challenging because it is notoriously noisy, requires cleaning before analysis and is difficult to interpret (e.g., discriminate the specific cause of the cognitive load changes) [3]. Moreover, there is no agreed upon way how to best measure cognitive load. To deal with these difficulties, a plethora of research aims to better understand cognitive load changes by analysing individual and combined modalities while using software artifacts [2, 8]. More, recently data-driven approaches based on machine learning became increasingly popular to link neurophysiological data and measures of interest [1]. Thus, extensible support for the processing of data is needed. Our second requirement is formulated as:

R2: A software architecture for neurophysiological experiments needs to be extensible in two ways: first, by supporting the ability to adopt external customized cleaning and analysis processes and second, by supporting the ability to include new devices and modalities.

Some software solutions (e.g., EEGLAB and OpenVibe) have emerged, that provide support for data analysis and enable researchers to incorporate their own cleaning and analysis processes [5, 14].

Transparent data analysis. Interlinked with the previous requirement of extensible data analysis, is the need for transparent data analysis. Transparent data analysis means to be able to trace how a particular measure was obtained from the raw neurophysiological measurements collected by the used neurophysiological tools. Thus, our third Requirement is formulated as follows:

R3: A software architecture for neurophysiological experiments needs to be transparent. The functionality of analysers and the flow of analysis processes should be visible to enable better understanding and optimization of these processes.

Current software solutions which support data analysis processes are domain and measurement specific. For example, EEGLAB and OpenVibe provide visibility and enable modifications of their analysis processes and their combinations [5, 14]. However, they primarily focus on EEG. On the contrary, other solutions (e.g., iMotions and CubeHX [10, 12]) offer more generic solution providing support for conducting multi-modal neurophysiological experiments. However, they are proprietary software solutions and it often remains opaque to the researcher how measures are derived from the raw neurophysiological measurements.

Repeatable data analysis. Deriving robust measures from raw neurophysiological measurements is often challenging because of the mapping of neurophysiological measures to theoretical constructs [6]. For instance, cognitive load can be measured using pupillary response data, but also using EEG or GSR data or combinations thereof [7, 8]. However, it is still unclear which (combination) of modalities and measures derived from them is the most robust and suitable for a particular setting. The application of data-driven methods based on machine learning to develop measures for theoretical constructs is getting increasingly popular [1, 8]. Therefore, we formulate our fourth requirement as follows:

R4: A software architecture for neurophysiological experiments needs to support repeatable data analysis, which allows data from previously collected neurophysiological experiments to be replayed and analysed with different analysis processes as well as applying pre-defined analysis processes on different data sets. Moreover, it needs to support the systematic testing and comparison of different analysis processes to develop more robust measures, even in near-realtime settings.

Software solutions like EEGLAB and OpenVibe support repeatable experiments by enabling the replay of data sets and the application of different data analysis processes including documentation of the process followed [5, 14]. However, the comparison between measures is non-automatic and, therefore determining a robust and suitable measure is non-trivial.

3 Proposed Software Architecture

We followed the design science research approach to develop the architecture [15]. To do that, we decomposed the architecture into subproblems based on the requirement list. To satisfy the list of requirements, we conducted a few engineering cycles and an architecture which is highly configurable and extensible, in the sense that a variety of components can be combined or included, has emerged.

Figure 1 shows the proposed architecture including its components which support researchers to define, deploy, execute and analyse experiments. The architecture

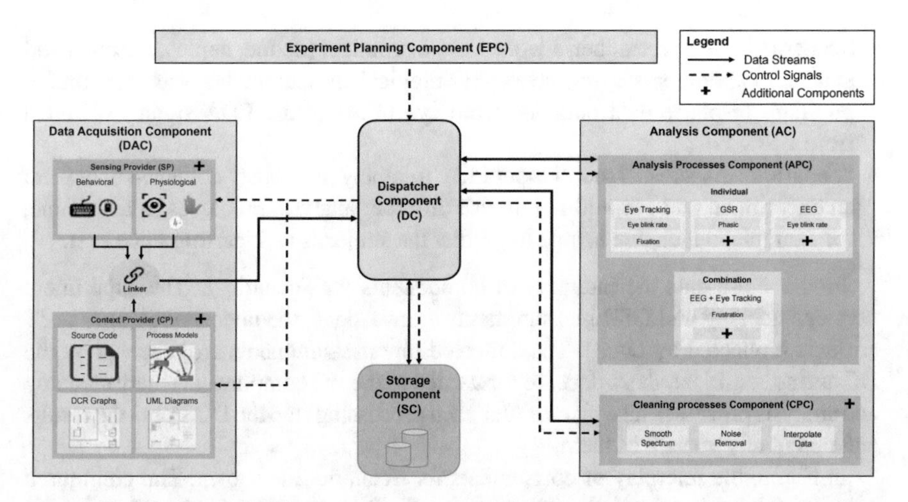

Fig. 1 The proposed architecture based on the selected requirements

consists of five main components: the Experiment Planning Component (EPC), the Data Acquisition Component (DAC), the Analysis Component (AC), the Storage Component (SC) and the Dispatcher Component (DC).

The EPC provides a user interface allowing researchers to define and deploy experiments including neurophysiological measurements. Moreover, the EPC allows researchers to specify reusable analysis pipelines. The DAC facilitates the collection of multi-modal measurements and the association of those with the task context, in a synchronised manner. The DAC consists of two providers: the sensing and the context provider. The sensing providers can cover user interactions (i.e. mouse clicks) and physiological measures (i.e. eye movements, skin conductance). The context providers, in turn, register events during the execution of an experimental task, e.g., events marking the start and end of sub-tasks (i.e. reading code).

The AC contains a vast amount of cleaning and analysis processes while new processes which extract new or existing measures can be incorporated. The AC enables modular analysis pipelines meaning that analysis pipelines can be easily set up, e.g., a pipeline to extract the eye blink rate from a stream of raw eye tracking data or to characterize pupillary responses to presented stimuli, and to replace cleaning and analysis processes in the pipeline to compare the behaviour of different processes.

The SC is responsible to store the collected raw data, the task context, the meta-data of the experiment (i.e., sampling rates and devices), the results of data analysis, and the applied analysis pipelines.

Finally, the DC is a scheduling component and is responsible to ensure that the experiment is executed according to its definition. To do that, it controls the data flow between the components (i.e., to retrieve, forward, and store data). Furthermore, the DC can instantiate multiple data analysis pipelines to run in a parallel manner.

Next, we will present the interplay of the different components using the following scenarios which are simple examples of how the software architecture can be useful.

- Scenario 1: As a researcher, I would like to conduct (define, deploy, execute, and analyse) experiments using neurophysiological measurements, e.g., combining pupillary response data obtained from eye tracking and EDA signals obtained from GSR.
- Scenario 2: As a researcher, I would like to replay previously collected data from an experiment and I would like to use different analysis processes to determine, for example, which one is able to predict the subject's task performance best.

Figure 2 presents the interplay of components for Scenario 1. The experiment uses eye tracking and GSR measurements to investigate the understandability of IT artifacts (collected by DAC). The collected raw measurements are received by the DC and stored. Next, these data are forwarded to the AC components where cleaning and analysis processes take place. After each processing step the DC stores the results and potentially visualizes them.

In Fig. 3, the interplay of components for Scenario 2 is shown. The configured pipeline represents an experiment where previously collected data (from Scenario 1) will be replayed to investigate the understandability of software artifacts, but in this case different analysis pipelines are used. For example, different analysis processes could be used to test which combination is better able to predict the subject's task performance.

The envisioned architecture supports researchers during all stages of the experimental cycle, i.e., experiment planning, experiment execution, and experiment analysis by fulfilling the requirements set in Sect. 2 [16]. The suggested architecture supports *Multi-modal measurements collection (R1)* by enabling the collection of neurophysiological measurements in a synchronised manner linked with the task context (cf. DAC). Moreover, it provides a modular architecture supporting *Extensible data analysis (R2)* with sub-components each representing a different data analyser (cf. AC). Additional components can be added without influencing the functionality

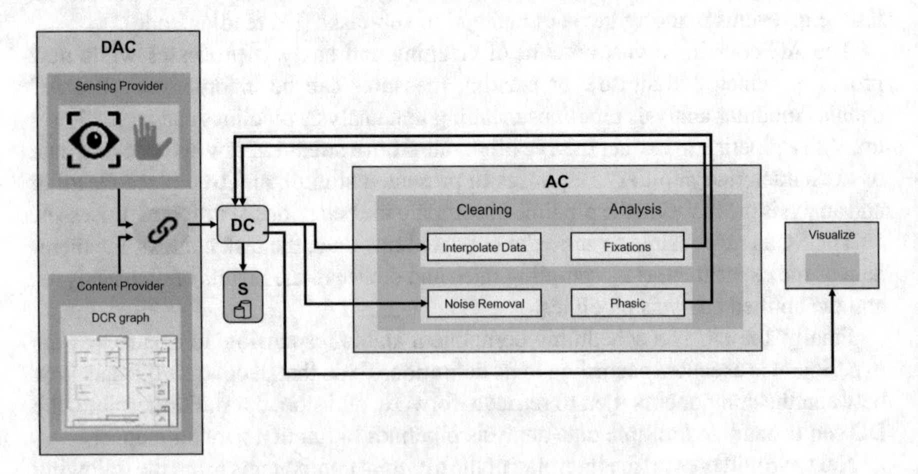

Fig. 2 Interplay of components: Scenario 1

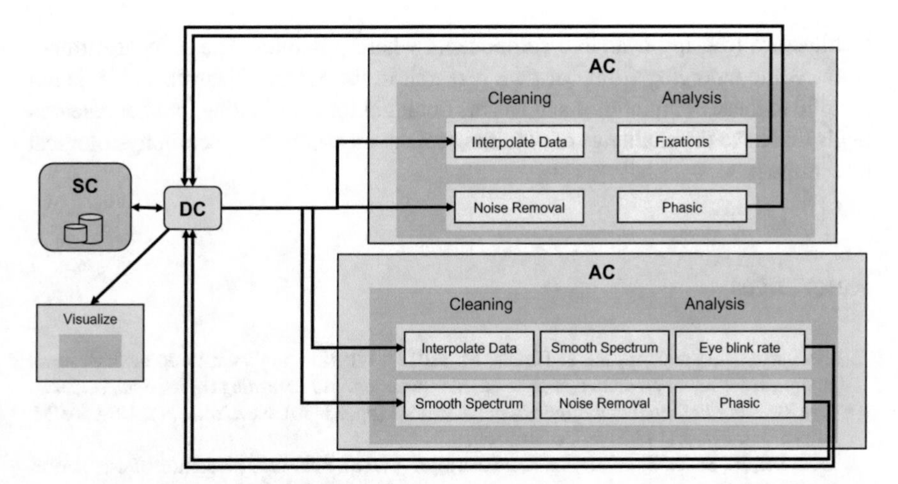

Fig. 3 Interplay of components: Scenario 2

of other components. The analysis components are transparent (*Transparent data analysis (R3)*) due to their open-source nature. Moreover, data analysis pipelines can be built that define the data analysis process in a reusable manner using the EDC and can be later traced since a history of those is maintained by the SC. Finally, the architecture supports *Repeatable data analysis (R4)* since data can be replayed (cf. DC) and experiments can be repeated using different analysis pipelines. Moreover, existing analysis pipelines can be reused for different data sets. Additionally, the DC ensures the capability of comparing analysis pipelines, because it is designed for instantiating several analysis processes and run those in a parallel manner. In conclusion, the proposed architecture supports the collection of neurophysiological measurements in a synchronised manner and supports the extensible, transparent, and repeatable analysis of data including the comparison of analysis processes.

4 Conclusions

This paper elaborates on the requirements and describes a preliminary software architecture that aids researchers conducting (defining, deploying, executing and analysing) neurophysiological experiments in an extensible, transparent and repeatable manner. While the collection and synchronisation of neurophysiological measurements is already well supported by existing software solutions, most software solutions only provide limited support to the extensible, transparent, and repeatable analysis of neurophysiological data. In particular, current solutions do not provide automatic means to systematically compare different data analysis pipelines to obtain robust measures. As a next step, we will start implementing a software framework for conducing neurophysiological experiments based on the proposed architecture (reusing available components where possible). The software framework, while emerging from our own research in the SE and NeuroIS fields, is not specific to these communities and has the potential for contributing in other research fields that have the challenge of obtaining robust measures from neurophysiological measurements.

References

1. Bednarik, R., Vrzakova, H., & Hradis, M. (2012). What do you want to do next: A novel approach for intent prediction in gaze-based interaction. In *Proceedings of the Symposium on Eye Tracking Research and Applications, ETRA'12* (pp. 83–90). New York, NY, USA: ACM. https://doi.org/10.1145/2168556.2168569.
2. Buettner, R., Sauer, S., Maier, C., & Eckhardt, A. (2018). Real-time prediction of user performance based on pupillary assessment via eye-tracking. *AIS Transactions on Human-computer Interaction,* 26–60. https://doi.org/10.17705/1thci.00103.
3. Chen, F., Ruiz, N., Choi, E., Epps, J., Khawaja, M., Taib, R., et al. (2012). Multimodal behavior and interaction as indicators of cognitive load. *ACM Transactions on Interactive Intelligent Systems (TiiS), 2*. https://doi.org/10.1145/2395123.2395127.
4. Chen, F., Zhou, J., Wang, Y., Yu, K., Arshad, S., Khawaji, A., & Conway, D. (2016). Robust multimodal cognitive load measurement. In *Human–computer interaction series*.
5. Delorme, A., & Makeig, S. (2011). Eeglab: An open source toolbox for analysis of single-trial eeg dynamics including independent component analysis. https://doi.org/10.1016/j.jneumeth.2003.10.009.
6. Dimoka, A., Banker, R. D., Benbasat, I., Davis, F. D., Dennis, A. R., Gefen, D., et al. (2012). On the use of neurophysiological tools in is research: Developing a research agenda for neurois. *Mis Quarterly: Management Information Systems, 36*(3), 679–702.
7. Duchowski, A. T., Krejtz, K., Krejtz, I., Biele, C., Niedzielska, A., Kiefer, P., et al. (2018). The index of pupillary activity: Measuring cognitive load vis-à-vis task difficulty with pupil oscillation. In *Proceedings of the 2018 CHI Conference on Human Factors in Computing Systems, CHI'18* (pp. 282:1–282:13). New York, NY, USA: ACM. http://doi.acm.org/10.1145/3173574.3173856.
8. Fritz, T., Begel, A., Müller, S. C., Yigit-Elliott, S., & Züger, M. (2014). Using psycho-physiological measures to assess task difficulty in software development. In *Proceedings of the 36th International Conference on Software Engineering, ICSE 2014* (pp. 402–413). New York, NY, USA: ACM. http://doi.acm.org/10.1145/2568225.2568266.
9. Hariharan, A., Lux, E., Pfeiffer, J., Dorner, V., Maller, M. B., Weinhardt, C., & Adam, M. T. P. (2017). Brownie: A platform for conducting NeuroIS experiments. *Journal of the Association of Information Systems, 18*(4), 264–296.
10. iMotions. (2018). https://imotions.com/.
11. Lee, S., Hooshyar, D., Ji, H., Nam, K., & Lim, H. (2017). Mining biometric data to predict programmer expertise and task difficulty. *Cluster Computing, 21*(1), 1–11.
12. Léger, P. M., Courtemanche, F., Fredette, M., & Sénécal, S. (2019). A cloud-based lab management and analytics software for triangulated human-centered research. *Lecture Notes in Information Systems and Organisation, 29,* 93–99.
13. Noldus, L. (1991). The observer—A software system for collection and analysis of observational data. *Behavior Research Methods Instruments and Computers, 23*(3), 415–429.
14. Renard, Y., Lotte, F., Gibert, G., Congedo, M., Maby, E., Delannoy, V., et al. (2010). Openvibe: An open-source software platform to design, test, and use brain-computer interfaces in real and virtual environments. *PRESENCE: Teleoperators and Virtual Environments, 19,* 35–53.

15. Wieringa, R. (2014). *Design science methodology for information systems and software engineering*. Berlin: Springer. https://doi.org/10.1007/978-3-662-43839-8.
16. Wohlin, C., Runeson, P., Höst, M., Ohlsson, M. C., Regnell, B., & Wesslé, A. (2012). Experimentation in software engineering. In *Experimentation in software engineering* (pp. 1–236). https://doi.org/10.1007/978-3-642-29044-2.

Machine Learning Based Diagnosis of Diseases Using the Unfolded EEG Spectra: Towards an Intelligent Software Sensor

Ricardo Buettner, Thilo Rieg and Janek Frick

Abstract In this research-in-progress work we sketch a roadmap for the development of a novel machine-learning-based EEG software sensor. In the first step we present the idea to unfold the EEG standard bandwidths in a more fine-graded equidistant 99-point spectrum to improve accuracy when diagnosing diseases. We use this novel pre-processing step prior to entering a Random Forests classifier. In the second step we evaluate the approach on alcoholism and epilepsy and demonstrate that the approach outperforms all benchmarks. The third step sketches a further improvement by replacing the hard-coded equidistant 99-point spectrum with a flexibly-grading spectrum. In the fourth step we combine the flexibly-grading EEG spectrum, the spatial locations of the EEG electrodes, and the EEG recording time to train an intelligent EEG software sensor using self-organizing feature mapping. Our work contributes to NeuroIS research by analyzing EEG as a bio-signal though a novel machine-learning approach.

Keywords Electroencephalography · Random forests · Spectral analysis · Machine learning

1 Introduction

Electroencephalography (EEG) is the second most dominant tool in NeuroIS research [1]. While the "*EEG reflects many thousands of simultaneously ongoing brain processes*" [2, p. 917], EEG scholars widely use fixed bandwidths to analyze this complex bio-signal (Fig. 1).

But the usage of very rigid and rough bandwidths substantially limits the capabilities when analyzing a complex signal. Müller-Putz et al. emphasized the following: "*Please note that the exact thresholds of these frequency bands vary in the scientific literature and, hence, are, at least to some degree, subject to debate*" [2, p. 917].

R. Buettner (✉) · T. Rieg · J. Frick
Aalen University, Aalen, Germany
e-mail: ricardo.buettner@hs-aalen.de

F. D. Davis et al. (eds.), *Information Systems and Neuroscience*,
Lecture Notes in Information Systems and Organisation 32,
https://doi.org/10.1007/978-3-030-28144-1_18

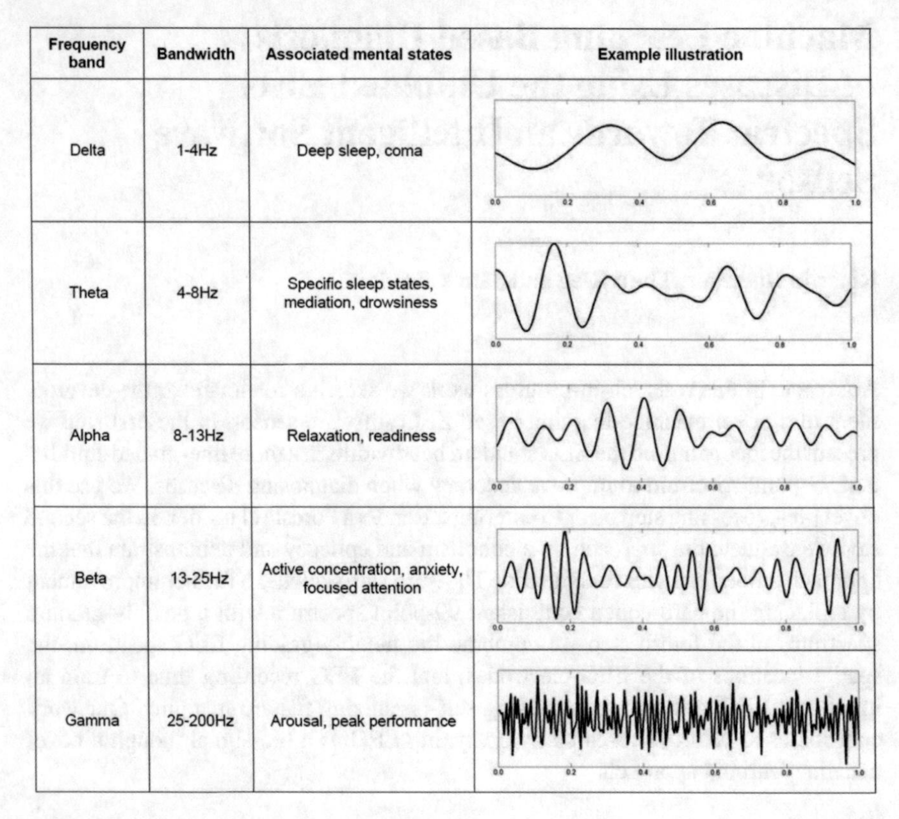

Frequency band	Bandwidth	Associated mental states	Example illustration
Delta	1-4Hz	Deep sleep, coma	
Theta	4-8Hz	Specific sleep states, mediation, drowsiness	
Alpha	8-13Hz	Relaxation, readiness	
Beta	13-25Hz	Active concentration, anxiety, focused attention	
Gamma	25-200Hz	Arousal, peak performance	

Fig. 1 Fixed EEG frequency bands from [2, p. 918]

In this work we relax the limitation of rigid and rough bandwidths by proposing an intelligent EEG software sensor that adapts to EEG information entropy. Instead of using fixed EEG bandwidths, the software sensor uses an adaptable fine-graded spectrum.

2 Methodology

The whole work is embedded in the Information Systems design science framework [3]. In order to clearly contribute to NeuroIS research and show strong methodological rigor, we followed the NeuroIS guidelines described by vom Brocke et al. [4].

We used two data sets for this work: An EEG data set with alcoholics collected by Henry Begleiter and an epilepsy data set from the University of Bonn. The data set on alcoholism comprised 122 subjects where brain activity was recorded at 256 Hz on 64 electrodes. The subjects completed 120 tests of different stimulation. The full alcoholism dataset is available from https://archive.ics.uci.edu/ml/machine-learning-databases/eeg-mld/eeg.html. For the epilepsy classification problem, four sets (A to D) of the epilepsy data from the University of Bonn were selected. Each set consists of 100 single channel EEG segments with duration of 23.6 s. These segments were obtained from multichannel EEG recordings and were visually inspected for artifacts triggered by eye movement or muscle activity. Sets A and B consist of EEG images of healthy people, A with eyes open, B with eyes closed. Sets C and D were measured on epileptics in the seizure-free state, D at the epileptogenic zone and C at the hippocampal formation of the opposite hemisphere. The sampling rate was 173.61 Hz. The full epilepsy dataset is available from http://epileptologie-bonn.de/cms/upload/workgroup/lehnertz/eegdata.html.

2.1 Step 1: Improve Random Forests Classification by Using the Fine-Graded Equidistant 99-Point EEG Spectrum

The Random Forests algorithm was proposed by Breiman [5]. It is a machine learning classifier which is based on an ensemble (a bag) of unpruned decision trees [6, p. 268]. Ensemble methods are related to the concept that an aggregated decision from various experts is often superior to a decision from a single system [7, p. 35]. The final classification decision is built on the majority vote.

In the first step we unfolded the EEG standard bandwidths in a more fine-graded equidistant 99-point spectrum, before entering this fine-graded spectrum into the Random Forests classifier (Fig. 2).

Before calculating the 99 equidistant fine-graded frequency power slices from 0.5 to 50 Hz at a step of 0.5 Hz, all regular preprocessing operations have to be conducted (including bandpass filtering, and ICA). Subsequently the power spectrum of each 0.5 Hz frequency band was calculated. For example, looking at the data of alcoholics, a matrix with 122 subjects and 99 features each was created. Based on this data, the Random Forests classifier was trained and classified according to the two classes "healthy" and "sick". In order to obtain a robust model and avoid overfitting, the 10-fold cross validation was applied. For evaluation purposes, 25% of the data were holding out. On these 25% of the data, which were not used for training the algorithm, the final model was evaluated.

2.2 Step 2: Evaluate the Improved Random Forests Classifier on Various Disease Classification Problems

In the second step we evaluated the improved Random Forests approach on various illnesses such as alcoholism and epilepsy.

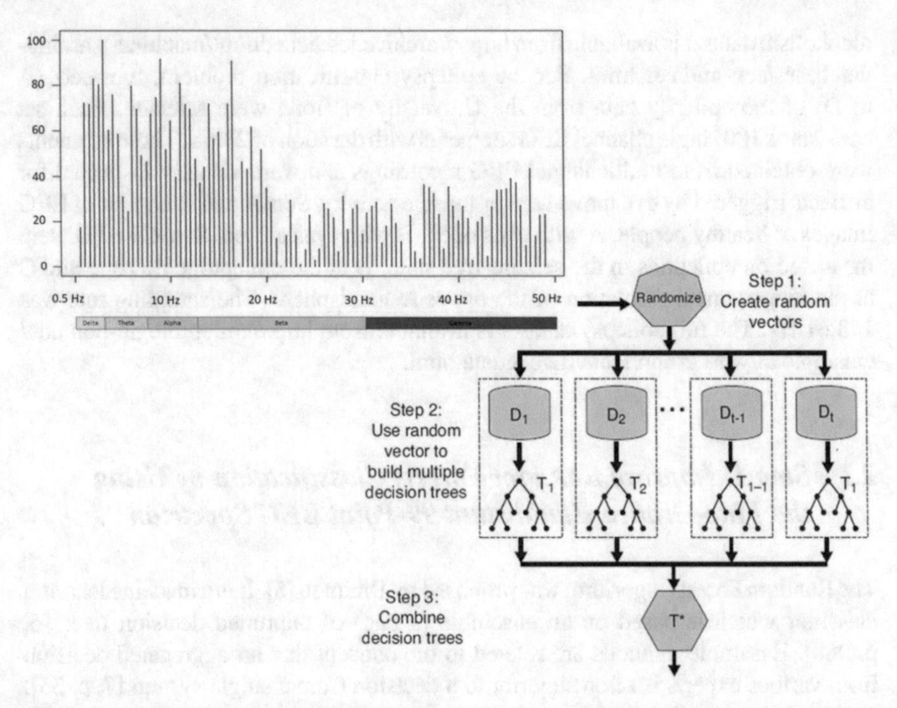

Fig. 2 Entering the fine-grade 99-point EEG spectrum data into the Random Forests classifier

Alcoholism is a globally relevant addiction problem. Regular and long-term alcohol consumption leads to a reduction of brain cells and shrinkage of brain mass. Alcoholism affects abilities such as coordination, regulation of body temperature, sleep, mood, cognitive abilities and memory [8], and its effects are reflected in EEG data [9].

Epilepsy is one of the most common disorders of the central nervous system and affects people of all ages, gender and ethnic backgrounds [10]. Epileptic seizures are characterized by an unpredictable occurrence pattern and transient dysfunctions of the central nervous system and excessive and synchronous abnormal locally neuronal activity in the cortex [11].

2.3 *Step 3: Transform the Hard-Coded 99-Point EEG Spectrum to a Flexibly-Grading Spectrum*

The third step sketches a further improvement by replacing the hard-coded equidistant 99-point spectrum by using a flexibly-grading spectrum based on information entropy analysis using decision trees. Since the conceptual idea underlying a decision tree is to recursively identify a predictor that allows the sample to be split in two subparts that are as homogeneous as possible with regard to the classification task at hand [12,

Table 1 Dataset-neutral comparison of diagnosis accuracies between our approach and the existing benchmark to date on diagnosing alcoholism and epilepsy

Disease	Benchmark to date	Our accuracy to date
Alcoholism	95.83% by Bajaj et al. [15]	97.40% (already published in [16])
Epilepsy	98.45% by Mursalin et al. [10]	99.23% (in preparation for publishing)

p. 537], decision tree analysis can be used to identify the importance of predictors (here: fine-graded bands). More important bands will then be sub-split again to see if these sub-splits are more effective compared to the original band.

2.4 *Step 4: Build an Intelligent EEG Software Sensor*

In the fourth step the flexibly-grading EEG spectrum (dimension 1), the spatial locations of the EEG electrodes (dimension 2), and the EEG recording time (dimension 3) are used to train an intelligent EEG software sensor combining these three dimensions. Technically we use self-organizing feature mapping [13]—an unsupervised learning approach mapping the input space (here: three dimensions partly with varying resolutions) to a low-dimensional (two-dimensional) sensor.

It was found that the natural representation of neural frequency sensors on the auditory cortex is nearly identical to the computational results of self-organizing feature mapping (e.g. for bats [14]).

3 Results

Here we report step 2 evaluation results on two diseases, i.e. alcoholism and epilepsy.

As shown in Table 1 our approach outperforms the current benchmarks (for the same datasets) and we nearly scrape the 100% line.

4 Discussion

Our first results demonstrate that, due to the massive aggregation of spectral data in EEG analyses using fixed bands, predictive information is lost. Based on our improved fine-graded equidistant 99-point EEG spectrum we can reliably diagnose globally relevant diseases and addictions such as epilepsy and alcoholism.

While our first results also challenge the medical—eventually outdated—EEG standard bandwidths, we expect to gain more insights from step 3 (flexibly-grading EEG spectrum).

We also expect new insights from step 4 since self-organizing feature maps are useful for the visualization of high-dimensional bio-data on low-dimensional views.

5 Limitation and Future Work

The main limitation is related to the fact that our ambitious work is a research-in-progress. Despite this we successfully evaluated our approach on two illnesses/addictions (epilepsy, alcoholism) with a very good level of accuracy; a more intensive evaluation of other diseases is needed. That is why in future work we will report results for dementia, brain tumors, strokes, autism, insomnia, and anesthesia as well as including step 3 and 4 results.

In addition we plan

- to re-evaluate previous studies on mental concepts such as cognitive workload [17–24], concentration [25], mindfulness [26, 27], and personality [28–30] with the improved Random Forests algorithm,
- to triangulate the intelligent EEG software sensor using other NeuroIS sensors [31, 32], and,
- to transfer our preprocessing step to convolutional neural network processing in completely other application domains [33–35].

References

1. Riedl, R., Fischer, T., & Léger, P. M. (2017). A decade of NeuroIS research: Status quo, challenges, and future directions. In *ICIS 2017 Proceedings: 38th International Conference on Information Systems*, December 10–13, 2017, Seoul, South Korea.
2. Müller-Putz, G. R., Riedl, R., & Wriessnegger, S. C. (2015). Electroencephalography (EEG) as a research tool in the information systems discipline: Foundations, measurement, and applications. *CAIS, 37,* 911–948.
3. Hevner, A. R., March, S. T., Park, J., & Ram, S. (2004). Design science in information systems research. *MISQ, 28*(1), 75–105.
4. vom Brocke, J., & Liang, T.-P. (2014). Guidelines for neuroscience studies in information systems research. *JMIS, 30*(4), 211–234.
5. Breiman, L. (2001). Random forests. *Machine Learning, 45*(1), 5–32.
6. Buettner, R. (2018). Robust user identification based on facial action units unaffected by users' emotions. In *HICSS-51 Proceedings* (pp. 265–273).
7. Buettner, R., Sauer, S., Maier, C., & Eckhardt, A. (2018). Real-time prediction of user performance based on pupillary assessment via eye-tracking. *AIS Transactions on Human-Computer Interaction, 10*(1), 26–56.
8. NIH National Institute on Alcohol Abuse and Alcoholism. (2010). *Beyond Hangovers—Understanding Alcohol's impact on your health* (Vol. 15, pp. 6–8). NIH Publication.

9. Mumtaz, W., Vuong, P., Xia, L., Malik, A., & Rashid, R. (2017). An EEG-based machine learing method to screen alcohol use disorder. *Cognitive Neurodynamics, 11*(2), 161–171.
10. Mursalin, M., Zhang, Y., Chen, Y., & Chawla, N. (2017). Automated epileptic seizure detection using improved correlation-based feature selection with random forest classifier. *Neurocomputing, 241,* 204–214.
11. Ngugi, A., Kariuki, S., Bottomley, C., Kleinschmidt, I., Sander, J., & Newton, C. (2011). Incidence of epilepsy a systematic review and meta-analysis. *Neurology, 77*(10), 1005–1012.
12. Buettner, R., Sauer, S., Maier, C., & Eckhardt, A. (2015). Towards ex ante prediction of user performance: A novel NeuroIS methodology based on real-time measurement of mental effort. In *HICSS-48 Proceedings* (pp. 533–542).
13. Kohonen, T. (1997). *Self-organizing maps*. Berlin: Springer.
14. Martinetz, T., Ritter, H., & Schulten, K. (1988). Kohonens self-organizing maps for modeling the formation of the auditory cortex of a bat. In R. Pfeifer (Ed.), *Connectionism in perspective* (pp. 403–412). Amsterdam: North-Holland.
15. Bajaj, V., Guo, Y., Sengur, A., Siuly, S., & Alcin, O. F. (2017). A hybrid method based on time–frequency images for classification of alcohol and control EEG signals. *Neural Computing and Applications, 28*(12), 3717–3723.
16. Rieg, T., Frick, J., Hitzler, M., & Buettner, R. (2019). High-performance detection of alcoholism by unfolding the amalgamated EEG spectra using the Random Forests method. In *HICSS-52 Proceedings* (pp. 3769–3777).
17. Buettner, R., Timm, I. J., Scheuermann, I. F., Koot, C., & Rössle, M. (2017). Stationarity of a user's pupil size signal as a precondition of pupillary-based mental workload evaluation. In *Information systems and neuro science: Gmunden Retreat on NeuroIS 2017*, June 12–14, 2017, Gmunden, Austria.
18. Buettner, R. (2016). The relationship between visual website complexity and a user's mental workload: A NeuroIS perspective. In *Information systems and neuro science* (Vol. 10 of LNISO, pp. 107–113), Gmunden, Austria.
19. Buettner, R. (2016). A user's cognitive workload perspective in negotiation support systems: An eye-tracking experiment. In *PACIS 2016 Proceedings*.
20. Buettner, R. (2015). Investigation of the relationship between visual website complexity and users' mental workload: A NeuroIS perspective. In *Information systems and neuro science* (Vol. 10 of LNISO, pp. 123–128), Gmunden, Austria.
21. Buettner, R. (2014). Analyzing mental workload states on the basis of the pupillary hippus. In *NeuroIS'14 Proceedings* (p. 52).
22. Buettner, R., Daxenberger, B., Eckhardt, A., & Maier, C. (2013). Cognitive workload induced by information systems: Introducing an objective way of measuring based on pupillary diameter responses. In *Pre-ICIS HCI/MIS 2013 Proceedings*, 2013, paper 20.
23. Buettner, R. (2013). Cognitive workload of humans using artificial intelligence systems: Towards objective measurement applying eye-tracking technology. In *KI 2013 Proceedings*, ser. LNAI (Vol. 8077, pp. 37–48).
24. Buettner, R., Daxenberger, B., & Woesle, C. (2013). User acceptance in different electronic negotiation systems—A comparative approach. In *ICEBE 2013: Proceedings of the 10th IEEE International Conference on e-Business Engineering*, September 11–13, Coventry, UK, 2013 (pp. 1–8). IEEE CS Press.
25. Buettner, R., Baumgartl, H., & Sauter, D. (2018). Microsaccades as a predictor of a user's level of concentration. In F. D. Davis, et al. (Eds.), *Information systems and neuroscience: NeuroIS retreat 2018*. Lecture Notes in Information Systems and Organisation (LNISO) (Vol. 29, pp. 173–177).
26. Sauer, S., Buettner, R., Heidenreich, T., Lemke, J., Berg, C., & Kurz, C. (2018). Mindful machine learning: Using machine learning algorithms to predict the practice of mindfulness. *European Journal of Psychological Assessment, 34*(1), 6–13.
27. Sauer, S., Lemke, J., Zinn, W., Buettner, R., & Kohls, N. (2015). Mindful in a random forest: Assessing the validity of mindfulness items using random forests methods. *Journal of Personality and Individual Differences, 81,* 117–123.

28. Buettner, R. (2017). Predicting user behavior in electronic markets based on personality-mining in large online social networks: A personality-based product recommender framework. *Electronic Markets: The International Journal on Networked Business, 27*(3), 247–265.
29. Buettner, R. (2016). Innovative personality-based digital services. In *PACIS 2016 Proceedings: 20th Pacific Asia Conference on Information Systems (PACIS)*, June 27–July 1, Chiayi, Taiwan.
30. Buettner, R. (2016). Personality as a predictor of business social media usage: An empirical investigation of XING usage patterns, In *PACIS 2016 Proceedings: 20th Pacific Asia Conference on Information Systems (PACIS)*, June 27–July 1, Chiayi, Taiwan.
31. Buettner, R. (2017). Asking both the user's brain and its owner using subjective and objective psychophysiological NeuroIS instruments. In *ICIS 2017 Proceedings: 38th International Conference on Information Systems*, December 10–13, 2017, Seoul, South Korea.
32. Buettner, R. (2013). Social inclusion in eParticipation and eGovernment solutions: A systematic laboratory-experimental approach using objective psychophysiological measures, In *EGOV/ePart 2013: Proceedings of the Joint Conference of IFIP EGOV 2013 & IFIP ePart 2013*, September 16–19, Koblenz, Germany, 2013 (Vol. P-221 of Lecture Notes in Informatics (LNI)—Proceedings, pp. 260–261).
33. Baumgartl, H., Tomas, J., Buettner, R., & Merkel, M. (2019). A novel deep-learning approach for automated non-destructive testing in quality assurance based on convolutional neural networks. In *Proceedings of the 13th International Conference on Advanced Computational Engineering and Experimenting*, July 1–5, 2019, Athens, Greece. Accepted.
34. Buettner, R., & Baumgartl, H. (2019). A highly effective deep learning based escape route recognition module for autonomous robots in crisis and emergency situations. In *HICSS-52 Proceedings*, January 8–11, 2019, Maui, Hawaii (pp. 659–666).
35. Baumgartl, H., Buettner, R., Bernthaler, T., Timm, I. J., Jansche, A., & Schneider, G. (2018). Colored micrographs significantly outperform grayscale ones in modern machine learning: Insights from a systematical analysis of lithium-ion battery micrographs using convolutional neural networks. In *Proceedings of the 13th Multinational Congress of Microscopy*, September 24–29, Rovinj, Croatia.

The Impact of Symmetric Web-Design: A Pilot Study

Aurélie Vasseur, Pierre-Majorique Léger and Sylvain Sénécal

Abstract In this research-in-progress paper we propose a pilot study to investigate the effect of symmetric web-design. Cognitive, affective, and behavioral impacts of symmetry are investigated using EEG, skin conductance, heart-rate measures, and a questionnaire on an experiment of 14 participants, each presented with 100 stimuli of symmetric (50) and asymmetric (50) conditions in a random order. We expect to observe a longer sustained posterior negativity in the Event-Related Potential (ERP), a higher emotional response, as well as a higher visual appeal for the symmetric condition. Results are currently being analyzed and will be presented at the conference.

Keywords Visual appeal · User experience · EEG · Event-related potential (ERP) · Symmetry

1 Introduction

In website design, the visual appeal (equally called *visual aesthetics* or *aesthetics* [1]) of a website is determined during the first 50 ms of viewing [1]. *Visual aesthetics* has been defined as a multidimensional construct with two main factors: Classic Aesthetics and Expressive Aesthetics [2]. One of the dimensions of the classic aesthetics is symmetry. Symmetry has also been correlated with visual appeal in a number of studies about website design [3–6].

However, little is known about the role of symmetric web design on users' first impressions. To our knowledge, there are no studies examining the affective and cognitive mechanisms involved in the relationship between symmetry and visual attractiveness. The literature on first impressions in human-computer interaction have

A. Vasseur (✉) · P.-M. Léger · S. Sénécal
HEC, Montréal, Canada
e-mail: aurelie.vasseur@hec.ca

P.-M. Léger
e-mail: pierre-majorique.leger@hec.ca

S. Sénécal
e-mail: sylvain.senecal@hec.ca

F. D. Davis et al. (eds.), *Information Systems and Neuroscience*,
Lecture Notes in Information Systems and Organisation 32,
https://doi.org/10.1007/978-3-030-28144-1_19

focused on higher level factors such as visual appeal [1], but have not considered specific dimensions of appeal such as symmetry as having an extensive role in users' judgement. Understanding the role of symmetry also has highly practical relevance and the findings could provide clear recommendations for web designers. If the effect of symmetry on visual attractiveness is found to be significant, web-designers could easily and virtually cost-free implement it to improve their visual appeal to their customer base.

In order to examine the impact of symmetry, we developed a one-factor (symmetric and control) within-subject experiment with 50 stimuli per condition. The pilot study recorded electroencephalography (EEG) measures of 14 participants. We conducted an event-related potential (ERP) analysis to extract information related to the first 800 ms after the presentation of the stimuli. Previous ERP studies have found that symmetry modulates later components of the ERP, more specifically, sustained negativity is observed after P1 and N1, typically from 220 ms after the onset of the stimulus [7]. Additionally, we recorded electrodermal activity (EDA) and heart rate (HR) measures to assess the emotional states (arousal and valence) of the participants.

This study provides preliminary results on the role of symmetry in visual appeal and on the mechanisms under which symmetry leads to better appreciation of a webpage, therefore indicating for which scenarios designers should opt for symmetry depending on their goals and audience.

2 Hypothesis Development

Symmetry has a long tradition of being synonymous with beauty and is typically studied in the artistic field. It is in that context that Leder et al. [8] developed their psychological model of aesthetic appreciation. They propose that symmetry, among other characteristics (complexity, contrast, order, grouping), impacts our aesthetic judgment through a continuous psychological evaluation that involves both implicit and explicit stages. When examining a piece of artwork, individuals make an aesthetic judgment: They first integrate the object of interest with their own experience through implicit memory integration, then try to understand it through cognitive processing. For Leder et al. [8], aesthetic judgment is the result of the evaluation of the artwork while aesthetic emotion is the byproduct of this processing. One is related to the cognitive aspect, the other to the emotional one, but both are involved when an individual is looking at a piece of artwork. We argue that the first appraisal of the visual aesthetics of websites follows the same psychological process as the appraisal of artwork.

Indeed, the design of a website has a dual purpose: being visually appealing and being functional. Since we are considering the first impressions of a user viewing a website in this study, the functional aspect will not be evaluated, only the visual aspect will be, and a parallel can be drawn between experiencing the visual of a webpage and visual art. Therefore, a potential mechanism under which symmetry impacts the visual appeal of a website would be through a change in cognitive processing and emotional state. One aspect of aesthetic judgment is understanding the image. Symmetry adds redundancy, and reduces ambiguity [8], the website becomes more predictable. The individual will be able to faster and more efficiently understand the image presented to him or her [8, 9]. This rapid understanding could lead to higher satisfaction and affect the emotional state of the individual. We already know that visual perception of symmetry alters brain rhythms [7] and activates higher-order regions of the human brain [10]. As mentioned, previous studies on symmetry have found that symmetry alters the ERP after 220 ms and results in sustained posterior negativity (SPN) [7]. Leder et al. [8] suggest an effect on the emotional state but no automatic emotional response to symmetry have been observed with neurophysiological tools so far [7]. However, behavioural measures suggest a positive affective response [11]. Overall, this led to the development of the following hypotheses:

H1: Symmetry positively impacts the perceived visual appeal of a webpage (vs. asymmetry).
H2: Symmetric webpage results in a longer SPN (vs. asymmetric).
H3: Symmetric webpage induces a higher emotional response (vs. asymmetric).

3 Event-Related Pilot Study

3.1 Symmetry and Asymmetry

Bilateral symmetry has been chosen due to its practical relevance in web design. Considering that most, if not all, websites allow the user to scroll down the page, it is difficult to consider a horizontally symmetric design for the purpose of this study. We refer to vertical symmetry in this study, other types of symmetry could be considered, however as this is the first study to examine this question, vertical symmetry has interesting properties in practice as mentioned.

In our experimental design, to be considered symmetric, the main component of the webpage has to be divided in the middle by an imaginary vertical line. Both sides of the line would have similar elements (i.e., buttons menu, text, images) that perform functions that have equal weight and importance to the user. For example, if an image has a form to fill only on one side and other types of content on the other side, the stimulus will not be considered symmetric (Fig. 1a, b). To test the validity of the stimuli, the full set was presented to three independent coders who had to sort the 100 stimuli under the symmetric or asymmetric condition. An average of 82% correct result was achieved and was considered salient enough to proceed with the pilot study. The stimuli are all screenshots of real websites. All have won awards in web-design in 2018 or were presented as an example of good design on specialized websites. This ensured that both conditions were of the same quality on average and limited bias in stimuli creation. They are all homepages with a limited amount of text to avoid inducing a high cognitive load through content.

(a)

Fig. 1 **a** Two examples of symmetric design. **b** Two examples of asymmetric design

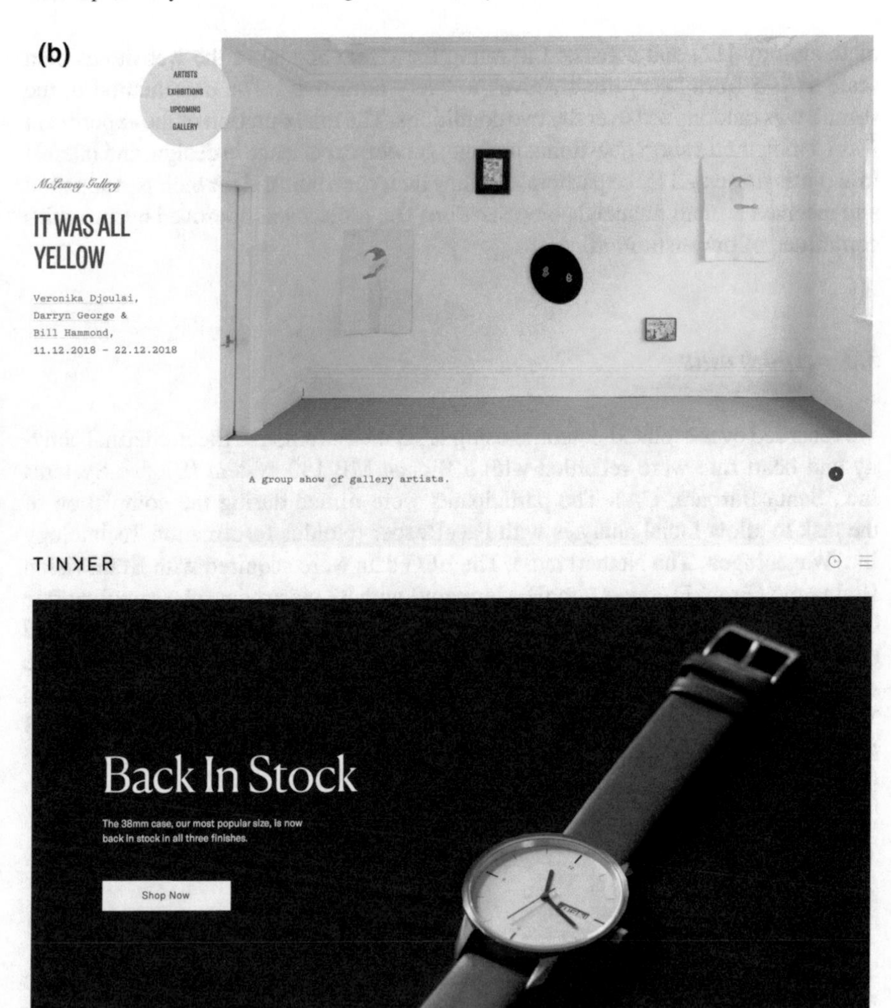

Fig. 1 (continued)

3.2 *Experimental Design*

An event-related study was conducted with 14 participants (6 male, 8 female; average age 23.4). The experimental task is illustrated in Fig. 1. One factor was tested over two conditions, symmetry and control (i.e., asymmetry). A total of 100 stimuli were presented to each participant in a within-subject experimental design (a total of 1400 ERP responses recorded). For each stimulus presented during 5 s, a perceptual question regarding the visual appeal of the page on a 5-point Likert scale was asked to the participant, with no time limit, then a fixation cross was displayed for 1 s before the presentation of the next stimulus. The perceptual question followed previous methodology [12] and consisted in rating the visual appeal of the website using a scale of 1–5 from "very unattractive" to "very attractive". The presentation of the stimuli was randomized over the two conditions. The total duration of the experiment was 15 min, then a short questionnaire (age, gender, experience in design, and interest in art) was presented to the participant. They then were thanked for their participation and received a small financial compensation. The project was approved by the ethics committee of our institution.

3.3 Apparatus

We recorded event-related potential using EEG measurements. Electrodermal activity and heart rate were recorded with a Biopac MP 150 system (Biopac Systems Inc., Santa Barbara, CA). The participants were filmed during the completion of the task to allow facial analysis with FaceReader (Noldus Information Technology Inc, Wageningen, The Netherlands). The EEG data were acquired with BrainVision BrainAmp (Brain Products CmbH, Germany) with 32 electrodes, at a sampling rate of 250 Hz. The software ePrime (eStudio 3.0, Psychology Software Tools) installed on a Windows 10 computer controlled the timing and presentation of stimuli. Time markers corresponding to the stimuli presentation were sent by ePrime to Brain Vision responsible for the recording of the EEG. Responses were given via a USB keyboard. The EEG data were reduced to segments of −200 ms to 800 ms around the stimulus onset for ERP analysis (Fig 2).

4 Ongoing Work

The results of the pilot study are currently being analyzed with SPSS 25.0.0.1 to test the three hypotheses. First, for the perceptual question regarding the relationship between symmetry and visual appeal, a repeated measure ANCOVA test will be run between the symmetric and the asymmetric conditions, with the results of the questionnaire as covariates. To test our second hypothesis, we will examine whether sustained posterior negativity is observed under the symmetric condition doing an ERP analysis. Last, we will test whether symmetry induces a higher emotional response by doing a repeated measure ANCOVA of emotional arousal on symmetry, with the results of the questionnaires as covariates. Additional post hoc analysis may include coding each of the stimuli under additional factors such as color, text quantity, composition rules (golden ratio, three tiers rules, etc.) which could potentially be impacting the visual attractiveness rating as well. We are also considering testing if there are different cognitive mechanisms underlying the visual attractiveness. We will be grouping the ERPs that correspond to a low to medium rating (1, 2, 3), and high rating (4, 5) to test if the cognitive mechanisms are the same or different under

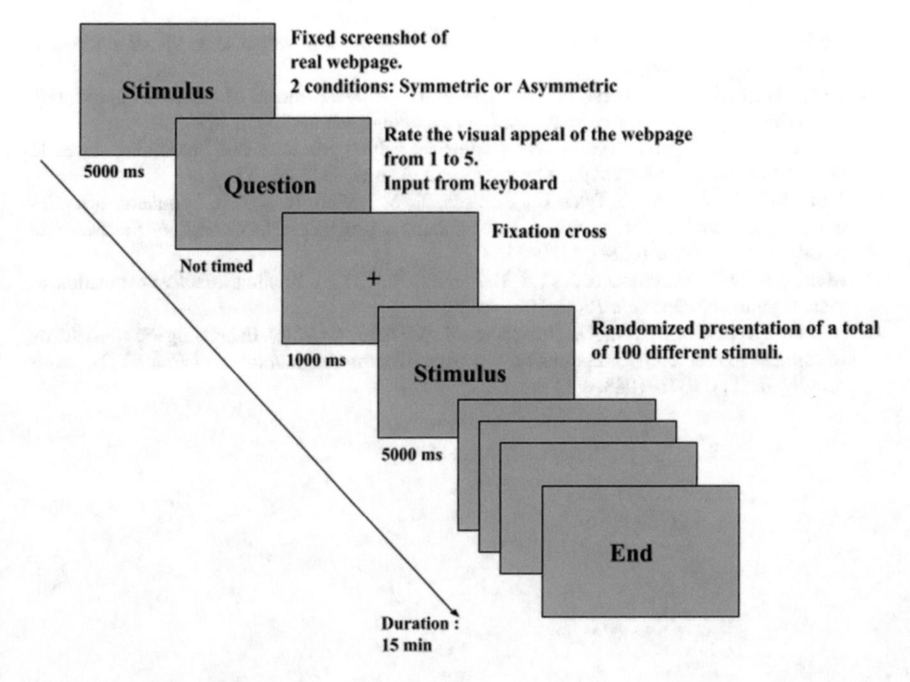

Fig. 2 Experimental design

the two categories. Preliminary results will be presented at the NeuroIS conference and feedback will help to prepare the main study.

References

1. Lindgaard, G., Fernandes, G., Dudek, C., & Browñ, J. (2006). Attention web designers: You have 50 milliseconds to make a good first impression! *Behaviour and Information Technology, 25*(2), 115–126.
2. Lavie, T., & Tractinsky, N. (2004). Assessing dimensions of perceived visual aesthetics of web sites. *International Journal of Human Computer Studies, 60*(3), 269–298.
3. Altaboli, A., & Lin, Y. (2011). Objective and subjective measures of visual aesthetics of website interface design: The two sides of the coin. *Lecture Notes in Computer Science (Including Subseries Lecture Notes in Artificial Intelligence and Lecture Notes in Bioinformatics), 6761 LNCS* (PART 1), 35–44.
4. Bauerly, M., & Liu, Y. (2006). Computational modeling and experimental investigation of effects of compositional elements on interface and design aesthetics. *International Journal of Human Computer Studies, 64*(8), 670–682.
5. Lai, C. Y., Chen, P. H., Shih, S. W., Liu, Y., & Hong, J. S. (2010). Computational models and experimental investigations of effects of balance and symmetry on the aesthetics of text-overlaid images. *International Journal of Human Computer Studies, 68*(1–2), 41–56.
6. Tuch, A. N., Bargas-Avila, J. A., & Opwis, K. (2010). Symmetry and aesthetics in website design: It's a man's business. *Computers in Human Behavior, 26*(6), 1831–1837.

7. Bertamini, M., & Makin, A. D. J. (2014). Brain activity in response to visual symmetry. *Symmetry, 6*(4), 975–996.
8. Leder, H., Belke, B., Oeberst, A., & Augustin, D. (2004). A model of aesthetic appreciation and aesthetic judgments. *British Journal of Psychology, 95*(489–508), 489–508.
9. Leder, H. (2013). Next steps in neuroaesthetics: Which processes and processing stages to study? *Psychology of Aesthetics, Creativity, and the Arts, 7*(1), 27–37.
10. Vanduffel, W., Sasaki, Y., Tyler, C., Knutsen, T., & Tootell, R. (2005). Symmetry activates extrastriate visual cortex in human and nonhuman primates. *Proceedings of the National Academy of Sciences, 102*(8), 3159–3163.
11. Makin, A. D. J., Pecchinenda, A., & Bertamini, M. (2012). Implicit affective evaluation of visual symmetry. *Emotion, 12*(5), 1021–1030.
12. Tractinsky, N., Cokhavi, A., Kirschenbaum, M., & Sharfi, T. (2006). Evaluating the consistency of immediate aesthetic perceptions of web pages. *International Journal of Human Computer Studies, 64*(11), 1071–1083.

Search Results Viewing Behavior vis-à-vis Relevance Criteria

Jacek Gwizdka and Yung-Sheng Chang

Abstract We conducted a lab-based experiment to investigate relationship between multiple criteria used in information relevance judgments and eye fixation behavior on search engine results. We collected eye-tracking data and conducted gaze-cued retrospective think-aloud (RTA). Data from RTA was coded with criteria used by participants in judging search results as relevant. The criteria were analyzed in relation to search engine result page (SERP) sequence and result rank on SERPs. The results of our study aligned with previous research, showing the effect of result rank on SERPs. Our results newly showed that *specific source* and *topicality* were the two most often used criteria for relevance judgments. *Specific source* was the most often used criteria initially but was then surpassed by *topicality* on subsequent SERPs and on lower result ranks. On first SERPs, fixation duration was significantly longer on results judged on *topicality* than on *specific source*. Pupils dilated significantly on the top ranked result on most SERP pages.

Keywords Information relevance · Eye-tracking · Retrospective think aloud

1 Introduction

The ways in which people access news have changed rapidly. Twenty years ago, only 12% of the U.S. adults accessed news online, while in recent years nearly 81% of news readers access them online. With the autonomy people have in selecting online news, it is vital to understand how they judge the relevance of online news information. In the present study, we investigate the relevance criteria participants used to select relevant news results on SERPs. We discuss how the relevance criteria judgment changes in search task session on subsequent SERPs (these could be re-

J. Gwizdka (✉) · Y.-S. Chang
Information Experience Lab, School of Information,
University of Texas at Austin, Austin, TX, USA
e-mail: neurois2019@gwizdka.com

Y.-S. Chang
e-mail: yscchang@utexas.edu

F. D. Davis et al. (eds.), *Information Systems and Neuroscience*,
Lecture Notes in Information Systems and Organisation 32,
https://doi.org/10.1007/978-3-030-28144-1_20

visits or new SERPs) and on results at different ranks on SERPs and how fixations and pupil dilation differ on them. We investigate the following research questions:

RQ-1: What are the criteria people use to make relevance judgments on online news search results?
RQ-2: What are the patterns of criteria use in making relevance judgments in task session in relation to the order of SERPs and result rank on SERPs?
RQ-3: What are the differences in eye-tracking measures on search result snippets (AOIs) between criteria and SERP sequential number and result rank?

2 Related Work

In information retrieval (IR), relevance is a vital but complex concept that aids an IR system to retrieve relevant information and, on the user side, relates to the process of human relevance judgment. Given its complexity, previous studies [2, 12, 13] had described two characteristics of relevance: multidimensionality and dynamics. Multidimensionality of relevance refers to the (potentially) many criteria used in relevance assessment, while dynamics refers to how the perception of relevance can change over time for the same user [2]. To research relevance multidimensionality, Zhang et al. [14] applied crowdsourcing and structural equation modeling to verify the five relevance criteria: reliability, topicality, scope, novelty, understandability. Gerjets et al. [5] conducted a search-task experiment by using preselected Google-like search result pages and analyzed five concepts of relevance: topicality, up-to-dateness, credibility, scope, and design. Past studies had also identified the criteria as either being positive or negative, and aggregated relevance criteria used per judgment on the spectrum from non-relevant, partial relevant, to relevant [6, 11]. Furthermore, users were found to use single criterion to judge non-relevant and partially relevant information while four or five criteria were used to evaluate relevant documents. Balatsoukas et al. [1] describe an extensive study of relevance and eye-tracking measures. They showed that users expend more cognitive effort (more frequent and longer fixations) on non-relevant document surrogates. Limitations of their study was the use of only a limited set of eye movement based features (fixation duration and number of fixations), a cut-off of fixations shorter than 200 ms, and the use of concurrent talk-aloud method that likely affected the fixation durations. The process of applying multiple relevance criteria has been theorized by da Costa Pereira et al. [3], who have proposed a mathematical model of how multiple relevance criteria can be hypothetically aggregated and weighted in their application by users. However, this proposal have not been validated with users.

In summary, there have been studies focusing on the multidimensional aspects of relevance, but there is scarcity of research with a focus on how readers apply relevance criteria to search results in relation to eye fixations on these results. To bridge this gap, we conducted a lab-based experiment and used eye-tracking and retrospective think-aloud to answer our research questions.

3 Method

40 participants recruited on university campus took part in the study (26 females; mean age 24 (sd = 8)). Native English speakers with no vision correction were recruited. Participants were additionally screened for their below medium familiarity with search topics. Since news information is time sensitive, we wanted participants to search for information related to recent news. Tobii TX-300 eye-tracker was used to capture participants' eye fixations and pupil size. The study was conducted in Information eXperience (IX) Lab at the School of Information, University of Texas at Austin.

Procedure. Participants received short instructions on the study, read and signed the consent form, and filled in demographic information. The experiment presentation was controlled and data collected by iMotions software. Each participant performed four assigned search tasks presented in randomized order on following topics: Zika virus, Brexit, 2016 Summer Olympics, and Bob Dylan Nobel Prize award. For each task, first, a search task scenario was displayed. Next, participants searched from Google home page. While performing search, participants were asked to save and annotate pages they thought were relevant. After each task an RTA interview was conducted by replaying session recording from iMotions and recording it using Camtasia. Participants were asked to recall their thoughts while viewing their search task sessions with overlaid eye-gaze. Thus, the researcher was able to point participants to SERP results they actually looked at. The experiment took 1.5–2 h to complete.

4 Data Analysis and Results

The RTA sessions were transcribed and coded (using content analysis) by one researcher using MAXQDA software. 19 codes were established. After the initial coding system was established, a second researcher verified 60% of the coded transcripts. The final coding system was reviewed by the first and an additional third researcher until agreement was reached. The analysis focused on relevance criteria used on SERP results selected as relevant, on the differences in criteria used on subsequent SERPs, on results by rank, and on eye fixations on individual search result AOIs on 1260 SERPs visited by participants. Dynamic AOIs were created manually in iMotions software on each individual search result that was viewed by participants producing 2988 AOIs. After considering eye-tracking data quality (bad quality was data marked by Tobii eye-tracker with validity > 1), this number was further reduced to 2460. These AOI were similar size, typically 4 or 5 lines of text (Fig. 1).

Relevance criteria used on SERPs. The percentage of relevance criteria used was calculated. Low frequency criteria (< 0.05%) were excluded from further analysis, leaving fourteen criteria codes. The top six criteria used when selecting a result on SERPs were: *specific source* (28%), *topicality* (27.8%), *rank* (9.4%), *credibility* (6%), *familiarity* (4.9%), and *recency* (4.3%). *Specific source* was defined as: the origin of

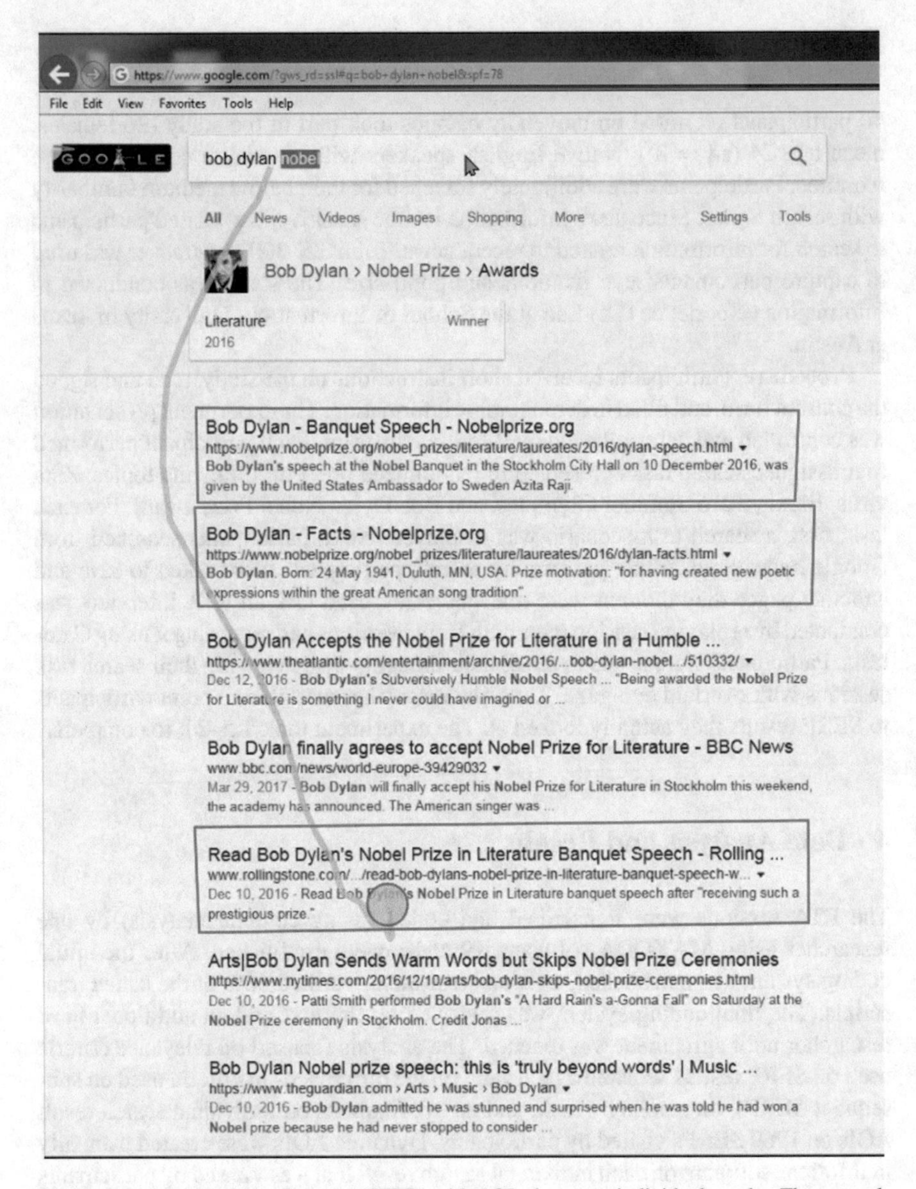

Fig. 1 Example search results page (SERP) with AOIs drawn on individual results. The example shows results for Bob Dylan Nobel Prize award task

the news information, such as BBC or CNN; *topicality:* the aboutness of the news content to the topic of a user's query [8]; *rank*: the order of all snippets (ads and news included) on the search engine results pages; *credibility*: the trustworthiness and expertise of the news source and its content [4]; *familiarity*: the subjective feeling of having the knowledge or past experience with the news source or news topic; *recency*: the closeness of the date when the news information was published in relation to the current date or the timeframe when the news events occurred. Figure 2 shows the criteria used by the order of SERP visits (left) and per result rank (right). *Specific source* and *topicality* were the two most often used criteria on irrespective of the order of SERP visit and result rank.

Relevance criteria and eye-tracking measures. During RTA sessions, participants referred to criteria used on whole SERPs or to individual results on SERPs. Our further analysis is focused on individual SERP results—432 manually created AOIs matched the criteria. The horizontal size of these AOI was the same, while the vertical size was relatively uniform (mean = 115 px (s.e. = 2.1); median = 104 px). A single criterion was used to make judgment on selected results in 61.6% of cases. Surprisingly, there was no difference in fixation durations or counts between the use of single criterion versus multiple criteria. To investigate differences in eye-tracking measures between individual relevance criteria we further consider individual search result AOIs with single criteria applied. There were 337 such AOI-single-criterion pairs. Kruskal-Wallis non-parametric test was used to test the differences. No difference were found on all SERP pages tested together. When examining individual orders of SERPs, significant difference was found on the first SERP page (Table 1), but not on other subsequent SERP pages. Post hoc tests (Wilcoxon rank sum test) revealed that the difference on first SERPs was due to significantly longer fixations on *topically* judged results compared with results judged by *recency* or *familiarity*. The difference between topical and *specific source* ($p = 0.059$) and rank was approaching significance. For fixation counts, *topicality* criterion had significantly more fixations associated with it than the *specific source* or *recency*.

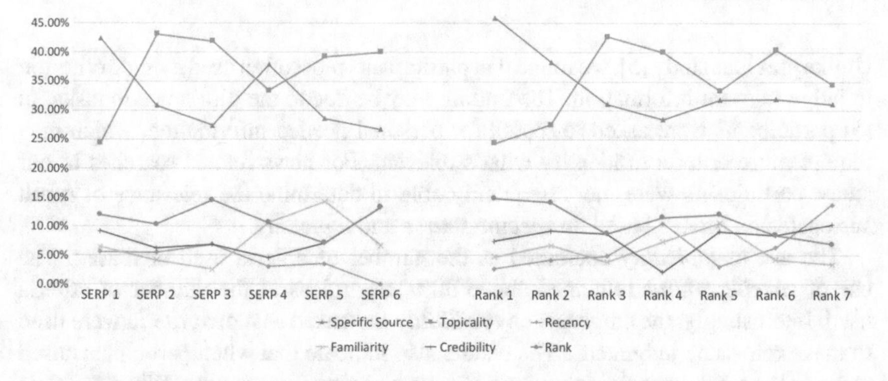

Fig. 2 Criteria [%] used by the order of SERP visits (left) and per result rank (right)

Table 1 Eye-tracking measures by relevance criterion (for first SERP visited)

	Specific source (S)	Topicality (T)	Recency (R)	Familiarity (F)	Credibility (C)	Rank (P)	K-W χ^2(df), p
Total fixation duration Mean(sd)	3055 (2127)	5333 (9523)	2713 (3954)	2238 (1076)	4031 (2706)	2608 (1700)	13.1 (5) $p = 0.02$
Fixation count Mean(sd)	13 (9.7)	24 (44)	10 (13)	11 (6)	16 (10)	12.(7)	12.2 (5) $p = 0.032$
Norm. pupil dilation Mean(sd)	0.0001 (0.058)	−0.019 (0.05)	−0.013 (0.03)	−0.003 (0.05)	−0.008 (0 .045)	0.005 (0.036)	6.9 (5) $p = 0.22$

Wilcoxon pairwise test significance levels: T&S, T&P fix dur $0.05 < p < 0.1$; T&R, T&F $p < 0.01$; fixation count T&R $p < 0.01$, T&S $p = 0.02$

We next investigated eye-tracking measures in relation to result ranks. Only normalized relative pupil change in diameter (using procedure described in [7]) was found to differ significantly on all SERPs (K-W: χ^2(df): 74.8(10), $p = 0.0001$) and on first and second SERPs. There were too few data points on lower ranked results and on subsequent SERPs to perform analysis. The normalized pupil dilation on top result across all SERPs was 0.0264 (sd = 0.046), while on results at rank 2–10 it ranged from −0.06 (at rank 10) to −0.0044 (at rank 2). All pairwise comparisons between rank 1 and lower result ranks were significant at $p < 0.0001$, except for rank 6 where $p = 0.006$. When considering SERP 1 and 2 separately, the same pattern was observed, pupil dilated significantly on the top ranked result.

5 Discussion

Unlike previous study [5], we found that participants most often used a single criterion to judge relevant information. The finding may be due to the difference in tasks. In [5] participants were asked to search for personal needed information, which may require more criteria to identify what's relevant. For news related searches in our study, participants were most frequently able to determine the relevance of result surrogates on SERPs based on *specific source* and *topicality*.

The use of *topicality* decreased as the number of criteria used increased. The use of *specific source* remained above 30%, regardless of the number of criteria used. Interestingly, the emphasis on *credibility* increased as more criteria were used to make relevance judgments. The results also indicate that when participants used more criteria, the *specific source* was always an important concern. When multiple criteria were used, *specific source* and *topicality* co-occurred most often.

The use of relevance criteria in relation to the order of SERP visits in search session and result rank on SERPs showed the highest use of *specific source* followed by *topicality* on the first visited SERPs and the top result ranks on SERPs. However, after the first visited SERPs in search session and on lower result ranks, the use of *topicality* increased and surpassed the use of *specific source*. Our results also coincided with previous studies [7, 14] showing the effect of result rank on information selection on SERPs: participants are inclined to select the top ranked results.

Longer fixation durations and counts on *topically* judged search results as compared with when other relevance criteria were applied indicate that results need to be considered more thoroughly in order to judge their *topicality* than to judge their *source*. That is intuitively understandable. One could speculate that searchers may be applying relevance criteria in parallel (possibly not realizing that they are doing so) and that criterion which is fastest to satisfy wins, and a decision is made. This speculation is supported by no significant difference in fixation duration between the use of single relevance criterion and multiple criteria. We believe that this is the first time such relationships were demonstrated empirically. Pupil dilation has been associated with mental workload [9] and interest [10]. Significantly more dilated pupils on top search results are a likely indication of more mental effort invested in processing the top result by searchers and, also, of more interest. Bias towards considering top search results is well known in information search and retrieval.

6 Conclusion

When assessing online news results on SERPs, *credibility* was, surprisingly, not the first criterion participants used to select news. Our results showed that the news' *specific source* and *topicality* were the two most often used criteria to make relevance judgment. This result is also shown when participants make judgment on subsequent SERPs and result rank on SERPs, with *specific source* most used initially but then replaced by *topicality* on subsequent SERPs and on lower results ranks. We found significant difference in fixation durations and counts between relevance criteria and in pupil dilation at top versus lower ranked results. We believe that this is the first study to examine relationship between relevance criteria and eye-tracking measures at the level of individual search results.

One of the limitations of the study is the use of assigned tasks and their relatively small number. In future work, we will further analyze the eye-movement data and relate participants' eye-movement patterns with their relevance judgments. The analysis is expected to provide a comprehensive understanding of how news readers judge online news information.

Acknowledgements This research has been funded, in part, by Portuguese Foundation for Science and Technology and the Digital Media Program at UT-Austin. We thank master's student Han Han for her help with manual creation of AOIs.

References

1. Balatsoukas, P., & Ruthven, I. (2012). An eye-tracking approach to the analysis of relevance judgments on the Web: The case of Google search engine. *Journal of the American Society for Information Science and Technology, 63*(9), 1728–1746. https://doi.org/10.1002/asi.22707.
2. Borlund, P. (2003). The concept of relevance in IR. *Journal of the American Society for information Science and Technology, 54*(10), 913–925. https://doi.org/10.1002/asi.10286.
3. da Costa Pereira, C., Dragoni, M., & Pasi, G. (2012). Multidimensional relevance: Prioritized aggregation in a personalized Information Retrieval setting. *Information processing & management, 48*(2), 340–357. https://doi.org/10.1016/j.ipm.2011.07.001.
4. Fogg, B. J., & Tseng, H. (1999, May). The elements of computer credibility. In *Proceedings of the SIGCHI conference on Human Factors in Computing Systems* (pp. 80–87). New York, NY, USA: ACM.https://doi.org/10.1145/302979.303001.
5. Gerjets, P., et al. (2011). Measuring spontaneous and instructed evaluation processes during Web search: Integrating concurrent thinking-aloud protocols and eye-tracking data. *Learning and Instruction, 21*(2), 220–231. https://doi.org/10.1016/j.learninstruc.2010.02.005.
6. Greisdorf, H. (2003). Relevance thresholds: A multi-stage predictive model of how users evaluate information. *Information Processing & Management, 39*(3), 403–423. https://doi.org/10.1016/S0306-4573(02)00032-8.
7. Gwizdka, J. (2018). Inferring web page relevance using pupillometry and single channel EEG. In *Information Systems and Neuroscience* (pp. 175–183). Springer, Cham.
8. Janes, J. W. (1994). Other people's judgments: A comparison of users' and others' judgments of document relevance, topicality, and utility. *Journal of the American Society for Information Science, 45*(3), 160–171. https://doi.org/10.1002/(SICI)1097-4571(199404)45:3%3c160:AID-ASI6%3e3.0.CO;2-4.
9. Kahneman, D., & Beatty, J. (1966). Pupil diameter and load on memory. *Science, 154*(3756), 1583–1585. https://doi.org/10.1126/science.154.3756.1583.
10. Krugman, H. E. (1964). Some applications of pupil measurement. *Journal of Marketing Research, 1*(4), 15–19.
11. Maglaughlin, K. L., & Sonnenwald, D. H. (2002). User perspectives on relevance criteria: A comparison among relevant, partially relevant, and not-relevant judgments. *Journal of the American Society for Information Science and Technology, 53*(5), 327–342. https://doi.org/10.1002/asi.10049.
12. Saracevic, T. (1996, October). Relevance reconsidered. In *Proceedings of the second conference on conceptions of library and information science* (*CoLIS* 2) (pp. 201–218). New York: ACM.
13. Schamber, L., et al. (1990). A re-examination of relevance: Toward a dynamic, situational definition. *Information Processing & Management, 26*(6), 755–776. https://doi.org/10.1016/0306-4573(90)90050-C.
14. Zhang, Y., Zhang, J., Lease, M., & Gwizdka, J. (2014, July). Multidimensional relevance modeling via psychometrics and crowdsourcing. In *Proceedings of the 37th international ACM SIGIR conference on Research & development in information retrieval* (pp. 435–444). ACM. https://doi.org/10.1145/2600428.2609577.

An Adaptive Cognitive Temporal-Causal Network Model of a Mindfulness Therapy Based on Humor

S. Sahand Mohammadi Ziabari and Jan Treur

Abstract In this paper the effect of a humor therapy is modeled based on a Network-Oriented Modeling approach. Humor therapy is a mindfulness therapy which has been used since many years ago, when Abu Bakr Muhammad ibn Zakariya al-Razi (https://en.wikipedia.org/wiki/Muhammad_ibn_Zakariya_al-Razi) as a Persian scientist who used humor theory to distinguish one contagious disease from another, to make stressed individuals more relaxed. The presented adaptive temporal-causal network model addresses the computational modeling of humor therapy for a person who in the first step triggers two incongruent beliefs in order to get the humor from a humor context to overcome an ongoing stressful event. This happens by showing a comedy movie. As a result, the stress level in the body reduces. Hebbian learning is incorporated to strengthen the effect of the humor therapy.

Keywords Cognitive temporal-causal network model · Hebbian learning · Extreme emotion · Humor therapy · Mindfulness

1 Introduction

To handle stress and its consequences for mental and physical health, often mindfulness therapies [8, 9] are considered. A wide variety of such therapies [33] working according to different mechanisms, is available, some of which have been analyzed by computational modeling; for example, see [30, 18–20]. The current paper addresses humor therapy. In the Oxford English Dictionary [26], humor is defined as 'That quality of action, speech, or writing which excites amusement; oddity, jocularity, facetiousness, comicality, fun'. Examples of contexts of humor are funny films, audio and videotapes of humorous songs, or reading materials [17]. Humor is considered when a driver like a comedy movie, triggers mental action involving cognitive and

S. Sahand Mohammadi Ziabari (✉) · J. Treur
Social AI Group, Vrije Universiteit Amsterdam, De Boelelaan 1105, Amsterdam, The Netherlands
e-mail: sahandmohammadiziabari@gmail.com

J. Treur
e-mail: j.treur@vu.nl

F. D. Davis et al. (eds.), *Information Systems and Neuroscience*,
Lecture Notes in Information Systems and Organisation 32,
https://doi.org/10.1007/978-3-030-28144-1_21

affective mental states, and often responses like mirth and laughter, which is the most common conducting expression of a jocular experience [12].

As has been described in [23], humor has psychological, social, emotional, behavioral, and cognitive components. Graceful emotional feeling are usually the result of humor and laughter [25]. Humor therapy has been considered as one of the distraction techniques, one of the remarkable uses of cognitive-behavioral techniques, in pain management and control [1, 14]. In [29] it has been noted that:

> In light of the light prevalence of chronic pain and its impact on physical and psychological perspectives among older people, the use of humor therapy as a means of reducing pain and loneliness as well as increasing happiness and life satisfaction is very appealing. [29], p. 3

Several minutes of powerful laughter generate results similar to exercising on a rowing machine or bicycle for about 10–15 min [5]. The reason is just because of releasing endorphins after an intensive laugh. In [10] the meaning of humor is defined.

The paper is organized as follows. In Sect. 2 the neuropsychological principles of the effects of stress and the parts of the brain which deal with stress are addressed, and the mechanisms by which humor can affect this. In Sect. 3 the adaptive temporal-causal network model is introduced. In Sect. 4 the simulation results of the model are discussed. Finally, Sect. 5 is a discussion Conclusion.

2 Neuropsychological Principles

In [12, 34] it has been illustrated that humor and laughter make lung capacity increase, abdominal muscles strengthen and also immunoglobulin A increment, the antibody which is generated by our immune system. Also, as it is declared in [1, 28] natural killer activity such as immunoglobulin G immunoglobulin M levels increases for 12 h since laughter or other humorous encounters and as stated in [1] the results of humor are also reduction in cortisol, increase in hormones and epinephrine and these changes are beneficial to the person's health. The continuous use of humor results in betterment in pain thresholds [13]. As stated in [3] the left amygdala is responsible for conscious and cognitively controlled emotional actions and the right amygdala is involved in unconscious and automatic emotional actions. Therefore in [21] it has been found that left amygdala activation will return the tendency of humor.

Pathways between Amygdala and other brain parts has been discovered: Amygdala and orbitofrontal cortex for discrimination of the valence, amygdala and cingulate cortex for computing an object's biological value, amygdala and anterior insula for emotional feelings, and amygdala and (colliculus and pulvinar) for filtering out a distractor driver [22]. Release of endorphins in the brain assists to control the pain as has been described in [7, 35]. In [4] it has been claimed that managing the pain accompanied with humor is more effective than managing the pain on itself. Also, in treatment of patients qualitative research described in [2, 9] supports the impacts of humor. As has been mentioned in [21] there are eight psychological benefits of humor based upon available quantitative and qualitive evidence in literatures discussed in [2]:

> 1. Humor reduces anxiety. 2. Humor reduces tension. 3. Humor reduces stress. 4. Humor reduces depression. 5. Humor reduces loneliness. 6. Humor improves self-esteem. 7. Humor restores hope and energy. 8. Humor provides a sense of empowerment and control. [2]

As mentioned in [5, 6] there are seven particular physiological gains that consists of central nervous, muscular, respiratory, circulatory, endocrine, immune, and cardiovascular systems. In much literature it has been claimed that there is a dual way architecture between sensory information and amygdala [15, 22].

In [27] it has been found that a positive emotional driver like pleasant taste or happy faces and a negative driver such as tremulous faces, sad faces or angry faces will activate the amygdala. There are differences in the experience of humor in relation to individual differences in personality, character strengths, age, gender, language, and culture. Research described in [10] has considered functional aspects of humor, such as its role in creativity, emotion regulation, and group cohesion. In [24] the incongruity and control state between them.

3 The Adaptive Temporal-Causal Network Model

First the Network-Oriented Modelling approach used to model this process is briefly explained. As discussed in detail in [31, Chap. 2, 29] this approach is based on temporal-causal network models which can be represented at two levels: by a conceptual representation and by a numerical representation. These three notions form the defining part of a conceptual representation of a temporal-causal network model:

- **Strength of a connection $\omega_{X,Y}$** Each connection from a state X to a state Y has a *connection weight value* $\omega_{X,Y}$ representing the strength of the connection, often between 0 and 1, but sometimes also below 0 (negative effect) or above 1.
- **Combining multiple impacts on a state $\mathbf{c}_Y(..)$** For each state (a reference to) a *combination function* $\mathbf{c}_Y(..)$ is chosen to combine the causal impacts of other states on state Y.
- **Speed of change of a state η_Y** For each state Y a *speed factor* η_Y is used to represent how fast a state is changing upon causal impact.

In Fig. 1 the conceptual representation of the temporal-causal network model is depicted. A brief explanation of the states used is shown in Table 1 (Table 2).

Next, the elements of the conceptual representation shown in Fig. 1 are explained in some more detail.

The described conceptual representation defines a numerical representation of the network model as follows [35, Ch 2]:

- at each time point t each state Y in the model has a real number value in the interval [0, 1], denoted by $Y(t)$

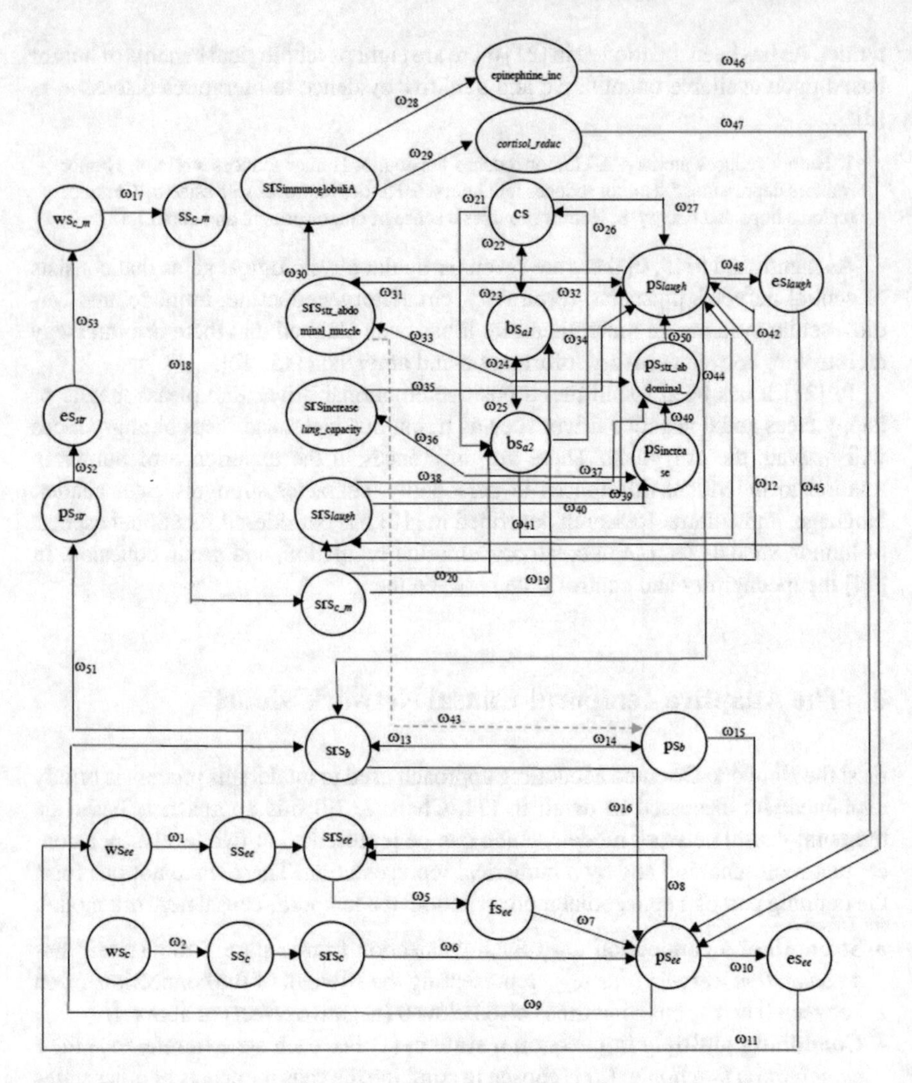

Fig. 1 Conceptual representation of the adaptive temporal-causal network model

- at each time point t each state X connected to state Y has an impact on Y defined as $\mathbf{impact}_{X,Y}(t) = \omega_{X,Y}\, X(t)$ where $\omega_{X,Y}$ is the weight of the connection from X to Y
- The *aggregated impact* of multiple states X_i on Y at t is determined using a *combination function* $\mathbf{c}_Y(..)$:

$$\begin{aligned}\mathbf{aggimpact}_Y(t) &= \mathbf{c}_Y\big(\mathbf{impact}_{X1,Y}(t), \ldots, \mathbf{impact}_{X_k,Y}(t)\big)\\ &= \mathbf{c}_Y\big(\omega_{X1,Y}X_1(t), \ldots, \omega_{X_k,Y}X_k(t)\big)\end{aligned}$$

Table 1 Explanation of the states in the model

X_1	ws_{ee}	World (body) state of extreme emotion *ee*	X_{15}	ss_{laugh}	Sensor state of laughing
X_2	ss_{ee}	Sensor state of body state for extreme emotion *ee*	X_{16}	$\text{ss}_{increase\_lung\_capacity}$	Sensor state of increasing the lung capacity
X_3	ws_{c}	World state for stress-inducing context *c*	X_{17}	$\text{ss}_{str\_abdominal\_muscle}$	Sensor state of stretching of abdominal muscle
X_4	ss_{c}	Sensor state for *c* (perceiving *c*)	X_{18}	$\text{srs}_{immunoglobulia\ A}$	Sensory representation of immunoglobulin A
X_5	srs_{ee}	Sensory representation state of body state for extreme emotion *ee*	X_{19}	Epinephrine_inc	Hormone increasing
X_6	srs_{c}	Sensory representation state of context *c*	X_{20}	Cortisol_reduc	Hormone reduction
X_7	fs_{ee}	Feeling state for extreme emotion *ee*	X_{21}	cs	Control state
X_8	ps_{ee}	Preparation state for response of extreme emotion *ee*	X_{22}	bs_{a1}	First belief
X_9	es_{ee}	Execution state (bodily expression) for response of extreme emotion *ee*	X_{23}	bs_{a2}	Second belief
X_{10}	srs_{b}	Sensory representation of body state *b*	X_{24}	ps_{laugh}	Preparation state for laughing
X_{11}	ps_{b}	Preparation state of body state *b*	X_{25}	$\text{ps}_{str\_abdominal}$	Preparation state for abdominal
X_{12}	ws_{c_m}	World state of comedy movie	X_{26}	$\text{ps}_{increase}$	Preparation state for increasing lung capacity
X_{13}	ss_{c_m}	Sensor state of comedy movie	X_{27}	es_{laugh}	Execution state of laughing
X_{14}	srs_{c_m}	Sensory representation state of comedy movie	X_{28}	ps_{str}	Preparation state of starting humor movie
X_{29}	es_{str}	Execution state of starting humor movie			

Table 2 States and their relations to domain literature

States	Principles	Quotation, References
srs_{ee}	Sensory representation of the body state for the extreme emotion	'*The dACC was activated during the observe condition. The dACC is associated with attention and the ability to accurately detect emotional signals.*' [16], p. 12
ws_{c_m}	World state of comedy movie	'*Humor can refer to a stimulus such as a comedy film, a mental process such as perception, or a response such as laughter and exhilaration*' [29], p. 2
ss_{c_m}	Sensor state of comedy movie	'*Humor can refer to a mental process such as perception, or a response such as laughter and exhilaration.*' [29], p. 4
srs_{laugh}	Sensory representation state of laughing	'*Humor and laughter are typically associated with a pleasant emotional feeling.*' [23], p. 610
$srs_{\text{increase } lung\_\text{capacity}}$	Sensory representation of increasing lung capacity	'*Humor has been shown to increase lung capacity, strengthen abdominal muscles, and increase immunoglobulin A, which is one of the major antibodies produced by the immune system.*' [12, 34, 29]
$srs_{\text{str\_abdominal\_muscle}}$	Sensory representation of strengthening of abdominal muscle	'*Humor has been shown to increase lung capacity, strengthen abdominal muscles, and increase immunoglobulin A, which is one of the major antibodies produced by the immune system.*' [12, 34, 29], p. 2

(continued)

Table 2 (continued)

bs_{a1} bs_{a2}	Two belief states for two incongruent interpretations as a basis for getting humor from the comedy movie	'*Cognitive theories typically analyze the structural properties of humorous stimuli or the way they are processed; sometimes these two levels are also mixed up. Perhaps beginning with Aristotle, incongruity was considered to be a necessary condition for humor. From this perspective, humor involves the bringing together of two normally disparate ideas, concepts, or situations in a surprising or unexpected manner.*' [24], pp. 24–25
cs	Control state for resolving the ingruency of the two beliefs	'*Only possible incongruities can be resolved completely while for an impossible incongruity only a partial resolution is possible, and a residue of incongruity is left. The definitions of incongruity ("… a conflict between what is expected and what actually occurs in the joke")*' [24], pp. 24–25
$srs_{immunoglobuliA}$	Sensory representation of immunoglobulin A	'*Humor has been shown to increase lung capacity, strengthen abdominal muscles, and increase immunoglobulin A, which is one of the major antibodies produced by the immune system.*' [12, 34, 29], p. 2
$es_{laughter}$	Execution state of laughing	'*Laughter is the most common behavioral expression of a humorous experience.*' [3, 29], p. 2
epinephrine_inc	Hormone	'*Humor causes reductions in cortisol, growth hormones, epinephrine.*' [1], p. 2
cortisol_reduc	Hormone	'*Humor causes reductions in cortisol, growth hormones, epinephrine.*' [1], p. 3

where X_i are the states with connections to state Y

- The effect of $\mathbf{aggimpact}_Y(t)$ on Y is exerted over time gradually, depending on speed factor η_Y:

$$Y(t + \Delta t) = Y(t) + \eta_Y\left[\mathbf{c}_Y\left(\mathbf{aggimpact}_Y(t)\right) - Y(t)\right]\Delta \mathrm{t}$$

$$\text{or} \quad \mathbf{d}Y(t)/\mathbf{d}t = \eta_Y\left[\mathbf{aggimpactY(t)} - Y(t)\right]$$

- Thus, the following *difference* and *differential equation* for Y are obtained:

$$Y(t + \Delta t) = Y(t) + \eta_Y\left[\mathbf{c}_Y\left(\omega_{X1,Y}X_1(t), \ldots, \omega_{X_k,Y}X_k(t)\right) - Y(t)\right]\Delta t$$
$$\mathbf{d}Y(t)/\mathbf{d}t = \eta_Y\left[\mathbf{c}_Y\left(\omega_{X1,Y}X_1(t), \ldots, \omega_{X_k,Y}X_k(t)\right) - Y(t)\right]$$

For states the following combination functions $\mathbf{c}_Y(\ldots)$ were used, the identity function **id**(.) for states with impact from only one other state, and for states with multiple impacts the scaled sum function $\mathbf{ssum}_\lambda(\ldots)$ with scaling factor λ, or the advanced logistic sum function $\mathbf{alogistic}_{\sigma,\tau}(\ldots)$ with steepness σ and threshold τ.

$$\mathbf{id}(V) = V$$
$$\mathbf{ssum}_\lambda(V_1, \ldots, V_k) = (V_1, \ldots, V_k)/\lambda$$
$$\mathbf{alogistic}_{\sigma,\tau}(V_1, \ldots, V_k) = \left[\left(1/\left(1 + \mathrm{e}^{-\sigma(V_1+\cdots+V_k-\tau)}\right)\right) - 1/(1 + \mathrm{e}^{\sigma\tau})\right](1 + \mathrm{e}^{-\sigma\tau})$$

The Hebbian Learning considered here makes that the strength ω of an adaptive connection between states X_1 and X_2 is adjusted using the following Hebbian Learning rule, taking into account a maximal connection strength 1, a learning rate $\eta > 0$ and a persistence factor $\mu \geq 0$, and activation levels $X_1(t)$ and $X_2(t)$ (between 0 and 1) of the two states involved. The first expression is in differential equation format, the second one in difference equation format:

$$\mathrm{d}\omega(t)/\mathrm{d}t = \eta[\mathrm{X}_1(\mathrm{t})\mathrm{X}_2(\mathrm{t})\,(1 - \omega(t)) - (1 - \mu)\omega(t)]$$
$$\omega(t + \Delta t) = \omega(t) + \eta[X_1(t)\mathrm{X}_2(t)(1 - \omega(t)) - (1 - \mu)\omega(t)]\Delta t$$

4 Example Simulation

An example simulation of this process is shown in Figs. 2 and 3. Table 3 shows the connection weights used, where the values for the Hebbian learning connection is the initial value as this weight is adapted over time. The time step was $\Delta t = 1$. The scaling factors λ for the states with more than one incoming connection are also

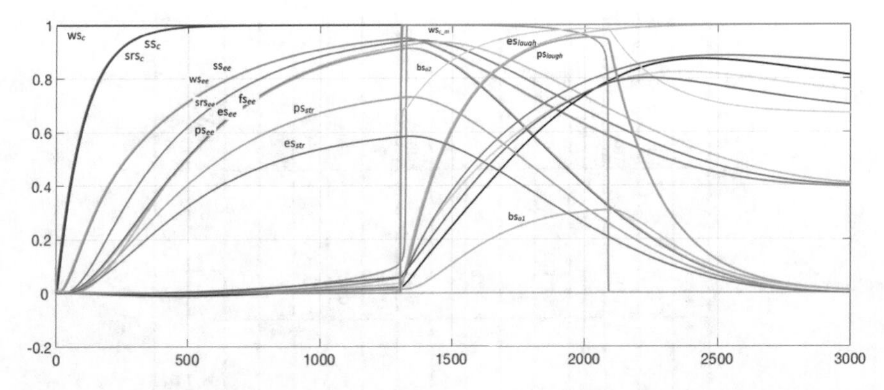

Fig. 2 Simulation results of the humor therapy

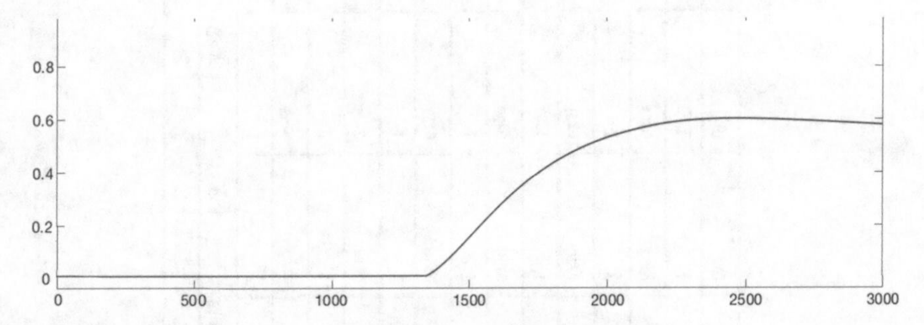

Fig. 3 Simulation results of adaptivity of humor therapy ($X_{17} - X_{11}$)

depicted in Table 3. In the scenario, the comedy movie is used as a basis for the humor therapy to decrease the level of the extreme emotion of the stressed individual.

The comedy movie gets a role from time around 1300 and finishes around 2200. After giving some time to the stressed individual to watch and sense the humor in the movie, the preparation and sensory representation of emotion starts to have a role and after internally being emotional she starts laughing from time around 1400 as an as-if loop from preparation state of watching movie and starts laughing (X_{15} and X_{16} and X_{17} as sensory representation states of laughing, ss_{laugh}, $ss_{increase\_lung\_capacity}$, $ss_{str\_abdominal\_muscle}$). The reduction of the stress level continues until the time around 3000 to become in the equilibrium level from 0.9 to just 0.4 (low-level of stress).

The results for adaptivity of the connection between sensory representation of stretching the abdominal muscle and the preparation state of a relaxed body state *b* has been shown in Fig. 3. As can been seen in Fig. 3 the adaptivity improves the effect of the Humor therapy and make it stable after finishing the therapy at time around 3000.

Table 3 Connection weights for the example simulation

Connection weight	ω_1	ω_2	ω_3	ω_4	ω_5	ω_6	ω_7	ω_8
Value	1	1	1	1	1	1	1	1
Connection weight	ω_9	ω_{10}	ω_{11}	ω_{12}	ω_{13}	ω_{14}	ω_{15}	ω_{16}
Value	−0.1	1	1	1	1	1	−1	−0.01
Connection weight	ω_{17}	ω_{18}	ω_{19}	ω_{20}	ω_{21}	ω_{22}	ω_{23}	ω_{24}
Value	1	1	1	1	1	1	1	−1
Connection weight	ω_{25}	ω_{26}	ω_{27}	ω_{28}	ω_{29}	ω_{30}	ω_{31}	ω_{32}
Value	−1	1	1	1	1	1	1	1
Connection weight	ω_{33}	ω_{34}	ω_{35}	ω_{36}	ω_{37}	ω_{38}	ω_{39}	ω_{40}
Value	1	1	1	1	1	1	1	1
Connection weight	ω_{41}	ω_{42}	ω_{43}	ω_{44}	ω_{45}	ω_{46}	ω_{47}	ω_{48}
Value	1	1	0.2	1	1	−0.1	−0.1	−0.9
Connection weight	ω_{49}	ω_{50}	ω_{51}	ω_{52}	ω_{53}			
Value	1	1	1	1	1			

State	X_5	X_8	X_{11}	X_{15}	X_{16}	X_{17}	X_{22}	X_{23}	X_{25}	X_{26}
λ_i	2	2	1.55	3	1	2	2	2	3	2

5 Conclusion

In this paper an adaptive cognitive temporal-causal network model of a mindfulness therapy based on humor to decrease the level of stress of individual with extreme stress was presented. Due to Hebbian learning the model is adaptive by which the influence becomes stronger over time. A variety of simulations were executed one of which was presented in the paper. Findings from Neuroscience and psychology were taken into account in the design of the adaptive cognitive model. This literature reports experiments and measurements of humor therapy for emotion-induced conditions as addressed from a computational perspective in the current paper.

This model can be used as the basis of a virtual agent model to get insight in such processes and to consider certain support or treatment of individuals and prevent some stress-related disorders that otherwise might develop.

References

1. Berk, R. A. (2001). The active ingredients in humor: Psychological benefits and risks for older adults. *Educational Gerontology, 27*(3–4), 323–339.
2. Cousins, N. (1976). Anatomy of an illness (as perceived by the patient). *New England Journal of Medicine, 295*(26), 1458–1463.
3. Dyck, M., Loughead, J., Kellermann, T., Boers, F., Gur, R. C., & Mathiak, K. (2010). Cognitive versus automatic mechanisms of mood induction differentially activate left and right amygdala. *NeuroImage, 54*(3), 2503–2513. https://doi.org/10.1016/j.neuroimage.2010.10.013.
4. Ferrell, B. R., Taylor, E. J., Grant, M., Fowler, M., & Corbisiero, R. M. (1993). Pain management at home: Struggle, comfort, and mission. *Cancer Nursing, 16*(3), 169–178.
5. Fry, W. F., Jr. (1992). The physiological effects of humor, mirth, and laughter. *Journal of the American Medical Association, 267*(4), 1857–1858.
6. Fry, W. F., Jr. (1986). Humor, physiology, and the aging process. In L. Nahemow, K. A. McCluskey-Fawcett, & P. E. McGhee (Eds.), *Humor and aging* (pp. 81–98). Orlando: Academic Press.
7. Haig, R. A. (1988). *The anatomy of humor*. Springfield. III, USA: Charles C. Thomas.
8. Holzel, B. K., Carmody, J., Vangel, M., Congleton, C., Yerramsetti, S. M., Gard, T., et al. (2011). Mindfulness practice leads to increases in regional brain gray matter density. *Psychiatry Research, 191,* 36–43. [PubMed:21071182].
9. Ljungdahl, L. (1989). Laugh if this is a joke. *Journal of the American Medical Association, 261*(4), 558.
10. Lopez, B. G., & Jysotna, V. (2017). Psycholinguistic approaches to humor. In *The Routledge handbook of language and humor* (pp. 267–281). Thousand Oaks: SAGE Publications.
11. Lundqvist, L. O., Carlsson, F., Hilmersson, P., & Juslin, P. N. (2009). Emotional responses to music: Experience, expression, and physiology. *Psychology of Music, 37*(1), 61–90.
12. Martin, R. A., & Dobbin, J. P. (1988). Sense of humor, hassles, and immunoglobulin A: Evidence for a stress-moderating effect of humor. *International Journal of Psychiatry in Medicine, 18*(2), 93–105.
13. Mahony, D. L., Burroughs, W. J., & Hieatt, A. C. (2001). The effects of laughter on discomfort thresholds: Does expectation become reality? *Journal of General Psychology, 128*(2), 217–226.
14. Melzack, R., & Wall, P. D. (1965). Pain mechanisms: A new theory. *Science, 150*(3699), 971–979.

15. McDonald, A. J. (1998). Cortical pathways to the mammalian amygdala. *Progress in Neurobiology, 55*(3), 257–332. https://doi.org/10.1016/S030301-0082(98)00003-3.
16. Murakami, H., Katsunuma, R., Oba, K., Terasawa, K., Motomura, Y., Mishima, K., et al. (2015). Neural networks for mindfulness and emotion suppression. *PloS One, 10*(6). https://doi.org/10.1371/journal.pone.0128005.eCollection2015.
17. Martin, R. A. (2000). Humor. In *Encyclopedia of psychology* (Vol. 4, pp. 202–204). USA: American Psychological Association.
18. Mohammadi Ziabari, S. S., & Treur, J. (2018). Cognitive modeling of mindfulness therapy by autogenic training. In *Proceedings of the 5th International Conference on Information Systems Design and Intelligent Applications, INDIA'18. Advances in intelligent systems and computing*. Berlin: Springer.
19. Mohammadi Ziabari, S. S., & Treur, J. (2018). An adaptive cognitive temporal-causal network model of a mindfulness therapy based. In *Proceedings of the 10th International Conference on Intelligent Human Computer Interaction, IHCI'18*. India: Springer.
20. Mohammadi Ziabari, S. S., & Treur, J. (2018). Integrative Biological, Cognitive and affective modeling of a drug-therapy for a post-traumatic stress disorder. In *Proceedings of the 7th International Conference on Theory and Practice of Natural Computing, TPNC'18*. Berlin: Springer.
21. Nakamura, T., Matsui, T., Utsumi, A., & Makita, K. (2017). The role of the amygdala in incongruity resolution: The case of humor comprehension. *Social Neuroscience, 13*(5), 553–565. https://doi.org/10.1080/17470919.2017.1365760.
22. Pessoa, L., & Adolphs, R. (2010). Emotion processing and the amygdala: Form a 'low roads' to 'many roads' of evaluating biological significance. *Nature Reviews Neuroscience, 11*(11), 773–783. https://doi.org/10.1038/nrn2920.
23. Ruch, W, & Ekman, P. (2001). The expensive pattern of laughter. In *Emotion, qualia, and consciousness* (pp. 426–443). Japan: World Scientific.
24. Raskin, V. (2008). *The primer of humor research*. Berlin: Mouton de Gruyter publishers (2008).
25. Ruch, W. (1993). Exhilaration and humor. In *Handbook of emotions* (pp. 605–616). New York, USA: Guilford Press.
26. Simpson, J. A., & Weiner, E. S. C. (1989). *The oxford English dictionary* (2nd ed.). Oxford, UK: Clarendon Press.
27. Sande, D., Grafman, J., & Zalla, T. The human amygdala: An evolved system for relevance detection. *Reviews in the Neuroscience, 14*(4), 303–316. https://doi.org/10.1515/revneuro.2003.14.4.303.
28. Takahashi, K., Iwase, M., & Yamashita, K. (2001). The evaluation of natural killer cell activity induced by laughter in a crossover designed study. *International Journal of Molecular Medicine, 8*(6), 645–650.
29. Tse, M. M. Y., Lo, A. P. K., Cheng, T. L. Y., Chan, E. K. K., Chan, A. H. Y., & Chung, H. S. W. (2010). Humor therapy: Relieving chronic pain and enhancing happiness for older adults. *Journal of Aging Research, 1*, 1–9. https://doi.org/10.4061/2010/343574.
30. Treur, J., & Ziabari, S. S. M. (2018). An adaptive temporal-causal network model for decision making under acute stress. In *Proceedings of the 10th International Conference on Computational Collective Intelligence, ICCCI'2018. Lecture Notes in Computer Science*. Berlin: Springer.
31. Treur, J. (2016). *Network-oriented modeling: Addressing complexity of cognitive, affective and social interactions*. Berlin: Springer.
32. Treur, J. (2016). Verification of temporal-causal network models by mathematical analysis. *Vietnam Journal of Computational Science, 3*, 207–221.
33. Verma, S. K., & Khanna, G. (2010). The effect of music therapy and meditation on sports performance in professional shooters. *Journal of Exercise Science and Physiotherapy, 6*(2).
34. Wanzer, M., Booth-Butterfield, M., & Booth-Butterfield, S. (2005). If we didn't use humor, we'd cry. Humorous coping communication in health care settings. *Journal of Health Communication, 10*(2), 105–125.

35. Weisenberg, M., Tepper, I., & Schwarzwald, J. (1995). Humor as a cognitive technique for increasing pain tolerance. *Pain, 63*(2), 207–212.

Neural Correlates of Dual Decision Processes: A Network-Based Meta-analysis

Ting-Peng Liang, Yen-Chun Chou and Chia-Hung Liu

Abstract It is well-received that human decision mechanism involves two processes: intuition and deliberation, which is also known as faster system 1 and slower system 2. A large volume of research has used this mechanism to interpret human decision behavior and the activation of associated bran regions in different scenarios. Recently, a trend of brain image research is to focus not on the role of individual brain areas but on the network of area connectivity. The purpose of this research is hence to explore how different brain regions are connected when these different decision processes are activated. In particular, we conduct a meta-analysis to build new knowledge on existing published primary research to construct neural networks associated with these dual processes. The social network analysis is used for this meta-analysis and results will be reported.

Keywords Neural correlation · Dual decision process · Brain network analysis

1 Introduction

Classical economic theory assumes that human decisions are rational. This, however, has been falsified as more and more evidence indicates that human behaviors are not totally rational. Instead, decisions are made by the collaboration of two brain mechanisms: one is faster, affective, and intuitive; while the other is slower, rational and deliberative. They are called "system 1" versus "system 2" [1]; "automatic" ver-

T.-P. Liang (✉)
National Sun Yat-Sen University, Kaohsiung, Taiwan
e-mail: tpliang@mail.nsysu.edu.tw

Y.-C. Chou · C.-H. Liu
National Chengchi University, Taipei, Taiwan
e-mail: yenchun@nccu.edu.tw

C.-H. Liu
e-mail: ga588950@gmail.com

F. D. Davis et al. (eds.), *Information Systems and Neuroscience*,
Lecture Notes in Information Systems and Organisation 32,
https://doi.org/10.1007/978-3-030-28144-1_22

sus "controlled" [2]; or "reflexive" (automatic system x) and "reflective" (controlled system c) [3]. System 1 is more automatic and heuristics-based, while system 2 is more deliberate and logical. This dual-process mechanism has also been adopted extensively in numerous research and conceptualized into a few major theories, such as the Elaboration Likelihood Model (ELM) [4, 5], the Heuristic-Systematic Model (HSM) [2, 6] and Heuristic-Analytic Model [7]. We also have empirical findings that show the existence of different brain regions are activated when solving different problems (e.g., [8, 9]). A number of prior studies have revealed that these two systems coexist and are employed in different decision tasks. Different brain regions are recruited and collaborated together to solve problems.

A large volume of papers in business and decision sciences have been published based on the dual process theories. However, most of them are behavioral in nature and draw interpretations from questionnaire survey or experimental data.

Recent development in cognitive neural science has allowed us to further examine how different brain areas are activated through the use of special instruments such as functional magnetic reasonance Imaging (fMRI) to better understand this dual systems model. Many research results under different contexts have been reported, but different experimental settings and the complexity of the human brains often lead to inconsistent observations that are hard to demonstrate the full picture.

The purpose of this study is to conduct a meta-analysis on published studies that adopted the dual systems theory to develop a better understanding of how these two subsystems work. We collected experimental results of published literature and aggregated their findings with the social network analysis, which is a data mining technique used for finding relationships among objects. The circuits of both subsystems are derived and evaluated. The result allows us to better understand the collaboration of brain areas in these two systems.

2 Literature Review

2.1 *The Dual-Process Accounts of Cognitive Processing*

Given the popularity of the dual-process theory, Evans [10] reviewed literature extensively to describe features from consciousness, age of evolution, functional characteristics, and individual differences. Key features associated with System 1 are automatic, experiential, heuristic, implicit, intuitive, holistic, reflexive, and impulsive; while key features associated with System 2 are controlled, rational, systematic, explicit, analytic, rule-based, and reflective. System 1 is the mechanism for fast reacting while System 2 is the slow system for deliberation and rational thinking.

While a large volume of empirical research has been published based on theories derived from this dual-process model, most of them are based on behavioral studies. Recently, the development of neuroscience, particularly the use of functional Magnetic Resonance Imaging, fMRI) instrument that allows for analyzing brain images, has motived many studies to examine whether different brain mechanisms actually exist in human decision making. For instance, Kuo et al. [8] found that the middle frontal gyrus, the inferior parietal lobule, and the precuneus were more activated in dominance-solvable games, while the insula and anterior cingulate cortex were more activated in coordination games that need more intuition.

In his review article, Lieberman [11] summarized brain regions associated with System 1 (X-System) as the amygdala, basal ganglia, ventromedial prefrontal cortex (VMPFC), lateral temporal cortex (LTC), and dorsal anterior cingulate cortex (dACC); while those associated with System 2 (C-System) as lateral prefrontal cortex (LPFC), medial prefrontal cortex (MPFC), lateral parietal cortex (LPAC), medial parietal cortex (MPAC), medial temporal lobe (MTL), and rostral anterior cingulate cortex (rACC). In fact, these two systems are closely coordinated. For example, Farrell et al. [12] reported that System 1 was not suppressed even under the performance-based contract that required economic reasoning.

2.2 *Analysis of Brain Networks*

Human brain is a complicated large network of neurons. Early research in cognitive neuroscience focuses on identifying brain regions associated with different cognitive processes such as risk assessment, rewards, trust, and choice. In reality, the mapping between brain regions and cognitive activities are many to many. Hence, it is often difficult to point out the exact role of individual brain regions. Furthermore, it is hardly found that only a single brain region is more activated significantly in solving a problem. Hence, a separate body of work has emphasized on determining how information exchange between distinct areas may give rise to cognitive processing. As described in Wig et al. [13]:

> Initial support for the idea that brain connectivity mediates cognition was largely based on observations obtained in animal models and careful examination of patient populations, wherein it was hypothesized that disrupted information exchange between brain areas could account for the behavioral disturbances that accompanied focal brain damage and psychiatric problems….the brain network, like numerous other systems, both man-made and biological, exhibits an underlying organization that characterizes and mediates its functions. A critical question is how does one go about untangling and making sense of this organized complexity?

Bassett and Sporns [14] argues that "approaching brain structure and function from an explicitly integrative perspective, network neuroscience pursues new ways to map, record, analyze and model the elements and interactions of neurobiological systems."

In recent research, we have also seen an increasing number of brain network analysis in neuroscientific papers. As an example, Van Den Heuvel and Pol [15] illustrates the co-activation of the resting-state fMRI functional connectivity.

3 Research Methodology

There are many ways to build a network of functionally linked brain regions that are connected when a task is processed. Brain network analysis may be at the anatomical or functional level. The common foundation is the graph theory that represents a network as a set of interconnected nodes and edges (e.g., [13, 16]). A brain network may be constructed from primary fMRI data or from published secondary data with meta-analysis, either at the voxel level or at the brain-regional level.

In this research, we use the social network analysis (SNS) technique common for analyzing human community to investigate how brain regions are coordinated when System 1 and System 2 are activated for processing tasks. The underlying assumption is that the brain is an organization of numerous actors who are recruited to be involved in processing a task when necessary. The procedures of our analysis include the following:

(1) Data collection
Search fMRI studies in decision making from the PubMed, Science Direct, Google Scholar, and Web of Knowledge using combinations of the following keywords: "heuristic," "emotional," "automatic," "logical," "rational," "analytical," plus "decision making" and "fMRI". This results in a total of 1258 papers.
(2) Data cleansing
We reviewed these papers by two researchers to keep those that have used fMRI for decision tasks that involves the dual processes. This results in a total of 113 papers.
(3) Sample construction
Finally, we chose those experiments with more than 10 subjects and the contexts are highly relevant such as "emotional versus non-emotional" or heuristic versus analytic." This results in our sample of 32 papers with 41 experimental reports, as some papers have more than one experiment.
(4) Coding and Network modeling
The selected experiments were coded by the same two researchers who have background in neuroscience based on the designed thinking process and reported brain region associations in these papers. The data were then analyzed with a social network software, GePhi, to determine their associated brain regions, connectivity, centrality, and subgroups.

SNS is a methodology for analyzing the structure of a community. It characterizes networked structures in terms of nodes (individual actors) and links (relationships or interactions between nodes, also called edges, ties or connections). There are a few properties that are important to SNS. First, nodes included in a networked graph show the involvement of the actor. In our analysis, each brain region is identified as a node. Since brain regions may be defined at different levels, we followed the term claimed by the authors in their papers and do not aggregate them. For example, one paper reports that dlPFC while another reports PFC was activated in the experiment, we coded them as two separate nodes. If we would like to examine a higher level connectivity, however, the instance of dlPFC will be included as an instance of PFC, but not vice versa.

When two brain regions are reported to be activated in the same experiment, we code them in the same record to assume that these regions had connectivity in processing the task during the experiment. Hence, when a study reported that Insula, inferior temporal gyrus (ITG), and ACC were significantly activated in an experiment that involved an emotional task, we assume that these three regions were connected as a network in the emotional decision making. That is, this study contributes one instance of three nodes and three connected edges (insular-ITG, insula-ACC, and ITG-ACC). The resulting brain network from our SNA report for emotional tasks (i.e., System 1) is the aggregation of all samples that involves emotional decision making. The more instances that report connectivity of two brain regions indicate "stronger" connectivity (i.e., the *strength* of a connection).

Another concept critical to SNS is the *centrality* of each node in the network. A node with higher centrality implies its importance or influence in the network. There are a few different types of centrality to represent the role of nodes from different perspectives. In this paper, we show the *weighted degree centrality* in our resulting network structures, which is a weighted connectivity measure of all connected nodes.

4 Preliminary Findings

Our analysis shows that System 1 and System 2 have both overlapped and distinct brain regions. Each is composed of more than one subgroups. Figures 1 and 2 show the functional network structure of System 1 and System 2, respectively. Larger notes indicate higher centrality in the decision process and wider edges indicate higher likelihood of connectivity. In Fig. 1, we know that Precuneus and ACC are the two most important brain regions when System 1 is involved in decision making and the top five brain regions are ACC, precuneus, Amygdala, PCC, insula, and Medial Frontal Gyrus. In other words, the network of these five regions are the key feature of decision making that involve System 1.

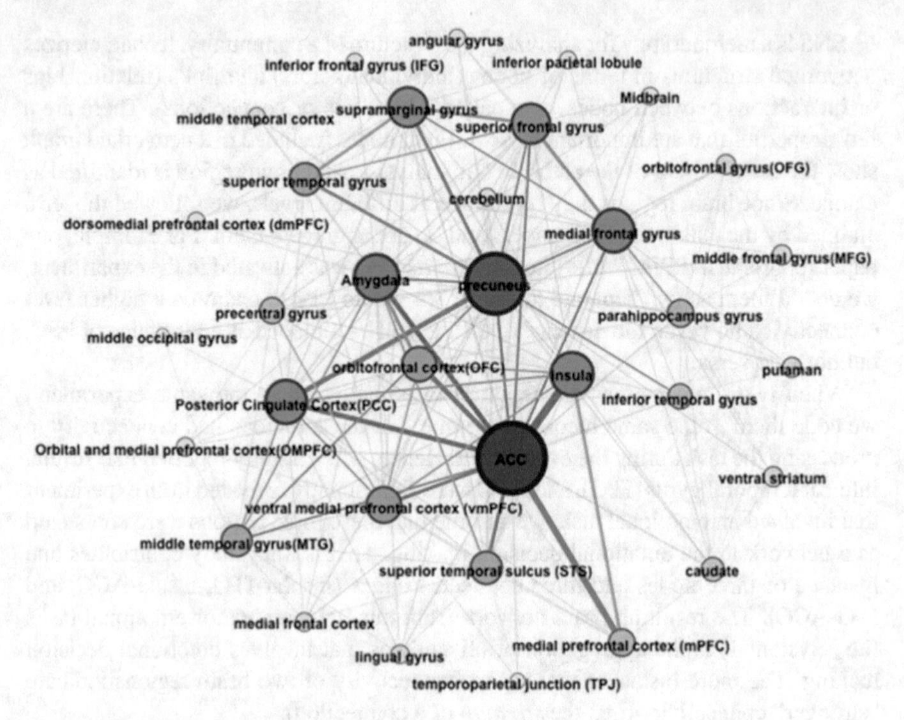

Fig. 1 Functional brain network of system 1

Figure 2 shows the weighted degree centrality of System 2, which indicates that dlPFC, Middle frontal gyrus, ACC, and inferior frontal gyrus are key regions in the System 2 network. ACC appears in both networks, which is understandable due to the nature that most decisions involve both processes and ACC is the bridge between them.

5 Concluding Remarks

The purpose of this research is to construct functional networks of the dual systems for cognitive information processing. We collected published research reports related to the thinking and intuition decision processes and coded their findings for our social network analysis. A few distinct brain regions and their connectivity are found. Our findings is interesting in that these identified brain networks can be used as neural markers for determining which brain subsystem is more involved in a decision making. Subgroup analysis and more implications from the social network analysis will be reported in future research.

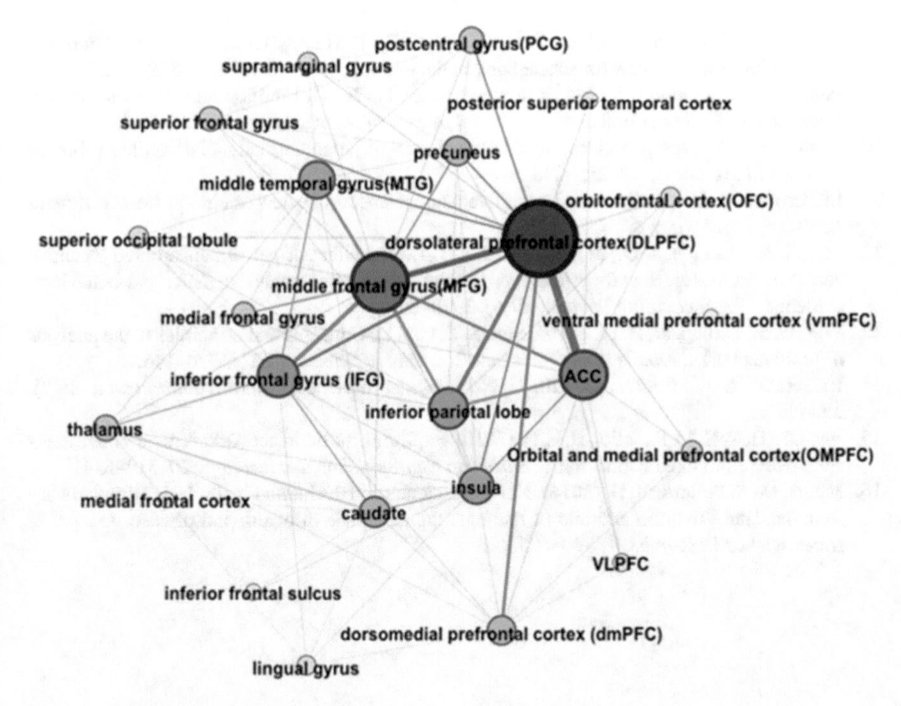

Fig. 2 Functional brain network of system 2

Acknowledgements This research was financially supported by grants from the Ministry of Science and Technology of ROC under the grant #103-2410-H-110-083 -MY2 and Research Center Grant from The Ministry of Education of ROC.

References

1. Kahneman, D. (2011). *Thinking, fast and slow*. New York, NY: Macmillan.
2. Chaiken, S., & Trope, Y. (Eds.). (1999). *Dual-process theories in social psychology*. New York, NY: The Guilford Press.
3. Lieberman, M. D., Gaunt, R., Gilbert, D. T., & Trope, Y. (2002). Reflection and reflexion: A social cognitive neuroscience approach to attributional inference. *Advances in Experimental Social Psychology, 34,* 199–249.
4. Petty, R. E., & Cacioppo, J. T. (1986). *Communication and persuasion: Central and peripheral routes to attitude change*. New York, NY: Springer.
5. Petty, R. E., & Cacioppo, J. T. (1986). The elaboration likelihood model of persuasion. *Advances in Experimental Social Psychology, 19,* 123–205.
6. Chaiken, S. (1980). Heuristic versus systematic information processing and the use of source versus message cues in persuasion. *Journal of Personality and Social Psychology, 39*(5), 752–766.
7. Evans, J. (1984). Heuristic and analytic processes in reasoning. *British Journal of Psychology, 75,* 451–468.

8. Kuo, W. J., Sjöström, T., Chen, Y. P., Wang, Y. H., & Huang, C. Y. (2009). Intuition and deliberation: Two systems for strategizing in the brain. *Science, 324*(5926), 519–522.
9. Phelps, E. A., Lempert, K. M., & Sokol-Hessner, P. (2014). Emotion and decision making: Multiple modulatory neural circuits. *Annual Review of Neuroscience, 37,* 263–287.
10. Evans, J. (2008). Dual processing accounts of reasoning, judgment, and social cognition. *Annual Review of Psychology, 59,* 255–278.
11. Lieberman, M. D. (2007). Social cognitive neuroscience: A review of core processes. *Annual Review of Psychology, 58,* 259–289.
12. Farrell, A. M., Goh, J. O., & White, B. J. (2014). The effect of performance-based incentive contracts on system 1 and system 2 processing in affective contexts: fMRI and behavioral evidence. *The Accounting Review., 89*(6), 1979–2010.
13. Wig, G. S., Schlaggar, B. L., & Petersen, S. E. (2011). Concepts and principles in the analysis of brain networks. *Annals of the New York Academy of Sciences, 1224,* 126–146.
14. Bassett, D. S., & Sporns, O. (2017). Network neuroscience. *Nature Neuroscience, 20*(3), 353–364.
15. Van Den Heuvel, M. P., & Pol, H. E. H. (2010). Exploring the brain network: A review on resting-state fMRI functional connectivity. *European Neuropsychopharmacology, 20,* 519–534.
16. Mears, D., & Pollard, H. B. (2016). Network science and the human brain: Using graph theory to understand the brain and one of its hubs, the amygdala in health and disease. *Journal of Neuroscience Research, 94,* 590–605.

Exploring the Neural Correlates of Visual Aesthetics on Websites

Anika Nissen

Abstract Perceiving beauty typically provides pleasure through which it becomes a general human need. In the context of online shopping, websites should be designed, bearing in mind that users prefer websites that are visually pleasing. In recent research, website aesthetics have mostly been explored with qualitative self-reports measurements, eye-tracking devices, or even mathematical models. Moreover, also neuroscientific methods have been utilized to investigate websites' aesthetics and beauty. Nevertheless there are only few studies that investigated the visual design of websites with neuroimaging tools, even though these neuroimaging studies might enable researchers to investigated unconscious cognitive processes. Against this background, this work in progress aims to open up fruitful avenues to measure website aesthetics and states hypotheses which brain regions are likely to be involved when it comes to the aesthetically pleasing perception of websites by users.

Keywords Visual design · Aesthetics · Websites · Appeal · Neural correlates · Brain imaging · fNIRS

1 Introduction

Visual design typically determines the perceived usefulness and ease of use, as well as the expected functionality of a website [1, 2]. This further determines whether users are likely to remain on the website and use it, for instance to make a purchase [3, 4]. As a consequence of this process from first visit to further use, it might be reasonable to consider two phases of visual aesthetics being (1) the beauty perceived *intuitively* which is in about the first 500 ms the user visits a website the first time, and (2) the one that is perceived *reflectively* after some time of actual use [5]. In other words, the design of websites needs to convince users to stay on the website in phase (1), but also needs to support the actual use of the website (2). Designing such visual appeal of graphical user interfaces (GUI) lies in the human-computer interaction

A. Nissen (✉)
University Duisburg-Essen, Essen, Germany
e-mail: anika.nissen@icb.uni-due.de

F. D. Davis et al. (eds.), *Information Systems and Neuroscience*, Lecture Notes in Information Systems and Organisation 32,
https://doi.org/10.1007/978-3-030-28144-1_23

(HCI) domain in which a variety of approaches has already been developed, that are often tightly connected to usability and user experience research [6–8]. But, in order to ensure that the design is as pleasing as intended, it needs to be measured in some way.

In respective research streams, a variety of metrics has been developed covering mostly self-reports, eye-tracking, and layout-based metrics [9, 10]. These metrics might assess the reflective aesthetics of websites and there have been several correlations found among them [10, 11]. For example, Altaboli and Lin [12] showed that there are significant correlations between the layout-based suggestions from Ngo et al. [13] and the self-reports of participants. However, we believe that self-reports come with several limitations. Firstly, as previously described, the perception of aesthetics might differ between intuitive and reflective appeal, from which the former might not be assessable with the given metrics. Reasons for this can be found in self-reports typically requiring conscious thought [14] which makes it impossible to capture anything intuitive with it. Furthermore, self-reports are prone to desirability biases and thus, they might not always reflect the *truth*. Consequently, we wonder whether there might be ways in which we can measure what people perceive without the need to ask them.

One established method is eye-tracking, which can tell us where people look and how long they focus on specific areas of interest (AOIs) [15]. This may allow us to draw conclusions regarding a GUI's attractiveness [16], however, it cannot tell us what the underlying *cognitive process* might be. A shortcoming that could be overcome by employing brain imaging methods that show neural activity. Consequently, this leads us to taking a look into neuroscientific literature upon which we found that aesthetics and perceptions of beauty are quite frequently researched, albeit not in the context of websites [17, 18]. Instead, it is explored for the beauties of music [19, 20], art [21], or human faces [22, 23] between which lie great differences in the recorded brain activity. Even when focusing on visual aesthetics only, concrete results from each study still vary depending on the stimulus material or the sample. However, some structures seem to be generally activated during aesthetic experience [24]. For instance, Reimann et al. [17] identified increased neural activity in the ventromedial prefrontal cortex (vmPFC) and striatum (particularly the right nucleus accumbens) for aesthetic product design. In contrast, Brown et al. [24] indicated in a meta-analysis of brain imaging experiments that visual aesthetics were located in the left inferior parietal lobule, fusiform gyri bilaterally, inferior frontal gyri, hypothalamus, caudate nucleus and the amygdala. Drawn to these results, mainly structures related to the limbic system and reward system show increased neural activity for aesthetic experience [24–26].

Taking together the shortcommings of the prior operationalized metrics for website aesthetics and the broad variability in active brain areas for the general perception of beauty, we question *which brain structures show a change in their activity when confronted with appealing versus less appealing website design?* Furthermore, it might be interesting in how far the manipulation of layout-based measures for aesthetic websites is reflected in specific brain activity.

As a result, this paper presents a future research design about how this research question could be adressed. The remainder of the paper is structured as follows. Firstly, we present some layout-based metrics taken from Ngo et al. [13] in the following chapter. These metrics are most often used in the literature we reviewed so far, and they offer a reasonable approach to preselect websites which might be perceived as pleasing. After that, we take a look into neuroaesthetic literature and review some studies in Sect. 3. From this, we derive hypotheses regarding which brain areas are likely to show activity for certain characteristics and describe how we plan to design the experiment in Sect. 4.

2 Layout-Based Measures of Website Aesthetics

Ngo et al. [13, 27] proposed 14 layout-based measures integrating the description of the size, ratios, and position of content elements on websites. These elements are based on mathematical calculations for balance, equilibrium, symmetry, sequence, cohesion, unity, proportion, simplicity, density, regularity, economy, homogeneity, rhythm and order/complexity. Some of these measures have been investigated in several studies before which validate their influence. For example, Altaboli & Lin [12] found for economy, density, balance, symmetry, sequence, simplicity, unity, and rhythm that they significantly correlated with the self-reported measures of the VisAWI (Visual Aesthetics of Website Inventory) developed by Moshagen & Thielsch [28]. Supporting these findings, another study identified that symmetry, sequence, balance, and unity significantly correlate with VisAWI items [29]. Except for unity, they were also identified by Zain et al. [30], who additionally included simplicity and rhythm measures. Moreover, Salimun et al. [31] also investigated the symmetry, unity, and sequence. As a result, recent research shows that the most important factors seem to be symmetry, sequence, balance, and unity.

Balance, it describes how evenly content elements are distributed on the screen [27]. For instance, Fig. 1 shows on the left side (a) an evenly distributed website, while (b) is less balanced as there are more content placeholders on the left than on the right site. As balance is calculated both horizontally and vertically, the red dotted

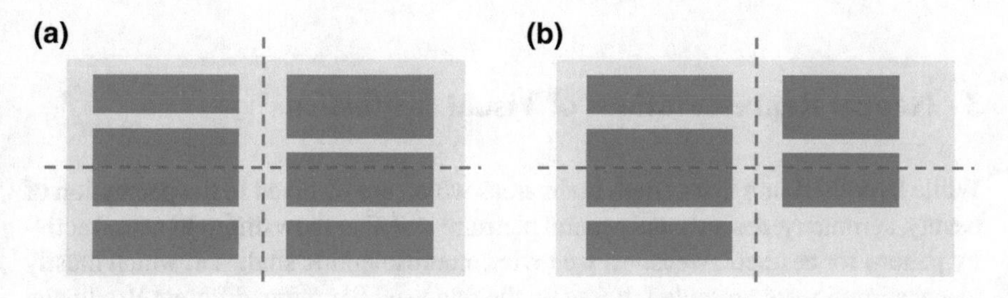

Fig. 1 Example for balance of content. Taken and adapted from [18]

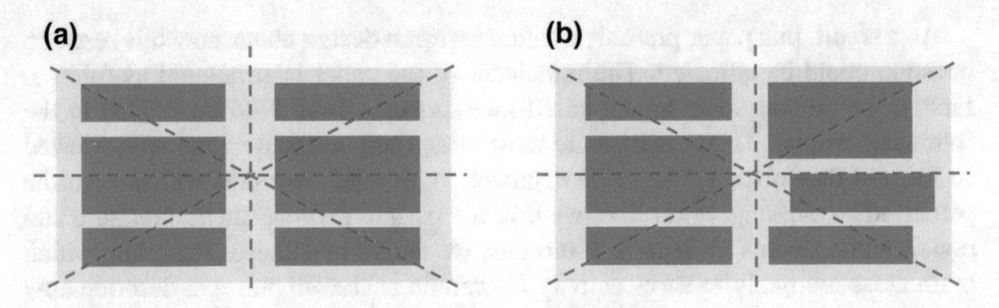

Fig. 2 Example of symmetry of content. Taken and adapted from [18]

lines are added for a better orientation. Finally, although (a) is not symmetric, and might also not fulfill the **sequence** criterium, it is still evenly filled with content in all four quarters.

The measure of **symmetry** describes that both sides of an axis have the same amount and forms of content elements. Symmetry does not only consider the horizontal and vertical axis, but also whether the content is diagonally balanced [27]. Analogous we have one version (Fig. 2a) in which all symmetries are considered and one (Fig. 2b) in which they are not considered.

Sequence "facilitates the movement of the eye" [13] which is for most people normally from the upper left to the lower right. As bigger sized objects also receive attention first in comparison to smaller objects, it might be reasonable to start a website with bigger sized elements and let the smaller ones follow. Finally, **unity** describes whether content elements are perceived as one entity or whether each element presents one unit on its own [13]. This can be realized using colors or putting elements that belong together.

In conclusion, if a website considers all these elements, it should be more likely perceived as aesthetic than a website that is highly unbalanced and/or asymmetric. Furthermore, websites that are not designed bearing in mind the way we look at things are more likely to confuse or irritate the user. Finally, and probably also in the sense of order and reducing complexity, elements that belong together should also be visually grouped together. Grounded on this work, the following section reviews neuroscientific literature that deals with visual aesthetics of paintings or computer-generated visuals.

3 Neural Representations of Visual Aesthetics

While hypothesizing that certain brain areas which are involved in the perception of beauty, symmetry, rewards and general pleasure will also show different neural activity pattern for aesthetic websites, we review neuroscientific studies in which mostly visual stimuli were presented. Based on the reviewed literature, different Brodmann areas (BA) which seem to be typically involved and might also show increased or decreased activity for beautiful versus ugly website designs are identified. Table 1 shows the resulting concept matrix of our brief review.

We are well aware that we cannot make any claims of completeness of our study, however it suffices to identify key brain regions that may show more activity than others. It should be mentioned that we assume that results from the other studies can be transferred to our own study, albeit having a different focus and including different stimuli material. In fact, we might also end up rejecting all of the identified results for our study, as websites afford active interaction of the user which might be different from passively perceiving a painting. As such, the results of this brief survey solely serve a starting point to suggest hypotheses.

In the table, brain areas that were used by about half of the included studies are marked. From this, we see that apparently BA 9/10, 32, and 47 seem to be most often activated (or observed to show activity) for aesthetic perception across different contexts and stimuli. BA 9/10/32 comprise the frontomedian and anterior cingulate cortex (ACC) [26, 32]. More precisely, parts of BA 10/32 and 47 make up what is referred to as medial to inferior orbitofrontal cortex (OFC) and ventromedial prefrontal cortex (vmPFC) [25, 33], with the latter belonging both functionally and anatomically to the OFC [34]. The included meta studies in our review being [24, 25, 35] let us suggest, that these areas are also supported in studies not included in this paper. All of these structures belong to the reward circuit which processes mostly pleasant stimuli [34].

4 Planned Experiment Design and Hypotheses

In the planned experiment, we aim to design a prototypical website that fulfills all of the four described and presented criteria versus a website that does not. Similarly, we vary the displayed criteria and investigate the neural reaction to it while keeping the overall design of the content elements the same. An example of potential stimulus material can be seen in the following figures; in Fig. 3 the left side (a) shows a nearly symmetrical website versus the same website as asymmetric on the right side (b). Figure 4 shows the same website in a version where the sequence criterion is fulfilled (a) versus a version where it is not fulfilled (b).

Several brain areas involved in the processing of visual website appeals are located in the prefrontal cortex (PFC), which makes them measurable with the innovative technology of mobile functional near-infrared spectroscopy (fNIRS) [40, 41]. fNIRS works by sending light with a wavelength of 830 nm and < 780 nm into the brain [42]. The brain tissue then absorbs or scatters the light which is received by detectors that allow to calculate the amount of oxygen running in the blood of the given brain tissue [43, 44]. Consequently, fNIRS measures the oxy-hemoglobin (O_2Hb) and deoxy-hemoglobin (HHb) in the blood; with increases in O_2Hb indicating increased activity of the corresponding brain structure [42].

Consequently, based on the capabilities of mobile fNIRS BAs of 9, 10, 47 might be observed [45]. However, brain areas such as the ACC (BA 32) might lie too deep

Table 1 Findings from studies focusing on perceived visual beauty

	Brodmann area															
Study	4	6	7	**9**	**10**	11	18	19	24	30	**32**	37	44	45	46	**47**
Boccia et al. [35]	X	X	X	X			X	X			X	X			X	
Brown et al. [24]								X			X	X	X			
Cela-Conde et al. [36]			X										X	X		X
Jacobs et al. [32]				X	X		X			X						
Jacobsen et al. [26]				X	X						X			X		X
Kirk et al. [33]					X	X										X
Kühn & Gallinat [25]					X	X			X	X	X					
Vartanian & Goel [37]					X		X				X	X				
Vartanian et al. [38]	X	X	X	X	X				X		X				X	
Zhang et al. [39]						X		X					X	X		X
Σ	2	2	3	**4**	**6**	3	3	3	2	2	**6**	3	3	3	2	**4**

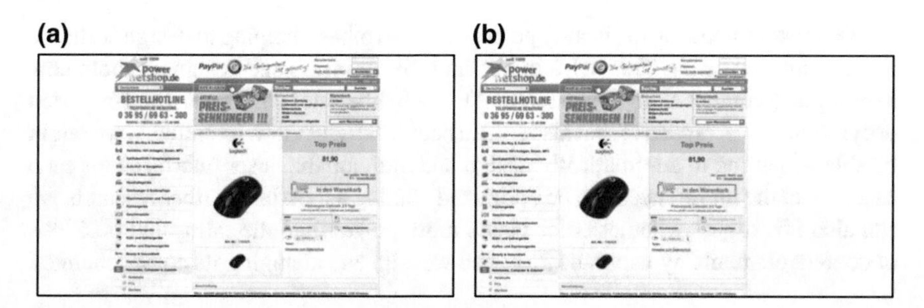

Fig. 3 Symmetric versus non-symmetric website design

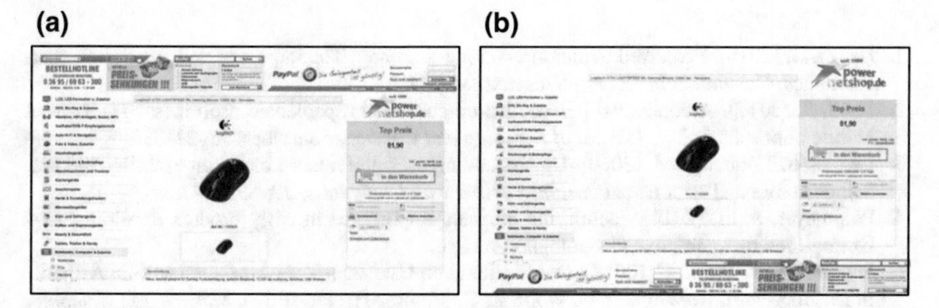

Fig. 4 Sequence fulfilled versus sequence not fulfilled

in the brain and thus, it is not possible to observe its activity because the fNIRS signal only reaches approximately 2–3 cm deep into the human brain [46]. Nevertheless, as most prefrontal cortex regions observed in previous studies could be observed using fNIRS, the planned experiment seems feasible.

Based on recent research, we derive the following hypotheses for our planned experiment:

H1 Aesthetic websites (AW) indicate an increase in O_2Hb in brain areas of the BA 9 in comparison to less aesthetic websites (NAW).

H2 AW indicate an increase in O_2Hb in brain areas of the BA 10 compared to the NAW.

H3 AW indicate an increase in O_2Hb in brain areas of the BA 47 compared to the NAW

In addition to neural data, we also assess self-reported measurements of beauty using the VisAWI [28], as this questionnaire has already showed its correlation to Ngo et al.'s formulas in several studies (see Sect. 2). This would leave us with 3 data types being (a) the layout-based calculations, (b) self-reported data from our participants, and (c) neural data from our participants.

Consequently, the planned study will receive information about whether the visual appearance of website designs is also manifested in brain activity related to aesthetics and pleasure. In line with this, we would investigate whether some design decisions have a greater impact on the beauty perception than others, helping to design aesthetic and effective websites. This helps to inform web engineers and designers about how to best place content elements in relation to each other; not only from a self-reported or eye-tracking perspective, but also from a neural activity view. Eventually, we might be able sometime to automatically design and individualize user interfaces for each user so that the preferences can be optimized. Taking a look into further research, we can also investigate whether color yields more power than the ratios and positions of content elements by using the designed website and changing its color scheme.

References

1. Hasan, B. (2016). Perceived irritation in online shopping: The impact of website design characteristics. *Computers in Human Behavior, 54,* 224–230.
2. Sohn, S. (2017). A contextual perspective on consumers' perceived usefulness: The case of mobile online shopping. *Journal of Retailing and Consumer Services, 38,* 22–33.
3. Lee, S., & Koubek, R. J. (2010). Understanding user preferences based on usability and aesthetics before and after actual use. *Interacting with Computers, 22,* 530–543.
4. Pohlmeyer, A. E. (2011). Identifying attribute importance in early product development—Exemplified by interactive technologies and age.
5. Schmidt, T., & Wolff, C. (2017). Der Einfluss von User Interface- Attributen auf die Ästhetik. In M. Burghardt, R. Wimmer, C. Wolff, & C. Womser-Hacker (Eds.), *Mensch und Computer 2017 - Tagungsband* (pp. 61–71). Regensburg: Gesellschaft für Informatik e.V.
6. Tractinsky, N. (2005). Does aesthetics matter in human-computer interaction. In C. Stary (Ed.), *Mensch & computer 2005: Kunst und Wissenschaft - Grenzüberschreitungen der interaktiven ART* (pp. 29–42). München: Oldenbourg Verlag.
7. Desmet, P. M. A., & Hekkert, P. (2007). Framework of product experience. *International Journal of Design, 1,* 57–66.
8. von Saucken, C., & Gomez, R. (2014). Unified user experience model enabling a more comprehensive understanding of emotional experience design. In *9th International Conference on Design and Emotion* (pp. 631–640). Bogotá.
9. Bargas-Avila, J. A, & Hornbæk, K. (2011). Old wine in new bottles or novel challenges? A critical analysis of empirical studies of user experience. In *Proceedings of the SIGCHI Conference on Human Factors in Computing Systems* (pp. 1–10).
10. Pappas, I. O., Sharma, K., Mikalef, P., & Giannakos, M. N. (2018). A comparison of gaze behavior of experts and novices to explain website visual appeal. In *Twenty-Second Pacific Asia Conference on Information Systems* 14.
11. Pappas, I., Sharma, K., Mikalef, P., & Giannakos, M. (2018). Visual aesthetics of E-commerce websites: An eye-tracking approach. In *51st Hawaii International Conference on System Sciences (HICSS), Big Island, Hawaii.*
12. Altaboli, A., & Lin, Y. (2011). Objective and subjective measures of visual aesthetics of website interface design: The two sides of the coin. In Jacko, J. A. (Ed.) *14th International Conference, HCI International 2011* (pp. 35–44). Orlando, FL, USA: Springer International Publishing.
13. Ngo, D. C. L., Teo, L. S., & Byrne, J. G. (2003). Modelling interface aesthetics. *Information Sciences (Ny), 152,* 25–46.
14. Riedl, R., Davis, F. D., Banker, R., & Kenning, P. H. (2017). *Neuroscience in information systems research.* Cham: Springer International Publishing.
15. Bojko, A. (2013). *Eye tracking the user experience: A practical guide to research.* Brooklyn, NY: Rosenfeld Media.
16. Al-Wabil, A., Alabdulqader, E., Al-Abdulkraim, L., & Al-Twairesh, N. (2010). Measuring the user experience of digital books with children: An eyetracking study of interaction with

digital libraries. In *International Conference for Internet Technology and Secured Transactions* (pp. 1–7). London, UK: IEEE.
17. Reimann, M., Zaichkowsky, J., Neuhaus, C., Bender, T., & Weber, B. (2010). Aesthetic package design: A behavioral, neural, and psychological investigation. *Journal of Consumer Psychology, 20,* 431–441.
18. Chatterjee, A., & Vartanian, O. (2014). Neuroaesthetics. *Trends in Cognitive Sciences, 18,* 370–375.
19. Brattico, E., Pearce, M. (2013). The neuroaesthetics of music. *Psychology of Aesthetics, Creativity, and the Arts, 7,* 48–61.
20. Salimpoor, V. N., & Zatorre, R. J. (2013). Neural interactions that give rise to musical pleasure. *Psychology of Aesthetics, Creativity, and the Arts, 7,* 62–75.
21. Vartanian, O., & Skov, M. (2014). Neural correlates of viewing paintings: Evidence from a quantitative meta-analysis of functional magnetic resonance imaging data. *Brain and Cognition, 87,* 52–56.
22. Zaidel, D. W., & Cohen, J. A. (2005). The face, beauty, and symmetry: Perceiving asymmetry in beautiful faces. *International Journal of Neuroscience, 115,* 1165–1173.
23. Zaidel, D. W., & Deblieck, C. (2007). Attractiveness of natural faces compared to computer constructed perfectly symmetrical faces. *International Journal of Neuroscience, 117,* 423–431.
24. Brown, S., Gao, X., Tisdelle, L., Eickhoff, S. B., & Liotti, M. (2011). Naturalizing aesthetics: Brain areas for aesthetic appraisal across sensory modalities. *Neuroimage, 58,* 250–258.
25. Kühn, S., & Gallinat, J. (2012). The neural correlates of subjective pleasantness. *Neuroimage, 61,* 289–294.
26. Jacobsen, T., Schubotz, R. I., Höfel, L., & Cramon, D. Y. V. (2006). Brain correlates of aesthetic judgment of beauty. *Neuroimage, 29,* 276–285.
27. Ngo, D. C. L., & Byrne, J. G. (2001). Application of an aesthetic evaluation model to data entry screens. *Computers in Human Behavior, 17,* 149–185.
28. Moshagen, M., & Thielsch, M. T. (2010). Facets of visual aesthetics. *International Journal of Human-Computer Studies, 68,* 689–709.
29. Mõttus, M., Lamas, D., Pajusalu, M., & Torres, R. (2013). The evaluation of user interface aesthetics. In: *MIDI'13, Warsaw*.
30. Zain, J. M., Tey, M., & Goh, Y. (2008) Probing a self-developed aesthetics measurement application (SDA) in measuring aesthetics of Mandarin learning web page interfaces. *IJCSNS International Journal of Computer Science and Network Security, 8*.
31. Salimun, C., Purchase, H. C., Simmons, D. R., & Brewster, S. (2010). Preference ranking of screen layout principles. In *Proceedings of the 24th BCS Interaction Specialist Group Conference* (pp. 81–87).
32. Jacobs, R. H. A. H., Renken, R., & Cornelissen, F. W. (2012). Neural correlates of visual aesthetics—Beauty as the coalescence of stimulus and internal state. *PLoS ONE, 7,* 1–8.
33. Kirk, U., Skov, M., Hulme, O., Christensen, M. S., & Zeki, S. (2009). Modulation of aesthetic value by semantic context: An fMRI study. *Neuroimage, 44,* 1125–1132.
34. Nadal, M. (2013). *The experience of art: Insights from neuroimaging*. Amsterdam: Elsevier B.V.
35. Boccia, M., Barbetti, S., Piccardi, L., Guariglia, C., Ferlazzo, F., Giannini, A. M., et al. (2016). Where does brain neural activation in aesthetic responses to visual art occur? Meta-analytic evidence from neuroimaging studies. *Neuroscience and Biobehavioral Reviews, 60,* 65–71.
36. Cela-Conde, C. J., Ayala, F. J., Munar, E., Maestu, F., Nadal, M., Capo, M. A., et al. (2009). Sex-related similarities and differences in the neural correlates of beauty. *Proceedings of National Academy of Sciences, 106,* 3847–3852.
37. Vartanian, O., & Goel, V. (2004). Neuroanatomical correlates of aesthetic preference for paintings. *NeuroReport Cognitive Neuroscience and Neuropsychology, 15,* 893–897.
38. Vartanian, O., Navarrete, G., Chatterjee, A., Fich, L. B., Leder, H., Modrono, C., et al. (2013). Impact of contour on aesthetic judgments and approach-avoidance decisions in architecture. *Proceedings of National Academy of Sciences, 110,* 10446–10453.

39. Zhang, W., Lai, S., He, X., Zhao, X., & Lai, S. (2016). Neural correlates for aesthetic appraisal of pictograph and its referent: An fMRI study. *Behavioural Brain Research, 305,* 229–238.
40. Neben, T., Xiao, B. S., Lim, E., Tan, C. W., & Heinzl, A. (2015). Measuring appeal in human computer interaction: A cognitive neuroscience-based approach. In *Gmunden retreat on NeuroIS 2015* (pp. 151–159). Vienna, Austria.
41. Krampe, C., Gier, N., & Kenning, P. (2018). The application of mobile fNIRS in marketing research—Detecting the 'first-choice-brand' effect. *Frontiers in Human Neuroscience*.
42. Scholkmann, F., Kleiser, S., Metz, A. J., Zimmermann, R., Mata Pavia, J., Wolf, U., et al. (2014). A review on continuous wave functional near-infrared spectroscopy and imaging instrumentation and methodology. *Neuroimage, 85,* 6–27.
43. Ferrari, M., & Quaresima, V. (2012). A brief review on the history of human functional near-infrared spectroscopy (fNIRS) development and fields of application. *Neuroimage, 63,* 921–935.
44. Funane, T., Atsumori, H., Katura, T., Obata, A. N., Sato, H., Tanikawa, Y., et al. (2014). Quantitative evaluation of deep and shallow tissue layers' contribution to fNIRS signal using multi-distance optodes and independent component analysis. *Neuroimage., 85,* 150–165.
45. Shimoda, K., Moriguchi, Y., Tsuchiya, K., Katsuyama, S., & Tozato, F. (2014). Activation of the prefrontal cortex while performing a task at preferred slow pace and metronome slow pace: A functional near-infrared spectroscopy study. *Neural Plasticity, 2014*.
46. Krampe, C., Gier, N., & Kenning, P. (2017). Beyond traditional neuroimaging: Can mobile fNIRS add to NeuroIS? *Lecture Notes in Information Systems and Organisation, 25,* 151–157.

Mitigating Information Overload in e-Commerce Interactions with Conversational Agents

Maria del Carmen Ocón Palma, Anna-Maria Seeger and Armin Heinzl

Abstract Information overload influences users' satisfaction and performance when completing a complex task. In e-commerce interactions, this has the effect that customers' decision making becomes confused, less accurate and less effective. For websites, numerous countermeasures to mitigate information overload have been presented, whereas not many attempts have been made to reduce cognitive load when conversational agents are used instead. Conversational agents are expected to increase the perceived overload due to the voice interface characteristics. In this pilot study, the cognitive load of subjects was measured during an online shopping task which required different custom shopping skills for Amazon Alexa. It was tested if the countermeasure filtered repetition can reduce subjects' perceived overload when using the voice assistant and which load differences can be found in comparison to a shopping website. To measure the mental load, the skin conductance level was recorded.

Keywords Information overload · Conversational agents · Skin conductance level

1 Introduction

The great convenience, large product range, and high amount of product-related information offered by online retailers ensures that an increasing number of customers use online channels for shopping. According to [1], the share of online shoppers in Germany has increased significantly over the last years. However, the vast amount of information and cognitive constraints of human information processing often cause

M. C. Ocón Palma (✉)
Camelot ITLab GmbH, Mannheim, Germany
e-mail: moco@camelot-itlab.com

A.-M. Seeger · A. Heinzl
University of Mannheim, Mannheim, Germany
e-mail: seeger@uni-mannheim.de

A. Heinzl
e-mail: heinzl@uni-mannheim.de

F. D. Davis et al. (eds.), *Information Systems and Neuroscience*,
Lecture Notes in Information Systems and Organisation 32,
https://doi.org/10.1007/978-3-030-28144-1_24

information overload (IO) which represents the limit of humans to gather and process information [2]. In e-commerce, IO has the consequence that decision making of consumers becomes confused, less accurate, and less effective [2]. For retailers, this issue is substantial, because confused customers are less likely to make rational buying decisions, to find satisfactory products, and to have an enjoyable shopping experience [3–5].

In e-commerce, different ways for reducing IO have been proposed. These options attempt to improve information quality and quantity as well as the format of websites [6]. However, one emerging technology which has been barley studied regarding IO issues are conversational agents (CAs). CAs are dialogue systems that make use of natural language processing to engage users in information-seeking and task-oriented dialogues [7]. The risk of affecting customers' performance due to IO is even more likely in interactions with voice-enabled CAs, i.e., spoken dialogue systems that are based on natural language. In an audio-linguistic context, consumers rely only on their auditory sense, making it a great challenge to process, understand, and memorize the same amount of information as in visual online shopping scenarios. Therefore, some fundamental modifications regarding the information provisioning as well as in the design of the user interface are required to mitigate IO.

In this research, a custom shopping skill for Amazon Alexa was developed following the design science research methodology. It integrates the theoretical-deduced feature filtered repetition to mitigate IO. The developed artifact was deployed in a pilot study to identify the differences in cognitive load when a CA was used for shopping items instead of the visual interactions with a shopping website. The study was conducted to test if subjects perceive less IO when the feature filtered repetition is enabled.

In the following section, a theoretical model will be developed. The third section outlines the experimental design and the results of our pilot study including the developed artifact. Finally, we discuss our contribution and summarize our findings.

2 Development of a Theoretical Model

According to [8], cognitive load can be classified into intrinsic, extraneous, and germane load. Intrinsic load is caused by the task-related complexity. In online shopping, the task complexity is determined by the number of products or product information and the extent of interaction that is required to search, compare, or purchase products [9]. Extraneous load is related to the design and organization of information, i.e., when information is presented in an inappropriate way, extraneous cognitive activity increases. Hence, extraneous load is additional load that does not help in solving the task. Finally, germane cognitive load describes the load required for the learning process. In contrast to extraneous and germane cognitive load, intrinsic cognitive load can rarely be reduced. According to [9], information processing can be improved with minimizing extraneous load and maximizing germane load.

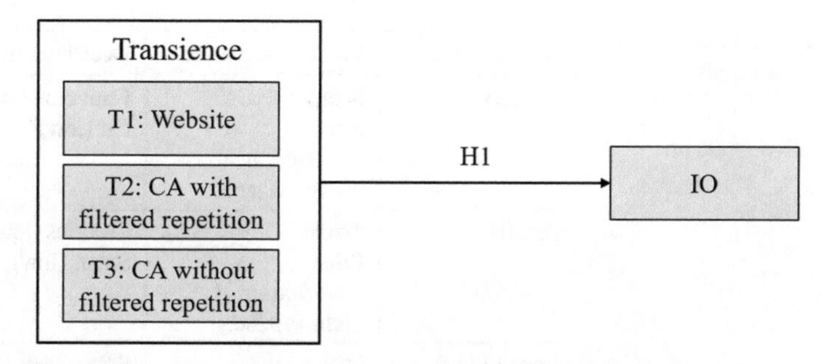

Fig. 1 The research model

In this study, it is argued that extraneous load is higher when using a CA instead of a website for online shopping. This statement is explained with the transient characteristics of voice user interfaces. Once users receive information by a CA, it vanishes, i.e., it becomes volatile, whereas the information cannot be revised like on a website interface. This characteristic is termed transient or when referring to fading information, transient information [10, 11]. Our research model is summarized in Fig. 1.

In the field of instructional design, transient information has already been studied. In this domain, spoken information causes loss of learning, especially when complex information is disseminated [12]. This occurrence is known as the transient information effect [13]. To reduce the transient information effect and, hence, to mitigate IO when using CAs, we propose the countermeasure filtered repetition. This feature refers to repeating a fraction of the received information whenever a user requests it. Reducing the information amount in the repetition is intended to help users memorize relevant information better. According to findings related to transient information, it has already been proven that segmented information causes less load [12, 14]. For the named reason, it is assumed that the feature filtered repetition will decrease the transience effect on IO. Therefore, we analyze three levels of transience (T1—website: low, T2—CA with filtered repetition: medium, T3—CA without filtered repetition: high) and investigate the following hypothesis:

H1 *The higher the level of transience, the higher the perceived IO.*

3 Experimental Design and Result

3.1 Participants, Materials and Procedure

To test our hypothesis, a pilot study was conducted. Overall, 20 native German speakers participated and most of the subjects had no previous experience with Amazon Alexa. The age ranged from 18 to 25 years. Due to some measurement

Table 1 The selected product sets and characteristics

Product category	Product attributes	Decision criteria
Yoghurt	Name Price Description Nutrition facts	Fruit content (high) Fat (low)
Cereals	Name Price Description Nutrition facts	Calories (low) Sugar (low)
Frozen Pizza	Name Price Description Nutrition facts	Price (low) Carbs (low)

problems with the skin conductance device that potentially distorted the results, only 12 out of the 20 subject's data were used for the evaluation.

In the experiment, our subjects were confronted with a set of food products of the same food category and with a high density level of information. Based on the product information and some predefined decision criteria, the subjects were asked to select one out of three presented products. For instance, participants were asked to chose the cheapest product with the lowest calories. The selected product sets as well as provided characteristics are listed in Table 1. The product categories were chosen because they belong to the products that are bought very often and are most probably well known by every subject. The shopping task had to be completed with the three constructs: website (T1), CA with filtered repetition (T2) and CA without filtered repetition (T3). For each construct, the density of information remained the same to create equal conditions. When chatting with Alexa, the subjects were additionally asked to sort a stack of paper. When using CAs, users typically also perform other, parallel tasks while conversing with the agent. For instance, they could be driving or conduct some work at home. In such constellations, the cognitive resources need to be shared between the subtasks which leads to fewer resources available for the primary task and a more demanded working memory [15, 16]. Regarding the cognitive load theory, this occurrence can also be explained with an increase in extraneous load [17]. In the present experiment, the parallel sorting task is considered to provide practical relevance and to induce a more demanded working memory when using CAs. With this condition, it is expected that the symptoms of IO are more visible.

It was decided to use a fix order in the experiment: At first, the subjects were asked to perform the task with website mock-ups, i.e., without filtered repetition and without a parallel task, afterwards with the developed artifact for Amazon Alexa without filtered repetition and with the sorting task (T3 + PT) and finally, with the shopping skill with filtered repetition and with the parallel task (T2 + PT). To avoid that any learning effect caused the subjects to perform better when using Amazon Alexa multiple times, an off-topic training session was added before T2 + PT and

T3 + PT. In the training tasks, the product information was replaced with animal information, but the developed Alexa skill versions remained the same.

To measure IO, the subject's skin conductance level (SCL) was recorded. The SCL was measured on the thenar and hypothenar of the subjects' non-dominant hand. In previous research, it has already been identified that electrodermal activity (EDA) is a suitable method for measuring cognitive load [18]. In particular, the method was selected because psychological arousal like emotions or stress cannot be controlled by humans and, thus, it allows to measure participants' reactions objectively. Additionally, the method is free of pain and does not require any physical burden. In this research, the mean SCL, the number of non-significant skin conductance response (NS.SCR), and the mean NS.SCR amplitude were calculated out of the SCL record. The NS.SCR was deemed to be appropriate because subjects were observed over a time interval and the occurrence of external stimuli was not explicitly induced. To prevent that stimuli such as speaking, deep breaths, or movements influence the NS.SCRs, the threshold level was set to 0.03 μS and the rejection threshold was set to 10%. Furthermore, the room temperature was monitored and kept between 22 and 24 °C. In addition to the EDA, two further measurements of IO were used for comparison: the NASA task load index (NASA-TLX) and the task completion time (TCT). The NASA-TLX is the subjective workload assessment deployed by the NASA [19]. It is a questionnaire which includes six subscales with a twenty-step bipolar rating scale. Schmutz et al., for instance, already used this method to analyze the effect of product listing page presentation on consumer's cognitive load [20]. Due to the similarities to this research, the method was used to capture the subjective perceived load during the treatments. The dimension physical demand was not added in the questionnaire because the experiment did not require physically challenging tasks.

3.2 Result

To validate the research model, a double repeated measure multivariate analysis of variance (MANOVA) considering all measurements of IO was performed. A within-subject design was chosen due to the individual differences in cognitive load which prevents comparing the subjects with each other. The results of the MANOVA test showed that there was a statistically significant difference between the three levels of transience, $F(10, 2) = 38.326$, $p = 0.026$. This means that the perceived overload changed over the course of the three shopping tasks. Therefore, the null hypothesis which states that the overload is the same at each level can be rejected. A pairwise comparison of each treatment using the Bonferroni post hoc test revealed that for the mean SCL, NASA-TLX score, and TCT, a significant difference exists between T1 and T2 + PT, and T1 and T3 + PT (see Table 2). No significant difference could be identified between the two CA versions for the named measurements. Regarding to the mean number of NS.SCR, a significant result was only achieved between T1 and T3 + PT ($p = 0.01$). The non-significant difference between the treatments

Table 2 Pairwise comparison using Bonferroni post hoc test (*p*-values)

Comparison	SCL	No. NS.SCR	NS.SCR amplitude	NASA-TLX	TCT
T1 versus T2 + PT	0.000*	0.333	0.557	0.000*	0.001*
T1 versus T3 + PT	0.000*	0.010*	0.526	0.003*	0.000*
T2 + PT versus T3 + PT	1.000	0.363	1.000	0.529	0.531

*significant, $p < 0.05$

T1 and T2 + PT could imply that the level of transience regarding both versions is similar which means, T2 + PT caused the same level of load than T1. The Bonferroni correction also revealed that the mean NS.SCR amplitude is the only variable showing no significant result for each pairwise comparison. Thus, no significant differences in the means cloud be identified between each level of transience.

4 Discussion and Conclusion

The aim of this study was to analyze if hypothesis H1 can be confirmed. We wanted to identify whether differences in IO exist between our treatments T1, T2 + PT as well as T3 + PT and whether IO can be reduced by the countermeasure filtered repetition. Apart from the results of mean NS.SCR amplitude which indicates that no differences exist between the three levels of transience, all other measurement methods have come to the same result: *They indicate that significant differences in load exist when using a CA for online shopping instead of a website.* This result is consistent with previous research in this area [12, 14]. Thus, it could be proven that the perceived IO is higher when using CAs compared to websites for online shopping. However, the second objective of our experiment was to find out whether differences between the two Amazon Alexa skills with and without filtered repetition exist. According to the gathered data, no significant difference could be found. Thus, H1 can only be proven partially. The major limitation of this pilot study is that both, T2 and T3 were tested with PT. Thus, it may be possible that the differences in IO in comparison to T1 are partly caused by the additional task. Since voice-enabled CAs are designed to be used in addition to other secondary tasks, we decided to focus on this practically more relevant setting.

In our study, one of the first attempts has been made to find a countermeasure to mitigate IO for CAs. The EDA, NASA-TLX, and TCT analysis allowed to gather an objective as well as a subjective view of subject's cognitive load during e-commerce interactions. The results of the pilot study provide evidence that users feel more overloaded when using CAs. In addition, it can be stated that the design of information provisioning needs to be adapted to reduce cognitive load. To further analyze the suggested countermeasure for IO, it is recommended to analyze user's feedback to identify whether other variables can explain the non-significant differences between the two Alexa skill versions. Furthermore, we recommend to conduct a study with more respondents and to let subjects test Amazon Alexa over a longer period of time to verify if the feature filtered repetition is able to reduce cognitive load.

References

1. Anteil der Online-Käufer an der Bevölkerung in Deutschland von 2000 bis 2016. https://de.statista.com/statistik/daten/studie/2054/umfrage/anteil-der-online-kaeufer-in-deutschland/.
2. Jacoby, J.: Information load and decision quality: Some contested issues. *Journal of Marketing Research*, 569–573.
3. Huffman, C., & Kahn, B. E. (1998). Variety for sales: Mass customization or mass confusion? *Journal for Retailing, 74*(4), 491–513.
4. Jacoby, J., & Morrin, M. (1998). "Not manufactured or authorized by …": Recent federal cases involving trademark disclaimers. *Journal of Public Policy & Marketing, 17*(1), 97–107.
5. Mitchell, V.-W., & Papavassiliou, V. (1999). Marketing causes and implications for consumer confusion. *Journal of Product & Brand Management, 8*(4), 319–342.
6. Eppler, M., & Mengis, J. (2004). The concept of information overload: A review of literature from organization science, accounting, marketing. *Information Society, 20*(5), 325–344.
7. Lester, J., Branting, K., & Mott, B. (2004). Conversational agents. In *The Practical Handbook of Internet Computing* (pp. 220–240).
8. Sweller, J. (1988). Cognitive load during problem solving: Effects on learning. *Cognitive Science, 12*(2), 257–285.
9. Schmutz, P. Heinz, S., Métrailler, Y., & Opwis, K. (2009). Cognitive load in eCommerce applications—Measurement and effects on user satisfaction. *Advances in Human-Computer Interaction*, 1–9.
10. Cohen M. J., Giangola J. P., Balogh, J. (2004). *Voice user interface design* (pp. 6ff). Addison-Wesley.
11. Schnelle, D., & Lyardet, F. (2006). Voice user interface design patterns. In *Proceedings of 11th European Conference on Pattern Languages of Programs EuroPlopm* (pp. 1–27).
12. Singh, A.-M., Marcus, N., & Ayres, P. (2012). The transient information effect: Investigating the impact of segmentation on spoken and written text. *Applied Cognitive Psychology, 26*(6), 848–853.
13. Leahy, W., & Sweller, J. (2011). Cognitive load theory, modality of presentation and the transient information effect. *Applied Cognitive Psychology, 25*(6), 943–951.
14. Wong, A., Leahy, W., Marcus, N., & Sweller, J. (2012). Cognitive load theory the transient information effect and e-learning. *Learning and Instruction*, 449–457.
15. Brünken, R., Plass, J. L., & Leutner, D. (2002). Direct measurement of cognitive load in multimedia learning. *Educationl Psychology, 38*(1), 53–61.
16. Vazquez-Alvarez, Y., & Brewster S. (2011). Eyes-free multitasking: The effect of cognitive load on mobile spatial audio interfaces. In *Proceedings of the 2011 Annual Conference on Human Factors in Computing Systems—CHI'11* (pp. 2173–2176).
17. Wood, E., Zivcakova, L., Gentile, P., Archer, K., De Pasquale, D., & Nosko, A. (2012). Examining the impact of off-task multi-tasking with technology on real-time classroom learning. *Computers & Education, 58*(1), 365–374.

18. Shi, Y., Ruiz, N., Taib, R., Choi, E., & Chen, F. (2007). Galvanic skin response (GSR) as an index of cognitive load. In *CHI'07 Extended Abstracts on Human Factors in Computing Systems* (pp. 2651–2656).
19. Hart, S. G., & Staveland, L. E. (1988). Development of NASA-TLX (task load index): Results of empirical and theoretical research. *Advances in Psychology, 52,* 139–183.
20. Schmutz, P., Roth, S. P., Sechler, M., & Opwis, K. (2010). Designing product listing pages—Effects on sales and users' cognitive workload. *International Journal of Human-Computer Studies, 68,* 423–431.

Positive Moods Can Encourage Inertial Decision Making: Evidence from Eye-Tracking Data

Yu-feng Huang and Feng-yang Kuo

Abstract We examine whether emotion can encourage inertial decision making, which is an emergent research topic in online shopping. Based on the information processing view, inertia is conceptualized as a decision process that involves repeated usage of a similar effortless information search pattern across multiple problems, and we propose that this conceptualization can be quantified using an eye-movement index based on the string-editing algorithm. We then examine whether positive moods, which have been shown to increase impulsive shopping, may promote inertia. Subjects, who either received positive moods priming or calculation (mood-suppressing) priming, participated in an eye-tracking experiment with multi-attribute decision tasks presented in a web map format like the Google Maps. The results showed that positive moods increased process inertia. We conclude that inertia can be quantified according to the information processing view, and that happy consumers tend to repeatedly use an effortless information search pattern to evaluate multiple products.

Keywords Decision making · Inertia · Eye movement · Moods · Online shopping

1 Introduction

One possible explanation to online impulsive buying is that consumers shop with inertia (or habit), which means that shoppers repeatedly adopt a similar decision process across multiple shopping tasks. In this study we identify, conceptualize, and quantify inertia according to the information processing view. Specifically, inertia is conceptualized as a decision-making process that repeatedly uses a similar effortless

Y. Huang (✉) · F. Kuo
National Sun Yat-sen University, Kaohsiung, Taiwan
e-mail: evanhuang@mis.nsysu.edu.tw

F. Kuo
e-mail: bkuo@mis.nsysu.edu.tw

F. D. Davis et al. (eds.), *Information Systems and Neuroscience*,
Lecture Notes in Information Systems and Organisation 32,
https://doi.org/10.1007/978-3-030-28144-1_25

information search pattern across multiple problems. While concepts related to decision inertia have been studied, those studies have focused on behavioral outcomes such as habitual choice or psychological states such as indecisiveness [1]. The examination of underlying cognitive process, which is critical to shape decision making theories [2], has been largely overlooked.

Theoretically, we integrate the effort-accuracy trade-off framework [3] with the theory of task-switch [4] to propose that inertia can be conceptualized as the decision makers' adoption of a similar effort-saving process across decision problems. Previously, the effort-accuracy trade-off framework has provided a good explanation for the choice of decision process within a decision problem. However, this framework has seldom been extended to explain the effort-saving behavior across problems. Based on the theory of task switching, we propose that retention of a previous decision process to the next problem is less effortful than process switching, and this retention contributes to inertial decision making.

In addition, we explore whether emotional factors such as positive moods can affect online shoppers' process inertia. Studies have shown that moods influence online shopping behavior [5] by promoting impulsive decision process. However, whether moods can promote process inertia is unknown. Here we empirically test the hypothesis that elicitation of positive moods leads to process inertia.

Methodologically, we capture process inertia with eye tracking in a multi-attribute decision making presented with a web map (such as Google Map, a common shopping environment). To examine the issue of strategy identification previous studies have been using process-tracing methods such as verbal report and mouse click [6]. More recently, eye tracking has been commonly used [7]. Nevertheless, although indices have been developed to associate eye movement patterns to decision strategies or processes [7, 8], such index for process inertia is still absent. We extend past eye-tracking studies of visual scanpath reliability [9] and process execution [10] to quantify process inertia in this study.

2 Background

2.1 Inertial Decision Making

Inertial phenomenon can be observed in individual decision making. During social interaction, for example, people spontaneously apply past interaction experience to categorize strangers into social groups, a behavior known as stereotyping [11]. Inertia can also refer to the status quo bias [12], a phenomenon that people prefer the default option and are reluctant to adopt a novel but possibly beneficial alternative. Inertia can also refer to state dependence in brand choice. That is, consumers tend to choose a brand that they have chosen before [13]. State dependence can result from the increase in product utility from past experience, which subsequently forms a psychological cost (brand loyalty) to discourage brand switch. However, these studies are based on consumers' behavioral outcomes or choices (hereafter referred to as choice inertia). Studies rarely examined inertia according to the information processing view (i.e., examination of the decision process, hereafter referred to as process inertia).

2.2 *Inertia: The Information Processing View*

The information processing view holds that decision makers adopt decision processes to collect and integrate information to reach a decision. This line of studies argues that choice behaviors are insufficient to fully explain human decision making; for example, distinct decision making processes may lead to a similar choice [2]. Generally speaking, these studies follow the accuracy-effort tradeoff framework. In this framework, cognitive effort is considered as a limited resource, and a strategy is considered more effortful when it requires conflict resolution and complete search of information [3]. Various psychophysiological tools have been used to measure effort; for example, using eye-tracing, a study [14] shows that pupil dilates when tasks become more effortful (difficult). However, studies of this line only consider effort associated with the one task currently being processed. When facing multiple tasks people may need additional effort to adapt to upcoming task characteristics. People might fail to adapt and become inertial by carrying the representation of the previous task characteristics to the current ones.

We extend the accuracy-effort tradeoff framework to include inertial decision behavior. We assume that an "inertial decision-making" is a strategy that effort-minimizing agents can use to reduce their cognitive effort when dealing with multiple choice tasks. We expect that when dealing with multiple tasks an effort-saving agent should not only try to save effort within the current task, but also seek to save effort between tasks. That is, our conceptualization of process inertia includes two constituents: the adoption of an effort-saving process within a task, and the consistency of applying the process between tasks. Once an effortless decision process is adopted, inertial decision makers are more likely to retain the process in mind and apply it to the subsequent decision problem other than to retrieve or construct a different one.

The argument that applying a similar process across tasks saves effort is based on the task-switching theory, which holds that switching between tasks generates extra effort (called the switching cost) in addition to the processing of the task itself [15]. Specifically, because the capacity of working memory is limited, to switch between tasks individuals have to 'save' the goals and procedures of the current task to long-term memory and 'load' the goals and procedures of the other task to the working memory [15]. A switch produces interference between task sets, and hamper decision performance in either task. Saving and loading costs additional effort and such effort cannot be completely eliminated even when the content of the upcoming task is predictable. Since decision makers estimate effort expenditure when predicting or scanning an upcoming decision problem [16], they may take into account the switching cost and refrain themselves from switching between decision processes if they seek to save effort for the upcoming decision task. In other words, decision makers may seek to save effort by simply applying the previous process to the next decision problem rather than retrieving a different process from a process pool or constructing a new one [17].

2.3 Moods and Process Inertia

Moods have been studied to examine their effects on shopping decisions and product evaluation [5]. Moods are the unspecific, transient and mild feelings that people integrate into decision processes to make decisions [18]. From the information processing view, it has been identified that mood protection is an integral part of choice of decision process [19]. According to the mood-managing theory, when positive moods are heightened people are more likely to enhance the goal to minimize the experience of negative emotion (i.e., protect their currently positive moods) [20]. Because effortful processing of decision information can induce negative mood [21], people with positive mood can select an avoidant behavior that prevent themselves from investing effort into a problem.

As discussed earlier, effort-saving can be achieved with the adoption of a less effortful process and apply such process to multiple problems. We hypothesize people with positive moods tend to adopt an effortless process and apply the similar process across problems. Conversely, mood-suppressed (or accuracy-oriented) decision makers tend to adopt a relatively effortful process. Given their willingness to invest effort into problems, they are less likely to rely on previous processes and demonstrate higher process dissimilarity between problems. In short, we hypothesize that:

H1 Positive moods leads to process inertia.

3 Methods

Based on the information processing view we measure inertia with the within-problem effort-saving process (search pattern) and across-problem process similarity (pattern similarity). Both are described below.

3.1 Search Pattern (Within-Problem Measurement of Effort-Saving)

Information search patterns can measure the degree of effort-saving of a decision process [8]. The search pattern is quantified by the formula (SB – SA)/(SB + SA), where SA is the frequency of within-attribute fixation pairs, and SB is the frequency of within-alternative pairs [8]. The range of search pattern is from 1 (completely alternative-based processing) to –1 (completely attribute-based processing). A person who uses attribute-based processing evaluates one attribute across all alternatives before going to the next attribute, while a person who uses alternative-based processing considers all attributes in one alternative before going to the next alternative. Because the attribute-based pattern has been usually associated to effort-saving processing [5, 22], a smaller score suggests that a more effort-saving processing is used.

3.2 Pattern Similarity (Across-Problem Measurement of Effort-Saving)

Process inertia is manifested as an application of a previous decision process to the upcoming problem. This suggests a high eye-movement pattern similarity between decision problems. We quantify across-problem pattern similarity with the string-editing algorithm, which calculates the difference (called Levenshtein Distance) between two sequences of patterns. Josephson and Holmes [9] adopted this algorithm to quantify scanpath reliability in their webpage repetition study. They calculated the differences between each pair of scan patterns, and averaged the differences of all possible combination of pairs as an index of an individual's similarity of scan pattern. Day [10] also adopted this algorithm to measure the difference between learned and executed processes. He showed that smaller distances indicated more successful executions of the learned process.

3.3 Data Collection

A total of 48 college students (20 males) with normal or corrected-to-normal vision participated our eye-tracking experiment. Participants finished four tasks, whose order was randomized, involving forced-preferential choice presented in a web map format. In each task, participants saw three screens (Fig. 1): first, the priming material, second, the task question, and third, the multi-attribute decision material. Participants performed the tasks at their own pace and chose one alternative they preferred by clicking the left mouse button on the chosen option.

In this between-subject design, participants were randomly assigned to the mood or the calculation priming condition. Hsee and Rottenstreich [23] show that mood

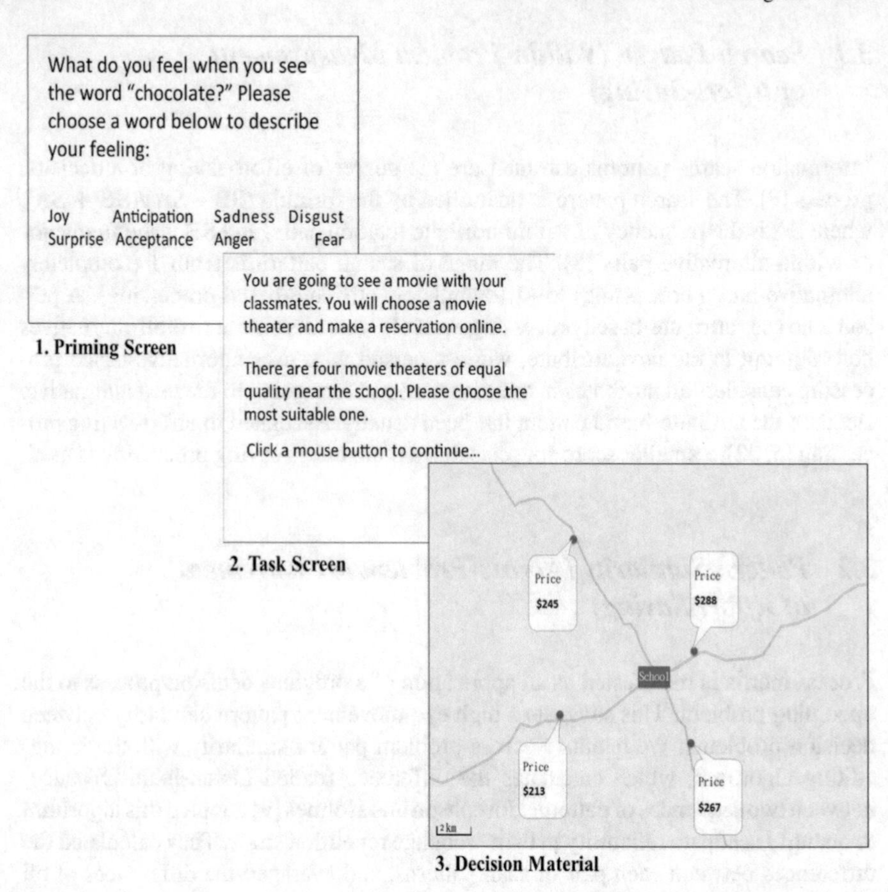

Fig. 1 The priming, task, and the forced-choice decision material utilized in the study

priming can increase participants' sensitivity in their feeling and calculation priming can promote the awareness of reasoning. In this study, four positive mood primes and four calculation primes were designed. To induce moods, participants were asked to reflect on concepts of 'love,' 'baby,' 'sunset' and 'chocolate' and reported their feelings. In contrast, math questions that require some degree of calculation served as the calculation primes (e.g., 'How much does an item of clothing, originally priced at NT$6000, cost after a discount of 35%?'). A pre-tested showed that the mood primes strengthened positive moods while the calculation primes suppressed the awareness of mood states ($p < 0.01$).

In the task screen, participants were asked to imagine that they were requesting services or buying products from four stores and would receive the services or products shortly. In the multi-attribute screen, each alternative was provided with price and distance attributes. For example, the dining task required participants to make an online reservation among four restaurants that varied with price and distance. The other three types of services were movies, karaoke, and book handling services.

Table 1 Mean and standard error of the study variables

Priming		Mean (Std. Err.)
Mood-priming (N = 24)	Search pattern	−0.3868 (0.0591)
	Pattern similarity	0.6384 (0.0115)
Calculation-priming (N = 24)	Search pattern	−0.1949 (0.0600)
	Pattern similarity	0.6534 (0.0102)

Participants completed the tasks with their eye movements recorded using the Eye-Link II system at 250 Hz sampling rate (SR Research, Mississauga, Canada) and its screen resolution was set to 800 × 600. After the tasks participants were debriefed and received their compensation (around 1.7 USD). Fixations were extracted from the raw recording data using the Data Viewer software (SR Research). Fixations were assigned to an option if they landed on its price information or around the lines connecting the origin to its locations. Fixating on the lines indicated that participants were attending on distance information.

4 Results

The descriptive statistics of the two eye movement measures are listed in Table 1. More importantly, Fig. 1 shows that participants who adopted an attribute-based pattern also tended to apply the similar pattern across multiple problems. Specifically, the score of search pattern is a continuum from −1 to 1, in which −1 indicates a complete attribute-based processing while 1 a complete alternative-based processing within a decision problem. The score of pattern similarity indicates the pattern similarity across decision problems. This similarity score ranges from 0 to 1, with lower score indicates higher similarity (less string editing distance). The correlation was significant for the mood-primed group ($r = 0.84$, $p < 0.01$) and calculation-primed group ($r = 0.50$, $p = 0.01$). Moreover, the correlation was stronger in the mood-priming group than in the calculation-priming group ($p = 0.02$), supporting H1 (Fig. 2).

5 Discussion

In this study we have conceptualized and quantified inertia from the information processing view. We have shown that the positive moods increase inertial decision making, evidenced in the strong relationship between within-problem effort-saving search pattern and across-problem pattern similarity. Our findings extend previous framework of choice of decision processes or strategies [3] by suggesting that people can save effort by repeatedly applying an effortless process across problems. Our quantification of inertia is useful for future studies to examine factors that can result

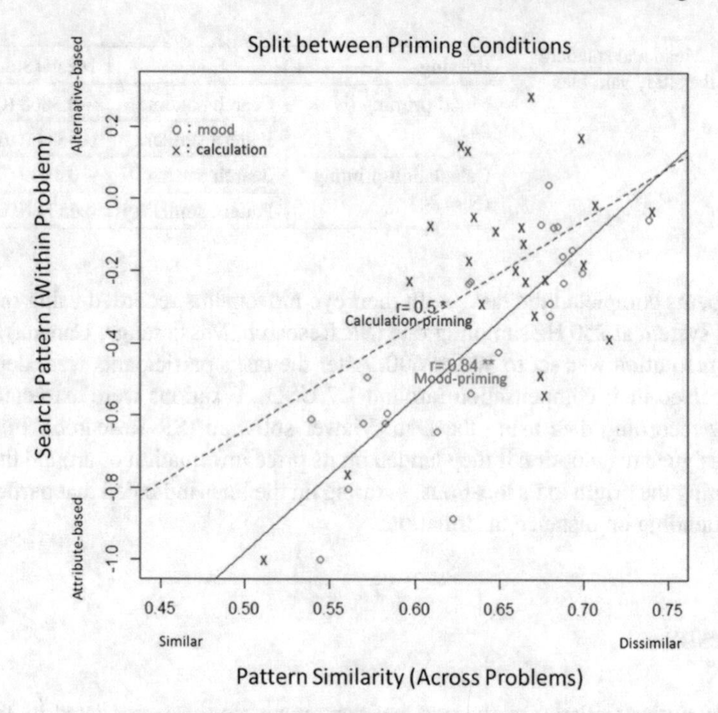

Fig. 2 Scatter plot of the two priming groups. $^{*}p < 0.05$, $^{**}p < 0.01$

in inertial decision making and to study whether and how people can detect and modify the previous process to adapt to contextual changes. Finally, given that the irrational online buying is alarming, we encourage future studies to apply our conceptualization and quantification to further explore factors affecting online shoppers' inertial decision behavior.

Practically, our results suggest that, for online shops, positive moods can boost sales more than they previously thought. Online consumers often buy multiple products, which can vary in their characteristics (e.g., utilitarian products vs. hedonic products). Our results suggest that happy consumers do not fully adapt to upcoming product characteristics and tend to keep using an effortless process to evaluate multiple products. In other words, happy consumers not only become more impulsive (effortless) in choosing one product but doing it across multiple products. Therefore, methods to promote moods, such as website aesthetics [24] and music [25], should play a more important role for online shops that include multiple products.

References

1. Polites, G. L., & Karahanna, E. (2012). Shackled to the status quo: The inhibiting effects of incumbent system habit, switching costs, and inertia on new system acceptance. *MIS Quarterly, 36*(1), 21–42.
2. Johnson, E. J., Schulte-Mecklenbeck, M., & Willemsen, M. C. (2008). Process models deserve process data: Comment on Brandstatter, Gigerenzer, and Hertwig (2006). *Psychological Review, 115*(1), 263–273.
3. Payne, J. W., Bettman, J. R., & Johnson, E. J. (1993). *The adaptive decision maker*. Cambridge, UK: Cambridge University Press.
4. Kiesel, A., et al. (2010). Control and interference in task switching—A review. *Psychological Bulletin, 136*(5), 849.
5. Huang, Y.-f., & Kuo, F.-y. (2012). *How impulsivity affects consumer decision-making in E-commerce. Electronic Commerce Research and Applications, 11*(6), 582–590.
6. Schulte-Mecklenbeck, M., Kühberger, A., & Johnson, J. G. (2011). *A handbook of process tracing methods for decision research: A critical review and user's guide*. Psychology Press.
7. Riedl, R., Brandstätter, E., & Roithmayr, F. (2008). Identifying decision strategies: A process- and outcome-based classification method. *Behavior Research Methods, 40*(3), 795–807.
8. Payne, J. W. (1976). Task complexity and contingent processing in decision making: An information search and protocol analysis. *Organizational Behavior and Human Performance, 16*(2), 366–387.
9. Josephson, S., & Holmes, M. E. (2002). Attention to repeated images on the world-wide web: Another look at scanpath theory. *Behavior Research Methods, Instruments, & Computers, 34*(4), 539–548.
10. Day, R. F. (2010). Examining the validity of the Needleman-Wunsch algorithm in identifying decision strategy with eye-movement data. *Decision Support Systems, 49*(4), 396–403.
11. Fazio, R. H. (1995). *Attitudes as object-evaluation associations: Determinants, consequences, and correlates of attitude accessibility*. In *Attitude strength: Antecedents and consequences*. Mahwah, NJ: Lawrence Erlbaum.
12. Samuelson, W., & Zeckhauser, R. (1988). Status quo bias in decision making. *Journal of Risk and Uncertainty, 1*(1), 7–59.
13. Dubé, J. P., Hitsch, G. J., & Rossi, P. E. (2010). State dependence and alternative explanations for consumer inertia. *The Rand Journal of Economics, 41*(3), 417–445.
14. Buettner, R., et al. (2018). Real-time prediction of user performance based on pupillary assessment via eye tracking. *AIS Transactions on Human-Computer Interaction, 10*(1), 26–56.
15. Monsell, S. (2003). Task switching. *Trends in Cognitive Sciences, 7*(3), 134–140.
16. Fennema, M. G., & Kleinmuntz, D. N. (1995). Anticipations of effort and accuracy in multiattribute choice. *Organizational Behavior and Human Decision Processes, 63*(1), 21–32.
17. Sénécal, S., et al. (2015). Consumers' cognitive lock-in on websites: Evidence from a neurophysiological study. *Journal of Internet Commerce, 14*(3), 277–293.
18. Pham, M. T. (1998). Representativeness, relevance, and the use of feelings in decision making. *Journal of Consumer Research, 25*(2), 144–159.
19. Bettman, J. R., Luce, M. F., & Payne, J. W. (1998). Constructive consumer choice processes. *Journal of Consumer Research, 25*(3), 187–217.
20. Andrade, E. B. (2005). Behavioral consequences of affect: Combining evaluative and regulatory mechanisms. *Journal of Consumer Research, 32*(3), 355–362.
21. Garbarino, E. C., & Edell, J. A. (1997). Cognitive effort, affect, and choice. *Journal of Consumer Research, 24*(2), 147–158.
22. Creyer, E. H., Bettman, J. R., & Payne, J. W. (1990). The impact of accuracy and effort feedback and goals on adaptive decision behavior. *Journal of Behavioral Decision Making, 3*(1), 1–16.
23. Hsee, C. K., & Rottenstreich, Y. (2004). Music, pandas, and muggers: On the affective psychology of value. *Journal of Experimental Psychology: General, 133*(1), 23–30.

24. Wells, J. D., Parboteeah, V., & Valacich, J. S. (2011). Online impulse buying: Understanding the interplay between consumer impulsiveness and website quality. *Journal of the Association for Information Systems, 12*(1), 32–56.
25. Day, R. F., et al. (2009). Effects of music tempo and task difficulty on multi-attribute decision-making: An eye-tracking approach. *Computers in Human Behavior, 25*(1), 130–143.

Application of NeuroIS Tools to Understand Cognitive Behaviors of Student Learners in Biochemistry

Adriane Randolph, Solome Mekbib, Jenifer Calvert, Kimberly Cortes and Cassidy Terrell

Abstract Cognitive load has received increased focus as an area that can be more richly explored using neuroIS tools. This research study presents the application of electroencephalography and eye tracking technologies to examine cognitive load of student learners in biochemistry. In addition to leveraging the Pope Engagement Index and eye tracking analysis techniques, we seek better understanding of the relationship that various individual characteristics have with the level of cognitive load experienced. While this study focuses on a particular STEM student population as they manipulate various learning models, it has implications for further studies in human-computer interaction and other learning environments.

Keywords Cognitive load · EEG · Eye tracking · Student learners · Individual characteristics

1 Introduction

In recent years, cognitive load has received increased focus as a construct of distinct interest that may be more richly explored using neuroIS tools [1, 2]. In particular, others have used neuroIS tools to examine the importance of engagement and cog-

A. Randolph (✉) · S. Mekbib
Department of Information Systems, Kennesaw State University, Kennesaw, GA, USA
e-mail: arandol3@kennesaw.edu

S. Mekbib
e-mail: shekbib@students.kennesaw.edu

J. Calvert · K. Cortes
Department of Chemistry and Biochemistry, Kennesaw State University, Kennesaw, GA, USA
e-mail: jcalver5@students.kennesaw.edu

K. Cortes
e-mail: klinenbe@kennesaw.edu

C. Terrell
University of Minnesota Rochester, Rochester, MN, USA
e-mail: terre031@r.umn.edu

F. D. Davis et al. (eds.), *Information Systems and Neuroscience*,
Lecture Notes in Information Systems and Organisation 32,
https://doi.org/10.1007/978-3-030-28144-1_26

nitive load in the areas of training and education [3] and shown their usefulness in understanding someone's full-body experience as they engage with technology [4]. Resulting, is a growing area of "neuro-education" [5, 6] to which we hope to contribute with our efforts.

In our ongoing study that is taking place as part of a federally-funded grant project in the United States, we use electroencephalography (EEG) and eye tracking technologies to assess cognitive load of student learners in biochemistry. Overall, the goals of this project are:

- to understand cognitive load as it impacts the development of undergraduate students' conceptual understanding of structure-function relationships in chemistry and biochemistry, and to
- refine the process for more effectively collecting and analyzing biometric data for mock classroom activities.

Use of neurophysiological tools such as EEG and eye tracking has been touted as complementary to traditional psychometric tools of survey and observation by providing increased understanding of human behavior [7], and we have found that to be the case here, as well. Further, although brain-computer interface (BCI) tools have typically been used to provide communication and environmental control to people with severe motor disabilities [8], they have also been used to more richly assess cognitive states such as cognitive load [9]. Here, we seek to use the concept of a passive BCI [10] to allow for enrichment of classroom-based interactions while students engage in various modeling exercises in support of learning biochemistry concepts. Passive BCI models have incorporated an EEG-based engagement index [11, 12] into their classifiers and we seek to do the same. A passive BCI represents a downline goal for this current three-year effort to collect and refine measurements of cognitive load.

In addition to measuring cognitive load, we are interested in how it relates to individual human characteristics. Understanding the relationship that various individual characteristics have with experienced cognitive load could help us better understand the pipeline for students engaging in science, technology, engineering and math (STEM) fields—of which information systems is considered a subset—and better provide support for students. While this study focuses on a particular STEM student population in biochemistry as they manipulate various learning models, it has implications for further studies with various student populations. Further, we may have more confidence when applying neuroIS tools to understand human-computer interaction phenomena, such as cognitive load, in seeing this case.

2 Methodology

The objective of the study is to evaluate the learning process and conceptual understanding of students in order to decrease their cognitive load. In the first year of this three-year study, more than sixty (60) students from a university in a metropolitan midwestestern city who are in the chemistry field have participated. Participants of the study were subdivided based on their stage of school year and four stages of curriculum. The classification categories included fall and spring General Chemistry, Organic Chemistry, and Biochemistry curriculum. Even though the potential study population consists of freshman to senior students, many of the actual participants in the first exercise were freshman students.

Students were asked to fill out a survey about their individual characteristics ranging from gender to level of athleticism. Individual characteristics of the study population were not limited to gender, race, and ethnicity, but also included differences in self-perceived levels of athleticism, dexterity, medication intake, smoking status, biometric tool use, and video game experience. This project is part of a larger study that will analyze the relationship between individual characteristics and various cognitive measures of spatial ability such as obtained using a Purdue Visual Rotation Test [13] and Hidden Figures Test [14].

The study is being conducted in a simulated learning environment where an instructor is present to explain the lesson and while the student works through exercises. Students' electrical brain activity is being measured using a 16-channel research-grade BioSemi ActiveTwo bioamplifier system (http://www.cortechsolutions.com/Products/Physiological-data-acquisition/Systems/ActiveTwo.aspx) running on a laptop. The electrode cap is configured according to the widely used 10–20 system of electrode placement [15]. Active electrodes are placed on the cap to allow for the recording of brain activations downsampled to 256 Hz using a Common Average Reference (CAR). The sixteen recorded channels are: frontal-polar (Fp1, Fp2), frontal-central (FC3, FCz, FC4), central (C3, Cz, C4), temporal-parietal (TP7, TP8), parietal (P3, Pz, P4), and occipital (O1, Oz, O2). Eye tracking data is being recorded using Tobii eye tracking glasses (www.tobii.com) while students are manipulating 2D, 3D, and virtual objects.

Afterward, data is being analyzed using the EEGLab plugin (https://sccn.ucsd.edu/eeglab/index.php) to Matlab to ascertain band powers and calculate cognitive load according to the Pope Engagement Index best represented by the calculation of (combined beta power)/(combined alpha power + combined theta power) [11].

3 Preliminary Results

Presently, data has been transcribed for the first year and cleaned with some initial analysis conducted. Statistical analysis will be used to assess the relationship between individual characteristics, spatial ability measures, and cognitive load as reflected by the Pope Engagement Index. The initial data indicates that the students are predominantly freshman, white females, and traditionally-aged ranging from 19 to 21 years.

Figure 1 starts to tell an interesting story of seven different student experiences based on EEG data that was able to be reliably captured and analyzed out of thirteen students in the first field visit. The y-axis in the figure represents the calculated

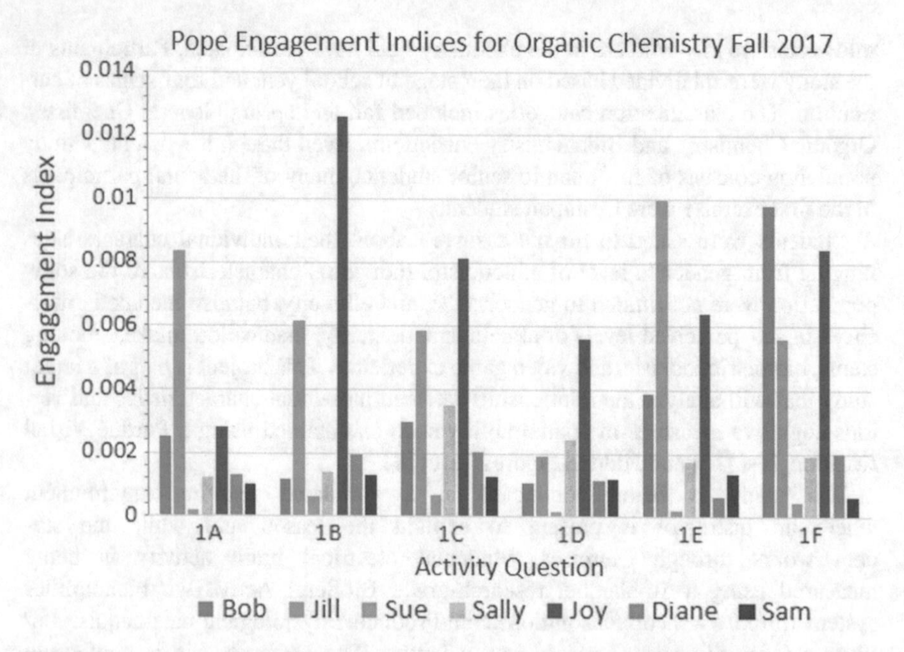

Fig. 1 Bar graph of Pope Engagement Indices calculated for Organic Chemistry students (Pseudonyms used to protect identities) across six classroom exercise questions

values of the Pope Engagement Index per question per student and serves as a reflection of cognitive load. It appears that Jill and Joy had a particularly difficult time with the classroom exercises whereas Sue and Diane did not necessarily have the same experience. This difference in cognitive load indicates that gender may not be the determining factor here. Data will be further analyzed to assess the relationship of individual characteristics, various spatial abilities, and cognitive measures to more fully understand student experiences. Already, later classroom exercises were modified based on preliminary understanding gained by reviewing general neurophysiological data, and there is early indication that cognitive load was able to be reduced for harder problems by providing better structure and scaffolding to solve these problems.

4 Conclusion

NeuroIS tools may be used to assess cognitive load of students while engaging in classroom learning activities and manipulating biochemistry models of varying types. There is a growing area of "neuro-education" research and use of neuroIS tools to assess training. Although the population of focus here is a student one in a particular subject area, this study has greater implications for future work and understanding the impact of individual characteristics on cognitive abilities. Further, this study presents an example of how we may inform passive BCI technologies and use them outside of a clinical setting typically reserved for patients with severe motor disabilities; hence, we may expand their use to a real-world, classroom-based setting to better understand cognitive ability.

Acknowledgements This work was funded by the National Science Foundation under Grant Number 1711425.

References

1. Riedl, R., Fischer, T., & Léger, P.-M. (2017). A decade of NeuroIS research: Status quo, challenges, and future directions. In *Thirty eighth international conference on information systems*. South Korea.
2. Fischer, T., Davis, F. D., & Riedl, R. (2019). NeuroIS: A survey on the status of the field. In *Information systems and neuroscience* (pp. 1–10). Springer.
3. Léger, P. M., et al. (2014). Neurophysiological correlates of cognitive absorption in an enactive training context. *Computers in Human Behavior, 34,* 273–283.
4. Courtemanche, F., et al. (2019). Texting while walking: An expensive switch cost. *Accident Analysis and Prevention, 127,* 1–8.
5. Charland, P., et al. (2017). Measuring implicit cognitive and emotional engagement to better understand learners' performance in problem solving. *Zeitschrift für Psychologie*.
6. Charland, P., et al. (2015). Assessing the multiple dimensions of engagement to characterize learning: A neurophysiological perspective. *Journal of Visualized Experiments: JoVE* 101.
7. Tams, S., et al. (2014). NeuroIS—Alternative or complement to existing methods? Illustrating the holistic effects of neuroscience and self-reported data in the context of technostress research. *Journal of the Association for Information Systems, 15*(10), 723–752.
8. McFarland, D. J., & Wolpaw, J. R. (2011). Brain-computer interfaces for communication and control. *Communications of the ACM, 54*(5), 60–66.
9. Randolph, A. B., et al. (2015). Proposal for the use of a passive BCI to develop a neurophysiological inference model of IS constructs. In F. D. Davis et al. (Eds.), *Information systems and neuroscience* (pp. 175–180). Springer.
10. Zander, T. O., & Kothe, C. (2011). Towards passive brain-computer interfaces: Applying brain-computer interface technology to human-machine systems in general. *Journal of Neural Engineering, 8*(2), 025005.
11. Pope, A. T., Bogart, E. H., & Bartolome, D. S. (1995). Biocybernetic system evaluates indices of operator engagement in automated task. *Biological Psychology, 40*(1), 187–195.
12. Demazure, T., et al. (2019). Sustained attention in a monitoring task: Towards a neuroadaptative enterprise system interface. In F. Davis, et al. (Eds.), *Information systems and neuroscience*. Cham: Springer.
13. Bodner, G. M., & Guay, R. B. (1997). The Purdue visualization of rotations test. *The Chemical Educator, 2*(4), 1–17.
14. French, J. W., Ekstrom, R. B., & Price, L. A. (1963). Manual for kit of reference tests for cognitive factors (revised 1963). Princeton NJ: Educational Testing Service.
15. Homan, R. W., Herman, J., & Purdy, P. (1987). Cerebral location of international 10–20 system electrode placement. *Electroencephalography and Clinical Neurophysiology, 66*(4), 376–382.

Task Switching and Visual Discrimination in Pedestrian Mobile Multitasking: Influence of IT Mobile Task Type

Pierre-Majorique Léger, Elise Labonté-Lemoyne, Marc Fredette, Ann-Frances Cameron, François Bellavance, Franco Lepore, Jocelyn Faubert, Elise Boissonneault, Audrey Murray, Shang-Lin Chen and Sylvain Sénécal

Abstract With the growing use of smartphones in our daily life, mobile multitasking has become a widespread (and often dangerous) behavior. Research on mobile multitasking thus far only focuses on a limited number of IT tasks that can be performed with a smartphone: talking, listening to music, and texting. Thus, we do not

---

P.-M. Léger (✉) · E. Labonté-Lemoyne · M. Fredette · A.-F. Cameron · F. Bellavance · E. Boissonneault · A. Murray · S.-L. Chen · S. Sénécal
HEC Montréal, Montréal, Canada
e-mail: pml@hec.ca

E. Labonté-Lemoyne
e-mail: elise.labonte-lemoyne@hec.ca

M. Fredette
e-mail: marc.fredette@hec.ca

A.-F. Cameron
e-mail: ann-frances.cameron@hec.ca

F. Bellavance
e-mail: francois.bellevance@hec.ca

E. Boissonneault
e-mail: elise.boissonneault@hec.ca

A. Murray
e-mail: audrey.murray@hec.ca

S.-L. Chen
e-mail: shang-lin.chen@hec.ca

S. Sénécal
e-mail: sylvain.senecal@hec.ca

F. Lepore · J. Faubert
Université de Montréal, Montréal, Canada
e-mail: franco.lepore@umontreal.ca

J. Faubert
e-mail: jocelyn.faubert@umontreal.ca

F. D. Davis et al. (eds.), *Information Systems and Neuroscience*, Lecture Notes in Information Systems and Organisation 32,
https://doi.org/10.1007/978-3-030-28144-1_27

know the extent to which these results generalize to other types of mobile multitasking behaviors such as reading while walking and gaming while walking. Also, we do not know the extent to which motor movement through physical space (i.e., walking vs. only standing) affects this phenomena. The current paper reports on an ongoing research that explores these questions. Our preliminary results suggest that mobile and standing multitasking leads to the inability to perceive incoming stimuli. Gaming appears to be the most dangerous mobile multitasking task for pedestrians.

Keywords Multitasking · Pedestrian · EEG · Texting while walking · Gaming while walking

1 Introduction

With the growing use of smartphones in our daily life, mobile multitasking has become a widespread (and often dangerous) behavior. We define *mobile multitasking* as the concurrent performance of one or more information technology (IT) tasks with a small computerized device (in most cases, a smartphone) while doing a motor movement such as walking. The behavior is increasingly common and can be seen almost anywhere. People are commuting to and from work, navigating the corridors of an office building, and even walking the halls of shopping malls, all while using their smartphone [1, 2].

While public safety research shows that mobile multitaskers are more cognitively distracted than non-mobile pedestrians [3] our team's recent work specifically measured this distraction using electroencephalography (EEG) and it is, to our knowledge, the first of its kind [4, 5]. Our results suggest that the influence of task-set inhibition on switch cost is more important when subjects are texting while walking. In other words, the more participants engage cognitively in texting while walking, the less attentional resources are available to attend to external (and potentially dangerous) stimuli.

Research on mobile multitasking thus far only focuses on a limited number of IT tasks that can be performed with a smartphone: talking, listening to music, and texting [3, 4, 6–9]. Thus, we do not know the extent to which these results generalize to other types of mobile multitasking behaviors such as reading while walking and gaming while walking. Also, we do not know the extent to which motor movement through physical space (i.e., walking vs. only standing) affects this phenomena. The current paper reports on an ongoing research program that explores these questions.

2 Related Work

Humans generally experience performance problems when multitasking [10]. *Multitasking* is defined as the concurrent performance of two or more distinct tasks [11]. Research clearly demonstrates that multitasking deteriorates performance as compared to performing tasks one at a time [11]. This deterioration is explained by theories underlying divided attention and dual-task performance, which have asserted that limitations in human multitasking are attributed to competition for processing resources (i.e., Multiple-resource theory), as well as to competition for processing mechanisms (i.e., Structural theory) [12].

Mobile multitasking is cognitively and perceptually complex [13]. While most dual-task research has been conducted in laboratory settings (e.g., [3], mobile multitasking is a daily activity which is arguably more complex than the experimental paradigms typically used to study the attentional mechanisms involved in dual-task interference. First, mobile multitasking involves one or more IT tasks on a smartphone, which require focused attention and fine motor control. Also, it involves gross motor control (during walking, cycling, and other physical activities involving motor movement through physical space). The mobile multitasker must divide his attention between the IT tasks and the dynamic visual scenes which necessitate a sustained vigilance to the external environment. Mobile multitasking in urban areas is even more demanding as it involves making spatial decisions in complex dynamic visual scenes (e.g., walking in a crowd where others are also moving, crossing streets, using public transit, and climbing stairs). Finally, the opportunity to become immersed in the IT task while walking is greater than for similar activities such as driving since individuals can engage in: (1) more numerous and complex range of IT tasks, (2) which may be sustained for longer periods of time (imagine a pedestrian slowly wandering through a crowd while staring at a mobile device, whereas the same person driving may quickly produce an accident or have other cars honking at them).

Recent studies confirm a significant increase in pedestrian injuries due to mobile phone usage between 2004 and 2010 [14], which coincides with the massive adoption of smartphones in urban areas. Numerous accidents involving pedestrians using phones have been reported with the majority of victims being less than 30 years of age [3]. Several articles on public safety report the unsafe and risky behaviour of mobile multitaskers [3, 6, 8, 9, 13, 14]. An observation study conducted at multiple high risk intersections in a metropolitan area revealed that more than 7% of pedestrians were mobile multitasking, and these individuals took significantly longer to cross the intersections [9]. Experimental studies using a virtual environment also show the prevalence of unsafe and risky behaviors [15]. In a virtual pedestrian environment, mobile multitaskers took more time to cross the street, missed several safe opportunities to cross, took longer to initiate crossing when a safe gap was available, looked left and right less often, spent more time looking away from the road, and were more likely to be hit or almost hit by an oncoming vehicle [3, 6]. This is not just an outdoor phenomena as mobile multitaskers are also at risk of accidents in an office environment [16]. Indeed, accidents are also increasing on work premises (e.g., falling down the stairs) and some organizations, such as General Motor, are now even prohibiting mobile multitasking inside company buildings [17].

Current research mostly focuses on texting while walking and we have little knowledge on the effect of other mobile multitasking behaviors. The device enables tasks with different levels of interactivity. Reading news on a mobile phone is unidirectional while texting is bidirectional. Writing an email usually entails slower communication speed than exchanging a short text message (SMS) which typically involve faster interactions. Some application like games may impose time constraints on the player (such as limited time to answer a question), which may have a consequence on the task switching behavior. Finally, some IT tasks are under the control of the user (such as scrolling through Facebook posts), while other events or alerts are not controlled by the user (e.g., pop-ups indicating new emails).

3 Methodology

Experimental design: We conducted a 2-factor within-subject experiment: Position (Standing vs. Walking) and Task type. Due to the complexity of the design, we conducted the project in two phases. In Phase 1 (standing condition only), we used 4 mobile tasks: (A) reading a document, (B) writing an email, (C) playing Tetris, and (D) group texting (i.e., texting with 2 individuals in the same conversation). In the second phase (walking condition only), 3 mobile tasks were used: (C) playing Tetris, (D) group texting and (E) individual texting (with one person).[1] In both phases, we also had a control group in which participants were only attending to the stimuli (F).

Participants: Thirty people (14 males, 16 females; ages 21–43, M = 25.6 years, SD = 5.9 years) participated in Phase 1 (standing) and 48 participated in Phase 2 (walking) (20 males, 28 females; ages 18–46, M = 25.5 years, SD = 5.5 years). All participants had normal or corrected-to-normal vision and were pre-screened for glasses, epilepsy, as well as health, neurological, and psychiatric diagnoses. This study was approved by the ethics committee of our institution. Participants provided written consent before participating and received a 40$ gift certificate as compensation upon experiment completion.

Stimuli and Apparatus: A dynamic point-light walker representation of a walking human form composed of 15 black dots was used as a biological motion stimulus. The dots, representing the head, shoulders, hips, elbows, wrists, knees, and ankles, were presented on a white background walking either leftward or rightward with a deviation angle of 3.0° (or −3.0°) from the participant. The point-light walker figure was displayed for 1000 ms with a resolution of 1280 × 1024 pixels using a projector (ViewSonic, Brea, California, United States). The walker stimulus had a height of 1.80 m and was displayed 4 m from participants, giving a 25° visual angle. Two speakers were located in front of participants which played a 1000 ms auditory stimulus cue with a random delay of ±500 ms before the presentation of the walker. Performance on point-like walker direction identification is strongly affected by divided attention in a dual-task paradigm, and the walker has ecological value in pedestrian safety research [4]. Thus, point-like walker stimulus is a suitable task for evaluating the switch cost of mobile multitasking in an authentic context. Mobile phone tasks

[1] In order to keep the number of conditions at a manageable level, two tasks which exhibited the smallest levels of dual task interference in Phase 1 (Tasks A and B) were excluded from Phase 2.

were performed using an iPhone 6s (Apple, USA). In phase 2, the threadmill used was the iMov iMovR's ThermoTread GT (iMovR, USA).

Instrumentation and Measures: EEG data was recorded from 32 Ag–AgCl preamplified electrodes mounted on the actiCap and with a brainAmp amplifier (Brainvision, Morrisville). The EEG signal was recorded using 32 electrodes with an acquisition sampling rate of 1000 Hz and analyzed with EEGLAB (San Diego, USA) and Brainvision (Morrisville, USA).

Procedure: While participants were standing (Phase 1) or walking (Phase 2), the point-like figure walker was presented shortly after the auditory cue stimulus. Participants were then asked to verbally identify the walker's direction by answering "left" or "right" according to the side on which they perceived the walker would pass them. Participants were also performing different mobile tasks for approximately 16–18 s per task. The experiment was composed of 5 blocks in Phase 1, one for each mobile tasks conditions and a control condition (5 × 22 trials), and 4 blocks in Phase 2, one for each tasks and a control condition (4 × 40 trials). The order of the blocks was counterbalanced and they were separated by a two-minute pause in which participants could sit on a chair while completing a short questionnaire. Prior to the first block, participants had a two-minute practice period to get used to the walker stimulus.

4 Preliminary Results and Ongoing Work

We recently completed the data collection for this project and the EEG analysis is currently underway. However, the behavioral results have already been analyzed. Linear regression with mixed model was performed to compare the least squares means (LSM) of performance across tasks. LSM was calculated using the model where we control for age, sex, and level of social use. Table 1 presents the performance (i.e., the proportion of walker directions correctly indicated by participants) by task. Unsurprisingly, the results show that mobile multitasking is risky.

Table 1 Behavioral results (least squares means of performance by task)

	Task name	Phase 1 standing	Phase 2 walking
A	Reading document	80.6%	–
B	Writing an email	76.6%	–
C	Playing tetris	66.6%[a]	80.8%[a, b]
D	Group texting	70.4%[a]	84.0%[a]
E	Individual texting	–	85.4%
F	Control	82.1%	89.1%

[a]Significantly lower than the control task
[b]Significantly lower than the individual texting task

In the standing position, there are significant differences between playing Tetris and the control task ($t(116) = -5.64$, $p < 0.0001$, Table 1) and between engaging in group texting and the control task ($t(116) = -4.26$, $p = 0.0003$) after adjustment for multiple comparisons [18].

In the walking condition, there are significant differences between playing Tetris and the control task ($t(141) = -5.11$, $p < 0.0001$, Table 1) and between engaging in group texting and the control task ($p = 0.0003$) with the same adjustment. In the walking condition, we even find that performance of playing Tetris is lower that individual texting ($t(141) = -2.85$, $p = 0.02$) after adjustment for multiple comparisons [18].

5 Discussion and Concluding Comments

Our preliminary results suggest that mobile and standing multitasking leads to the inability to perceive incoming stimuli. While the impacts on behavioral performance might seems small percentage wise, in real life only a single instance of not attending to an external stimuli can lead to physical injuries or have life threatening consequences. Our results also suggest that walking might not be the main contributor in reducing the behavioral performance. Even in the standing position, some tasks like playing a game and engaging in a group text might prevent people from noticing external events in a public environment. Finally, gaming appears to be the most dangerous mobile multitasking task for pedestrians.

This research contributes to a better understanding of the impact of mobile multitasking on user behaviors. It contributes to the literature by exploring and comparing a larger number of IT tasks performed on smartphones. Preliminary results suggest that playing a game and engaging in group texting (tasks that were absent from the mobile multitasking literature) diminish individuals' performance compared to individuals not using a smartphone.

These preliminary findings suggest that smartphone application designers and smartphone manufacturers should be careful in developing mobile apps, especially those related to games and group texting, by considering features related to user security. Given that some smartphone games are especially designed for mobility (e.g., Pokemon Go), we believe that these results should raise awareness of this very dangerous behavior in a pedestrian context.

As with any experimental studies, our research has limitations. We used a relatively young sample, so potentially very good at multitasking and do have the cognitive performance declines of older adults (e.g. [19]). Thus, research might not be representative for the whole population, especially an issue as large-scale adoption of devices such as smartphones spreads to older age groups. Also, the walker required attention, but the impact of errors is not near as significant as in real life (imagine an oncoming car). Thus, it is conceivable that a pedestrian in a real-world situation may perform differently.

References

1. Taylor, R. (2014). Can't walk the walk? Stop texting! *The Wall Street Journal.*
2. Whitehead, N. (2015). Texting while walking: Are you cautious or clueless? *NPR.*
3. Schwebel, D. C., et al. (2012). Distraction and pedestrian safety: How talking on the phone, texting, and listening to music impact crossing the street. *Accident Analysis and Prevention, 45,* 266–271.
4. Courtemanche, F., et al. (2019). Texting while walking: An expensive switch cost. *Accident Analysis and Prevention, 127,* 1–8.
5. Courtemanche, F., et al. (2014). Texting while walking: Measuring the impact on pedestrian visual attention. In *Proceedings of the Gmunden Retreat on NeuroIS.*
6. Byington, K. W., & Schwebel, D. C. (2013). Effects of mobile Internet use on college student pedestrian injury risk. *Accident Analysis and Prevention, 51,* 78–83.
7. Hyman, I. E., Jr., et al. (2010). Did you see the unicycling clown? Inattentional blindness while walking and talking on a cell phone. *Applied Cognitive Psychology, 24*(5), 597–607.
8. Stavrinos, D., Byington, K. W., & Schwebel, D. C. (2011). Distracted walking: Cell phones increase injury risk for college pedestrians. *Journal of safety research, 42*(2), 101–107.
9. Thompson, L. L., et al. (2013). Impact of social and technological distraction on pedestrian crossing behaviour: An observational study. *Injury prevention, 19*(4), 232–237.
10. Sanbonmatsu, D. M., et al. (2013). Who multi-tasks and why? Multi-tasking ability, perceived multi-tasking ability, impulsivity, and sensation seeking. *PLoS ONE, 8*(1), e54402.
11. Strayer, D. (2013). Multitasking and human performance. In H. Pashler (Eds.), *Encyclopedia of the mind* (pp. 543–546). London: Sage.
12. Folk, C. (2010). Attention: Divided. In E. B. Goldstein (Eds.), *Encyclopedia of perception* (pp. 85–88). London: Sage.
13. Stavrinos, D., & Byington, K. W. (2009). Effect of cell phone distraction on pediatric pedestrian injury risk. *Pediatrics, 123*(2), e179-e185.
14. Nasar, J. L., & Troyer, D. (2013). Pedestrian injuries due to mobile phone use in public places. *Accident Analysis and Prevention, 57,* 91–95.
15. Licence, S., et al. (2015). Gait pattern alterations during walking, texting and walking and texting during cognitively distractive tasks while negotiating common pedestrian obstacles. *PLoS one, 10*(7).
16. Léger, P. M., et al. (2013). Travailler à l'extérieur des frontières de l'organisation: Un modèle pour étudier les effets des multitâches technologiques en contexte piétonnier. In *Congrès de l'Association des Sciences Administratives du Canada (ASAC).* Alberta.
17. Vorano, N. (2018). *GM has banned cellphone use while walking in its offices.* [cited March 13th 2019; Available from https://driving.ca/buick/auto-news/news/gm-has-banned-cellphone-use-while-walking-in-its-offices.
18. Holm, S. (1979). A simple sequentially rejective multiple test procedure. *Scandinavian Journal of Statistics,* 65–70.
19. Clapp, W. C., Rubens, M. T., Sabharwal, J., & Gazzaley, A. (2011). *Deficit in switching between functional brain networks underlies the impact of multitasking on working memory in older adults.*

Interpersonal EEG Synchrony While Listening to a Story Recorded Using Consumer-Grade EEG Devices

Nattapong Thammasan, Anne-Marie Brouwer, Mannes Poel and Jan van Erp

Abstract Interpersonal EEG synchrony derived from the hyperscanning technique has the potential to reveal brain mechanisms beyond the border of traditional analysis within an individual subject. However, the inter-brain connectivity has not been fully investigated using wearable consumer-grade EEG devices which can enable a variety of application in a real-world scenario. In this study, we investigate interpersonal synchrony by capturing EEG signals using wearable EEG devices, from multiple participants ($N = 6, 7, 15$) who simultaneously listened to a novel being read to them. The results show that similar power-spectral patterns from neural responses evoked by perceiving the same auditory stimuli exhibit the synchrony, which is likely to have a transient characteristic rather than being stationary.

Keywords Inter-brain synchrony · Electroencephalogram · Hyperscanning

1 Introduction

Hyperscanning, a neuroimaging technique that simultaneously measures the brain activity from multiple subjects, has demonstrated its usefulness in neuroscience studies beyond the analysis within an individual brain [1]. Inter-brain correlation of electroencephalography (EEG) has been recently proposed a marker of attention, engagement, and emotion [2]. Despite its potential, such brain correlates were mainly discovered in laboratory settings, where the reproducibility of EEG

N. Thammasan (✉) · M. Poel · J. van Erp
University of Twente, Enschede, The Netherlands
e-mail: n.thammasan@utwente.nl

M. Poel
e-mail: m.poel@utwente.nl

J. van Erp
e-mail: jan.vanerp@tno.nl

A.-M. Brouwer · J. van Erp
TNO, The Hague, The Netherlands
e-mail: anne-marie.brouwer@tno.nl

F. D. Davis et al. (eds.), *Information Systems and Neuroscience*,
Lecture Notes in Information Systems and Organisation 32,
https://doi.org/10.1007/978-3-030-28144-1_28

synchrony in real-world scenarios has yet to be fully explored. Further, most of the previous studies in neurophysiological synchrony were conducted using sophisticated high-cost EEG devices, therefore, limiting the number of participants. The recent development of wearable EEG enables a possibility to explore inter-brain synchrony among a higher number of individuals using portable wireless EEG in naturalistic environments. Here, we investigate the synchrony of EEG signals recorded from multiple participants who simultaneously listened to a reading-out-loud sound of a novel story in a group setting with minimal control, primarily aiming to monitor synchrony between the emotional experiences of the listeners.

2 Methodology

2.1 Data Collection

Data was collected in three different sessions from the informed participants who attended a science exhibition and volunteered to involve in this study (Fig. 1). Brainwaves were simultaneously recorded at the sampling frequency of 220 Hz from 4 electrodes (TP9, Fp1, Fp2, and TP10) of Muse-2014 wearable EEG headbands.[1] After putting on the EEG headband and ensuring the position and impedance appropriateness, the main experimenter started to read out loud the first two chapters of a novel written by Arnon Grunberg at a naturalistic speed. Participants were encouraged to minimize body movements to avoid artifacts. The story lasted for 28.13 min on average across three sessions. Upon data quality inspection and discarding data of participants who left the experiment before it ended, the numbers of remaining participants for each session are 14, 6, and 7.

2.2 Data Preprocessing

Data were trimmed at the arbitrarily selected starting and ending points. The data compression module of MUSE inevitably led to the jitters in the EEG timestamps, which is crucial information to temporally align EEG signals from all participants. To alleviate the issue, a linear regression technique together with a sliding window technique was exploited in order to reduce the deviation of the timestamps. In particular, timestamps within one sliding window were fitted to a linear curve and the corrected timestamps were acquired from this linear model; in addition, overlapping technique was also employed to mitigate a sharp increment or decrement of timestamps at the end of the window. The corrected timestamps were yielded by taking the average of overlapped windows. Afterwards, EEG data were re-sampled at new equal-bin

[1]More specification can be found in http://developer.choosemuse.com/.

Fig. 1 Experiment set-up

timestamps of 200 Hz, which was applied to all participants in the same session, using an interpolation technique. Finally, signals of interesting have a duration of 24.09 min (3.04 min from the start and 1.03 min before the end of the task).

Subsequently, a Butterworth band-pass filter were applied to acquire signal within the frequency of interest between 1 and 40 Hz. To tackle adverse effects of artifacts, an artifact removal technique based on Independent Component Analysis (ICA), implemented in EEGLAB [3] was employed. Specifically, ICs were computed using *infomax* algorithm and then visually evaluated. The artifactual ICs were then removed from the source space, and the remained ICs were used to project back to original space to generate artifact-free EEG data. Normalization by subtracting mean values was then applied to obtain zero-mean EEG signals.

2.3 Synchrony Measurement

The EEG synchrony of each dyad in the same session was assessed by computing the correlation of sliding power spectra. Specifically, power spectral density (PSD) was computed from each EEG channel in each particular window whose size was 5 s (20% overlapping). In this study, a multi-taper PSD technique implemented in Chronux toolbox [4] was used to obtain PSD; the taper bandwidth was set to 2 Hz in each 1-s sliding window, and the number of tapers was set to 16. Afterwards, spectral power levels within particular frequency bands were averaged to yield power band of theta (4–8 Hz), alpha (8–12 Hz), beta (12–20 Hz), and gamma (20–30 Hz).

Upon acquiring PSD series, the synchrony of EEG between participants x and y in the same session was measured by computing the correlation (r_{xy}) of PSD in a specific band and a particular channel using the following formula:

$$r_{xy} = \frac{\sum_{i=1}^{k}\left(PSD_i^x - \overline{PSD^x}\right)\left(PSD_i^y - \overline{PSD^y}\right)}{\sqrt{\sum_{i=1}^{k}\left(PSD_i^x - \overline{PSD^x}\right)^2 \sum_{i=1}^{k}\left(PSD_i^y - \overline{PSD^y}\right)^2}} \quad (1)$$

where PSD_i^x represents PSD of a signal from subject x at the i-th window from all of the k windows, and $\overline{PSD^x}$ represents its average taken from all windows. As it is reasonable to assume that a pair of participants might not have a synchrony of PSD throughout the entire experiment but rather have intermittent synchronization due to fluctuating mental states, k is defined as the sliding window size for calculating correlation in the region of interest rather than using all data from the whole experiment. In this study, k is arbitrarily set as 20 and there is overlapping between consecutive windows; given the defined step size of PSD sliding window as 1 s, this means we are interested in finding a series of correlation of power spectra within a 20-s time-frame.

2.4 *Statistical Testing*

To test the null hypothesis that the power spectra time series of two different participants in the same session are not correlated, the non-parametric surrogate data method proposed in [5] was exploited. This approach randomizes PSD windows differently to build surrogate PSD time series and computes surrogate correlation. This process is repeated for 1000 times to produce distributions in which the null hypothesis holds i.e., the chance-level of cross-participant PSD correlation. The original, non-permuted data are then compared to the surrogate distribution to obtain p-values which, in this study, were later compared with a significance threshold of 0.05.

3 Results

3.1 *Transient Correlation*

The correlation coefficients of power-spectra time series from each pair of participants were averaged within the experimental session, separately by channel-frequency band. Then the values were aggregated from all pairs of participants to derive the grand-average, the 25th and 75th percentiles, the minimum and the maximum (Fig. 2a, b, c). In addition, the coefficient at a particular window from a pair of participants, whose value is over 0.75 and its original PSDs significantly correlate, is also shown. In general, the averaged correlation coefficients are relatively low (between around 0.2 and 0.3) compared to an intermittent coefficient from a window. The results suggest that the synchronization of EEG power spectra is transient rather than stationary throughout the whole experiment. It is also noticeable that the

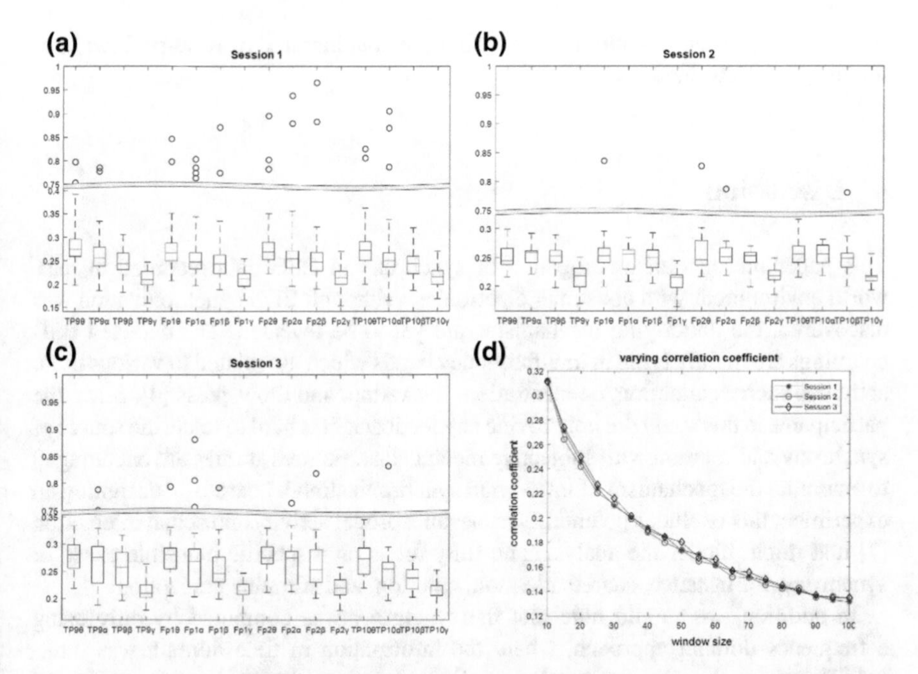

Fig. 2 The resulted correlation coefficient of power spectra averaged across all pairs of participants in the same session (session 1, 2, and 3). A correlation coefficient that is derived from significantly correlated PSDs at a particular window is shown as a circle. The decline of coefficients by increasing the size of the sliding window is displayed in sub-figure (**d**)

correlations in theta and alpha are relatively higher than in beta and gamma, which could be owing to that the listening task might homogenously enhance the level of concentration and relaxation of participants [6].

3.2 *Decline of Correlation by the Increased Size of Sliding Window*

It is sensible to assume that the size k of the sliding window for calculating correlation may affect the obtained results as it refers to the minimum duration of synchronous PSD. Henceforth, another investigation was conducted by varying the size of k from 10 to 100 s with an increment of 5 s. The results, depicted in Fig. 2(d), suggest that correlation coefficients decrease when increasing the size of the sliding window. From this it can be deduced that the synchronization might not last long and the enlarged window might encompass the epochs where EEG signals are not synchronized. However, this phenomenon should be investigated further in a follow-up study. Moreover, one should investigate the possible increase of false-positive rates for small window sizes.

4 Discussion

This paper presents an investigation of synchrony in EEG data recorded in real-world environment with consumer electronics. Although EEG synchronization was discovered, the underlying mechanisms are yet to be revealed. The detected PSD couplings are mainly lying in low-frequency bands which are related to various brain activities such as attention, concentration, relaxation, and drowsiness [4]. Since the participants in this study did not provide any feedback, it is hard to relate the source of synchrony and relation with subjective mental state. Futures studies are encouraged to elaborate the mechanism of inter-brain synchronization by carefully designing an experiment that excludes potentially irrelevant cortical activity and social interaction [7] that might hinder the analysis and fully focus on a specific plausible cause of synchrony, for instance, shared attention, emotion, and stimulus perception.

In addition, we should note that the synchrony was computed by only using a frequency-domain approach, where the information in time domain was omitted. Therefore, the size and overlap of sliding window play an important role and might cause high variance. Our future work will include applying time-frequency approach, such as wavelet analysis, of inter-brain synchrony to gain further insights and employing statistical analysis to validate the detected synchrony. Further, the synchrony measurement should go beyond dyadic analysis and toward discovering a similar pattern among a group of multiple brains [8].

Acknowledgements This work was supported by The Netherlands Organization for Scientific Research (NWA Startimpuls 400.17.602).

References

1. Dmochowski, J. P., Bezdek, M. A., Abelson, B. P., Johnson, J. S., Schumacher, E. H., & Parra, L. C. (2014). Audience preferences are predicted by temporal reliability of neural processing. *Nature Communication, 5,* 4567.
2. Dikker, S., Wan, L., Davidesco, I., Kaggen, L., Oostrik, M., McClintock, J., et al. (2017). Brain-to-brain synchrony tracks real-world dynamic group interactions in the classroom. *Current Biology, 27*(9), 1375–1380.
3. Delorme, A., & Makeig, S. (2004). EEGLAB: An open source toolbox for analysis of single-trial EEG dynamics including independent component analysis. *Journal of Neuroscience Methods, 134*(1), 9–21.
4. Mitra, P., & Bokil, H. (2007). *Observed brain dynamics*. Oxford University Press.
5. Canolty, R. T., Edwards, E., Dalal, S. S., Soltani, M., Nagarajan, S. S., Kirsch, H. E., et al. (2006). High gamma power is phase-locked to theta oscillations in human neocortex. *Science, 313,* 1626–1628.

6. Weisz, N., & Obleser, J. (2014). Synchronisation signatures in the listening brain: A perspective from non-invasive neuroelectrophysiology. *Hearing Research, 307,* 16–28.
7. Dumas, G., Nadel, J., Soussignan, R., Martinerie, J., & Garnero, L. (2010). Inter-brain synchronization during social interaction. *PLoS ONE, 5*(8), e12166.
8. Palva, S., & Palva, J. M. (2012). Discovering oscillatory interaction networks with M/EEG: Challenges and breakthroughs. *Trends in Cognitive Sciences, 16*(4), 219–230.

Using Eye-Tracking for Visual Attention Feedback

Peyman Toreini, Moritz Langner and Alexander Maedche

Abstract In the age of big data, decision-makers are confronted with enormous amounts of information coming from various resources at high velocity. However, humans have limited cognitive capabilities such as attentional resources. Inappropriate attentional resource allocation can lead to severe losses in performance. Nowadays, the usage of eye-tracking devices brings the opportunity to design neuro-adaptive information systems that support users in better managing their limited attentional resources. In this study, we investigated the design of an attentive information dashboard which provides visual attention feedback (VAF) as live biofeedback. Later, we examined how three different VAF types assist decision makers in their visual attention allocation (VAA) performance and focused attention while conducting a data exploration task. The results show that providing an individualized VAF as live biofeedback using real-time gaze data supports users in managing their attention better than general VAFs.

Keywords Attention · Live biofeedback · Information dashboard · Eye-Tracking

1 Introduction

The deployment and usage of Business Intelligence (BI) systems has massively increased in recent years [1]. A critical capability of BI systems is presenting analytical results from different sources on information dashboards. Information dashboards are graphical user interfaces that visualize information from different sources on a single display [2]. This enables decision makers to get an overview of the business at

P. Toreini (✉) · M. Langner · A. Maedche
Karlsruhe Institute of Technology (KIT), Institute of Information Systems and Marketing (IISM), Karlsruhe, Germany
e-mail: peyman.toreini@kit.edu

M. Langner
e-mail: moritz.langner@student.kit.edu

A. Maedche
e-mail: alexander.maedche@kit.edu

F. D. Davis et al. (eds.), *Information Systems and Neuroscience*,
Lecture Notes in Information Systems and Organisation 32,
https://doi.org/10.1007/978-3-030-28144-1_29

a glance and helps them to make timely business decisions [3]. Generally, decision makers use such dashboards in two common modes, exploring the business status in data explorations mode or finding specific information in focused search mode [4]. To accomplish data exploration tasks properly, there is a need to allocate attention efficiently among the overabundance of information [5–7]. However, when there is high information density on dashboards, decision makers fail to allocate their attention to all the information. Missing important information may be the consequence of high attentional demand or an inappropriate attention allocation [8]. Therefore, there is a need to receive support for managing attention while exploring dashboards to avoid such failures.

Attention is known as a limited resource of humans [9] and the huge amount of information can create a poverty of attention [7]. In the digital environment, researchers called for designing attentive user interfaces (AUI) [10, 11], attention-aware systems [12], attention-management systems in affective computing [13, 14] and attention management support features that preserve users from attentional breakdowns [15, 16]. Eye-tracking devices are known as the main tools for designing such systems [10, 17, 18] since according to the eye-mind hypothesis, where users are fixating is underlying their attention allocation [19]. Also, in the information systems (IS) community, researchers called for designing neuro-adaptive IS that recognize the physiological state of users in real-time and adapt based on that information [20]. Therefore, these devices can be used to design attentive dashboards as a type of neuro-adaptive IS.

The attentional process in the visual field is known as visual attention [21] and visual attention allocation (VAA) is the set of processes enabling and guiding the selection of incoming perceptual information [12, 22]. Tracking the VAA of users can be used for designing live biofeedback [14, 23–25] that informs users about their attentional state. Moreover, AUIs that provide visual attention feedback (VAF) are known as being supportive to recover from attentional breakdowns and improve the performance in several contexts [13, 26–28]. Although the NeurosIS community called for the usage of eye-tracking devices to design and evaluate innovative systems [18, 20, 29], there is limited research on using eye activity for designing live biofeedback [25]. Also, to the best of our knowledge, designing VAF to support users exploring information dashboards is not investigated so far. Thus, this study aims to design an attentive information dashboard that supports users in improving their attention management:

RQ: What type of VAF supports users in their VAA performance and focused attention while exploring information dashboards?

To answer this question, we investigated the effects of three different VAF types as part of a design science research (DSR) project. In this DSR project, we used eye trackers in both the design and evaluation process as suggested by researchers in the NeuroIS field [29]. In the design phase, we used them for developing an attentive information dashboard as a neuro-adaptive IS and also for designing individualized VAFs as a live biofeedback [24]. In the evaluation phase, we analyzed users' eye-movement data to investigate impacts of such feedback on the VAA performance and the focused attention while revisiting the dashboard.

2 Research Methodology and Instantiation of VAF Types

This DSR project is structured following the approach of Kuechler and Vaishnavi [30]. As the first step, we conducted an exploratory study in addition to a literature review to extract difficulties of users in managing their attention while exploring information dashboards [31]. Moreover, we identified that neuro-adaptive dashboards and live biofeedback are not studied in this context so far. This study is focused on the next steps of the first cycle of the DSR project. In the suggestion phase, we proposed to use eye-trackers to design attentive information dashboards and different VAF types that can support users in managing their limited attention. Later, we developed the suggestions by integrating Tobii 4C eye-tracker as apparatus with the appropriate license for conducting research studies. For the evaluation phase, we designed a controlled lab experiment to investigate effects of suggested VAF types by analyzing users' eye-movement data. To better understand the impact of each VAF, we designed a distinct dashboard that increases the internal validity of the experiment. As it can be seen in Fig. 1 (left), this dashboard is designed to minimize the influence of external factors on the VAA. The dashboard includes six charts while each is acknowledged as an area of interest (AOI). We tried to design each chart with the same complexity by having the same format, amount of information chunks, size, no color, etc. With having the same complexity, we suppose that a proper VAA is close to an even

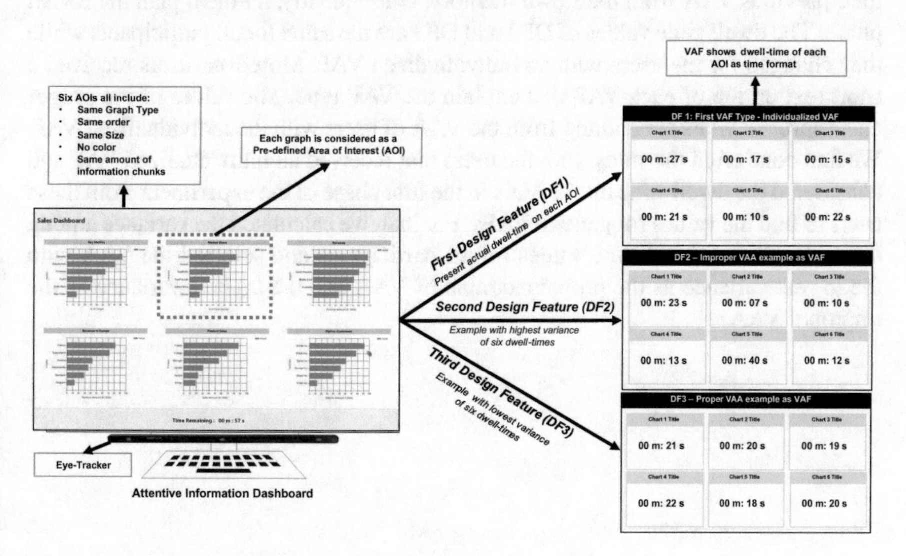

Fig. 1 Designed attentive information dashboard and three design features as VAF types

distribution of attention on all six AOIs. Therefore, having lower variance among six dwell-times for each user shows that the user had a proper VAA and having high variance indicates that the attention is not distributed properly among these six charts.

Figure 1 (right) shows three ***design features (DF)*** as VAFs that were implemented for this study. Feedback is known as a constructive element that sends back information about what action has been done while allowing the user to continue with the activity [32]. Therefore, a proper design of VAF for data exploration tasks should serve as a memory aid and facilitate users to recognize their previous VAA. Also, it should appears during the data exploration task and let the user resume it after processing the feedback. This can be done either by presenting the actual VAA as a form of live biofeedback, or by giving general feedback as a hint for improving VAA within next steps.

First, we designed the live biofeedback as an individualized VAF by considering the user's eye-movement in real-time. ***DF1*** in Fig. 1 represents an example of this VAF type. This type of VAF displays the ***actual VAA*** by presenting the dwell-time on each chart of the dashboard as a time format. We assume that having such information assists users to remember their previous VAA. Consequently, it helps in selecting a suitable data exploration strategy within the revisit phase. The second VAF type is a general VAF that presents an example of improper VAA. ***DF2*** in Fig. 1 shows the design of this VAF type which is equivalent to the individualized VAF and only the duration values are different in order to provide an example of an ***improper VAA***. The third VAF, ***DF3*** in Fig. 1, is again a general VAF with the same design. In contrast to DF2, the values in this VAF type expose an example of a ***proper VAA***. We assume that both DF2 and DF3 as general VAFs give a hint to the participants for recalling their previous VAA from their own memory, consequently, let them plan the revisit phase. The dwell-time values of DF2 and DF3 are the same for all participants while they changed for the users with an individualized VAF. Moreover, users received a short text on top of each VAF that explain the VAF type. The values of the proper and improper VAFs are coming from the VAA of users with the individualized VAF. We first conducted the study with the users that received an individualized VAF and considered the dwell-time on six AOIs in the first phase of the experiment from these users to find the values for general VAFs. For that, we calculated the variance among the collected six dwell-time values for all participants and selected the VAA with the lowest variance as the proper example of VAA and the highest variance as the improper VAA.

3 Experiment and Preliminary Results

3.1 Experimental Design

To test impacts of VAF types, we executed an eye-tracking experiment in a controlled lab environment. The experimental design was a 2 (with and without VAF) by 3 (VAF types) mixed design in which with or without VAF was manipulated within subjects and VAF types were manipulated between subjects. In this experiment, each subject had to conduct data exploration tasks in two rounds on two different information dashboards, which both had the same design but different content. For each data exploration task, participants had three minutes to explore the dashboard in total. After two minutes they were interrupted for 30 s and later resumed the task for one more minute. During the first round, the interruption phase counted as a break and participants were asked to wait for 30 s before continuing. During the second round, each participant got one of the three VAF types presented in the previous section.

The experiment procedure started by calibrating the eye-tracker for each participant individually with the software provided by Tobii. In the second step, screen-based instructions were given to explain the experimental steps and illustrated the concepts used in the dashboard. These instructions were followed by control questions to ensure the common understanding of concepts on the dashboard. After that, the main part of the experiment started and can be seen in Fig. 2. First, participants conducted the first round (without VAF) of the experiment in which they explored the dashboard for two minutes (first phase) and then they had a break for 30 s. Next, they had the opportunity to resume the data exploration task for one more minute and finish the first round. Then, the calibration status was checked again, and in case of any errors the system enforced the execution of a recalibration. Also, a rest phase was included for two minutes to control for carryover effects from the first round. Next, the second round (with VAF) of experiment is started. In this round, users had a data exploration task similar to the first round but this time with a new dashboard. In this round, participants received one of the designed VAF types after two minutes instead of having a break. As time was controlled in each step, the participants got a timer in the footer of the screen that displayed the remaining time in all phases of the experiment. At the end, demographic questions were asked as a survey.

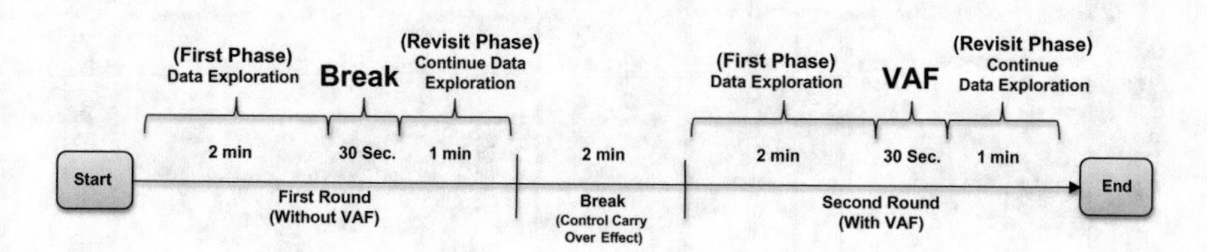

Fig. 2 The main part of the experiment

3.2 Data Analysis Results

In total 29 university students (8 female, 21 male) with an average age of 25.03 years (SD = 2.32) participated in this experiment. After checking the collected data, we removed two participants since the total collected dwell-time was less than 3/4 of the time assigned to each phase. This can be because these users ignored some part of the task or an error in the calibration. The remaining 27 participants were distributed across three groups (G). 11 participants were assigned to G1 with the individualized VAF, 8 participants to G2 with the improper example of VAA (general VAF) and 8 participants to G3 with the proper example of VAA (general VAF).

The data analysis is focused on comparing the VAA performance and focused attention of the participants as the dependent variables (DVs) in the revisit phase of both rounds (with and without VAF). The revisit phase is considered as an opportunity to improve the VAA by focusing on previously low-attended charts of the first phase. Thus, for each round, we detected the three low attended charts in the first phase and measured the dwell-time on these charts in the revisit phase. Also, we captured the focused attention of users by tracking the number of transitions between the AOIs. Focused attention is defined as the centering of attention on a limited stimulus field in IS [33]. Here, having a lower number of transitions is considered as having a higher focused attention. Figure 3 presents the VAA performance and focused attention of all three groups in the first and second round (with or without VAF conditions).

To test the difference between groups in these two rounds, we conducted a mixed design ANOVA test with groups as between subject and the DVs as within subject. We did this test for each DV separately and the results did not show any significant difference among the three groups for both DVs. We assume that these results occurred because of the low sample size in this pilot study. To investigate the effects of VAFs in more details, we performed a within-subject analysis for each group separately by conducting a paired-sample t-test. For the group with an individualized VAF (G1), the results confirm that there was a significant difference in the VAA performance of the first (M = 45.69, SD = 14.08) and second (M = 64.95, SD = 16.38) round of

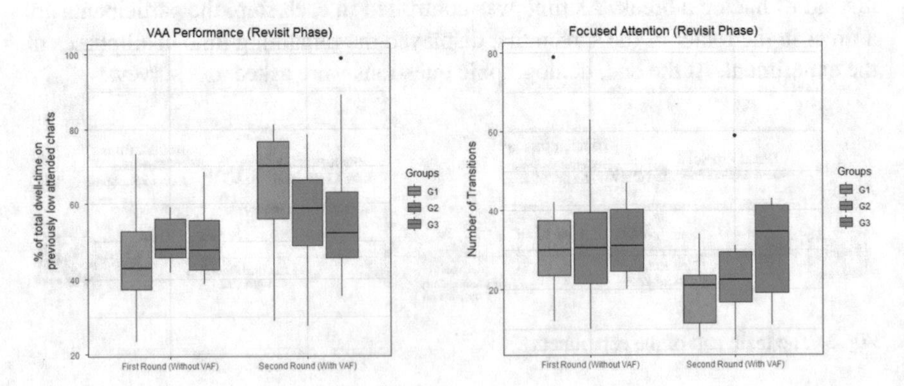

Fig. 3 VAA performance and focused attention during the revisit phase

the experiment ($t(10) = -3.773$, $p = 0.003$). Although Fig. 3 shows that the VAA performance improved in the second round for all groups, the paired-sample t-test for G2 and G3 does not show a significant difference. Therefore, we can infer that the individualized VAF helped participants to find previously low attended charts in a better way than the other two general VAFs.

Moreover, after checking the normality assumptions for transitions, we used the paired sample Wilcoxon test to investigate the effect of the individualized VAF on focused attention. For G1, the results reveal that there was a weak significant difference in focused attention for the first ($M = 32.82$, $SD = 17.86$) and second ($M = 19.55$, $SD = 9.05$) round of the experiment ($v = 56.5$, $p = 0.040$). Also, for G2, the results of a paired sample t-test show a weak significant difference for the first ($M = 32$, $SD = 16.14$) and second ($M = 25$, $SD = 16.06$) round ($t(7) = 2.3287$, $p = 0.05$). However, for G3 the results do not show any significant difference. We can infer that, although the focused attention improved for all three VAF types, the effect of an individualized VAF and an improper VAA example was stronger than the effect of a proper VAA example.

4 Discussion and Next Steps

Results from the previous section reveal that individualized VAF as a live biofeedback supported users to improve both VAA performance and the focused attention in comparison with proper or improper examples of VAA as general VAFs. Providing actual values of the VAA supports users to recall this information from their memory in a better way than receiving the general hints. These results confirm that users with general VAFs had difficulty to recall their previous VAA and to plan for the revisit phase. Regarding the focused attention, we found that giving the individualized and improper example of VAA as VAFs support users to focus better on the task. However, having a proper VAA example influences neither the VAA performance nor focused attention. The results explain that having the improper VAA example as a hint helps users to better focus during the revisit phase than the proper VAA example.

Although the existing results provided valuable findings, our study has some limitations that can be covered in future investigations. First, this study is considered as an exploratory study to get preliminary results of using the three suggested VAFs. Next, there is a need to differentiate the effects of these VAFs types in a larger scale for robust theorization and confirmatory studies. Second, the Tobii 4C eye-tracker is used for both designing and evaluating parts. However, this device is mainly used for designing interactive systems based on gaze position rather than evaluation purposes. Therefore, there is a need to evaluate this VAFs with a more accurate eye-tracker and analyze other eye-movement data such as fixations or pupil size to measure the workload [34]. In addition, there is a need to investigate other design suggestions for the individualized VAF than providing it in a time format. The attention distribution can be visualized as heatmap, scan paths, etc. Also, there is a need to study the role of working memory capacity of the users to check how such VAF types affect different users with different capabilities. Previous studies showed a close relationship between the working memory capacity and the control of attention [35]. Furthermore, besides designing live biofeedback for data exploration tasks, other tasks such as search tasks, resumption support, shared attention in meetings, etc. can be investigated.

5 Conclusion

In the age of big data, decision-makers are confronted with an enormous amount of information coming from various resources. To ease processing and extracting value, this information is visualized and presented in a form of information dashboards. However, humans have difficulties to process all information since they have limited attentional resources. In this study, we focused on the design of an attentive information dashboard as a neuro-adaptive IS and VAF as live biofeedback to support users managing their limited attentional resources. For that, eye-tracking devices are used for both designing and evaluating this systems [20, 29] and this DSR project contributes to the field of NeuroIS by enhancing user capabilities [18] in managing their limited attentional resources. Based on the list of possible contributions in the field of NeuroIS provided by Riedl and Léger [24], our design is related to the eighth contribution, using NeurosIS tools and delivering an IT artifact which tracks and adapts to the user's attentional state. Moreover, we contribute to the ninth contribution by providing VAF as a live biofeedback that assists users to better control their limited attentional resources. Also, based on Gregor and Hevner [36] this DSR study contributes as an "improvement" since we addressed an existing problem (attention management) by providing the design of a new solution (individualized VAF). By evaluating three suggested VAF types, we identified that providing individualized VAF as live biofeedback supports users in both VAA performance and focused attention more than having a proper or improper example of a VAA as general VAFs.

References

1. Chen, H., Chiang, R. H. L., & Storey, V. C. (2012). Business intelligence and analytics: From big data to big impact. *MIS Quarterly, 36,* 1165–1188.
2. Few, S. (2006). *Information dashboard design: The effective visual communication of data.* O'Reilly Sebastopol, CA.
3. Yigitbasioglu, O. M., & Velcu, O. (2012). A review of dashboards in performance management: Implications for design and research. *The International Journal of Accounting Information Systems, 13,* 41–59.
4. Vandenbosch, B., & Huff, S. L. (1997). Searching and scanning: How executives obtain information from executive information systems. *MIS Quarterly, 21,* 81.
5. Davenport, T. H., & Völpel, S. C. (2001). The rise of knowledge towards attention management. *Journal of Knowledge Management, 5,* 212–222.

6. Proctor, R. W., & Vu, K. L. (2006). The cognitive revolution at age 50: Has the promise of the human information-processing approach been fulfilled? *International Journal of Human—Computer Interaction, 21,* 253–284.
7. Simon, H. A. (1971). Designing organizations for an information-rich world. *Computers, Communications, and the Public Interest, 72,* 37.
8. Roda, C. (2011). Human attention and its implications for human–computer interaction. *Human Attention* in *Digital Environments*, 11–62.
9. Chun, M. M., Golomb, J. D., & Turk-Browne, N. B. (2011). A taxonomy of external and internal attention. *Annual Review of Psychology, 62,* 73–101.
10. Bulling, A. (2016). Pervasive attentive user interfaces. *IEEE Computer, 49,* 94–98.
11. Vertegaal, R., & Shell, J. S. (2008). Attentive user interfaces: The surveillance and sousveillance of gaze-aware objects. In *Social Science Information* (pp. 275–298).
12. Roda, C., & Thomas, J. (2006). Attention aware systems: Theories, applications, and research agenda. *Computers in Human Behavior, 22,* 557–587.
13. D'Mello, S., Olney, A., Williams, C., & Hays, P. (2012). Gaze tutor: A gaze-reactive intelligent tutoring system. *International Journal of Human-Computer Studies, 70,* 377–398.
14. Allanson, J., & Fairclough, S. H. (2004). A research agenda for physiological computing. *Interacting with Computers, 16,* 857–878.
15. Bailey, B. P., & Konstan, J. A. (2006). On the need for attention-aware systems: Measuring effects of interruption on task performance, error rate, and affective state. *Computers in Human Behavior, 22,* 685–708.
16. Davenport, T. H., Völpel, S. C., & Vo, S. C. (2005). The rise of knowledge towards attention management. *Journal of Knowledge Management.*
17. Majaranta, P., & Bulling, A. (2014). Eye tracking and eye-based human–computer interaction. In *Advances in physiological computing* (pp. 39–65).
18. Dimoka, A., Davis, F. D., Pavlou, P. A., & Dennis, A. R. (2012). On the use of neurophysiological tools in IS research: Developing a research agenda for NeuroIS. *MIS Quarterly, 36,* 679–702.
19. Just, M. A., & Carpenter, P. A. (1976). Eye fixations and cognitive processes. *Cognitive Psychology, 8,* 441–480.
20. Riedl, R., Hevner, A., & Davis, F. (2014). Towards a NeuroIS research methodology: Intensifying the discussion on methods, tools, and measurement. *Journal of the Association for Information Systems, 15,* I–XXXV.
21. Bundesen, C. (1990). A theory of visual attention. *Psychological Review, 97,* 523–547.
22. Eriksen, C. W., & Yeh, Y.-Y. (1985). Allocation of attention in the visual field. *Journal of Experimental Psychology: Human Perception and Performance, 11,* 583–597.
23. Astor, P. J., Adam, M. T. P., Jerčić, P., Schaaff, K., & Weinhardt, C. (2014). Integrating biosignals into information systems: A NeuroIS tool for improving emotion regulation. *Journal of Management Information Systems, 30,* 247–278.
24. Riedl, R., Léger, P.-M. (2016). Fundamentals of NeuroIS. In *Studies in Neuroscience, Psychology and Behavioral Economics* (p. 127).
25. Lux, E., Adam, M. T. P., Dorner, V., Helming, S., Knierim, M. T., & Weinhardt, C. (2018). Self and foreign live biofeedback as a user interface design element: A review of the literature. *Communications of the Association for Information Systems, 34,* 555–606.
26. Sharma, K., Alavi, H. S., Jermann, P., & Dillenbourg, P. (2016). A gaze-based learning analytics model. In *Proceedings of the Sixth International Conference on Learning Analytics & Knowledge—LAK '16* (pp. 417–421). New York, New York, USA: ACM Press.
27. Sarter, N. B. (2000). The need for multisensory interfaces in support of effective attention allocation in highly dynamic event-driven domains: The case of cockpit automation. *The International Journal of Aviation Psychology, 10,* 231–245.
28. Otto, K., Castner, N., Geisler, D., & Kasneci, E. (2018). Development and evaluation of a gaze feedback system integrated into eyetrace. In *Proceedings of the 2018 ACM symposium on eye tracking research & applications—ETRA '18* (pp. 1–5). New York, New York, USA: ACM Press.

29. vom Brocke, J., Riedl, R., & Léger, P.-M. (2013). Application strategies for neuroscience in information systems design science research. *Journal of Computer Information Systems, 53,* 1–13.
30. Kuechler, B., & Vaishnavi, V. (2008). Theory development in design science research: Anatomy of a research project. *European Journal of Information Systems, 17,* 489–504.
31. Toreini, P., & Langner, M. (2019). Designing user-adaptive information dashboards: Considering limited attention and working memory. In *Proceedings of European Conference on Information Systems (ECIS 2019)*, Stockholm, Sweden, June 08–14, 2019.
32. Sharp, H., Rogers, Y., & Preece, J. (2007). *Interaction design: Beyond human-computer interaction*. Wiley.
33. Hong, W., Thong, J. Y. L., & Tam, K. Y. (2004). Does animation attract online users' attention? The effects of flash on information search performance and perceptions. *Information Systems Research, 15*.
34. Perkhofer, L., & Lehner, O. (2019). Using gaze behavior to measure cognitive load. In F. D. Davis, R. Riedl, J. vom Brocke, P.-M. Léger, & A. B. Randolph (Eds.), *Information systems and neuroscience* (pp. 73–83). Cham: Springer International Publishing.
35. Kane, M. J., & Engle, R. W. (2003). Working-memory capacity and the control of attention: The contributions of goal neglect, response competition, and task set to Stroop interference. *Journal of Experimental Psychology: General, 132,* 47–70.
36. Gregor, S., & Hevner, A. R. (2013). Positioning and presenting design science research for maximum impact. *MIS Quarterly, 37,* 337–355.

Perturbation-Evoked Potentials: Future Usage in Human-Machine Interaction

Jonas C. Ditz and Gernot R. Müller-Putz

Abstract Brain-computer interfaces (BCIs) can be used to improve human-machine interactions (HMIs) by providing implicit information about the mental state. We introduce a brain activity, perturbation-evoked potentials (PEPs), that was not yet investigated in the context of BCIs although it has the required properties. An experimental setup for studying PEPs is proposed and validated and two possible use cases for this brain activity are introduced.

Keywords Passive brain-computer interface · Perturbation-evoked potential · Human-machine interaction · Rehabilitation · Assistive device

1 Introduction

A brain-computer interface (BCI) is a system that translates brain activity into control commands for computers or machines. During the last decades, BCIs were successfully used to compensate for lost neural functionality and restore several key abilities in paralyzed patients, e.g. communication [1], motor control [2], and locomotion [3]. Additionally, BCI systems were used in rehabilitation medicine to improve the recovery rate of patients [4].

Brain signals can be obtained using different recording methods. Typically, non-invasive techniques like electroencephalography (EEG) [5], functional magnetic resonance imaging (fMRI) [6], or near-infrared spectroscopy (NIRS) [7] are used. However, several BCI systems using invasive recording techniques for acquiring brain activity have been proposed. More prominent methods used are electrocorticography (ECoG) [8] and multi-unit recordings [9]. Invasive methods offer the advantage of excellent spatial and temporal resolution of obtained signals as well as a better signal-to-noise ratio at the cost of inflicting physical trauma to the brain and/or the

J. C. Ditz · G. R. Müller-Putz (✉)
Graz University of Technology, Graz, Austria
e-mail: gernot.mueller@tugraz.at

J. C. Ditz
e-mail: jonas.ditz@tugraz.at

F. D. Davis et al. (eds.), *Information Systems and Neuroscience*,
Lecture Notes in Information Systems and Organisation 32,
https://doi.org/10.1007/978-3-030-28144-1_30

skull [10]. Acquisition of brain signals is the first step in realizing a brain-computer interface. In order to generate control commands, brain activity has to be mapped onto user intentions. For this purpose, BCI systems rely on activity patterns that can be actively modulated by users and therefore hold information about intention. Patterns that are often used for active BCI control are event-related potentials (ERPs, e.g. P300) [11], movement-related cortical potentials (MRCPs) [12], and steady-state evoked potentials (SSEPs) [13].

BCI systems using one of the brain signals mentioned above are often referred to as active brain-computer interfaces (aBCIs). This term reflects the intention of these systems to provide users with autonomous control over a target computer or machine, in other words, the user can actively trigger a response. Over the last decade, a second type of BCI system, whose purpose was not to provide users with active control over a machine or computer, but instead to enrich human-machine interaction with implicit information about the state of a user, was suggested by different research groups. These systems are often referred to as passive brain-computer interfaces (pBCIs) [14]. The main difference between pBCIs and aBCIs is the brain activity that is used. A pBCI system relies on activity patterns that are not voluntarily modulated by users but automatically occur due to changes of internal or external parameters. Proposed systems used, among others, mental workload [15], error-related potentials (ErrPs) [16], and band power [17]. The implicit information about a user's mental state encoded in these activity patterns is used to evoke a reaction of the controlled machine or computer, e.g. Zander et al., used a pBCI to implicitly control movement of a cursor based on workload [15].

This work focuses on possible use cases for perturbation-evoked potentials (PEPs) to enrich HMI. Such a PEP is released by the brain whenever a person loses balance. PEPs belong to the group of event-related potentials. Therefore, they have similarities with other ERPs such as time-locking to an event. Especially error-related potentials seem to be closely related with PEPs. However, they offer different information about users. While ErrPs reflect a general mismatch between user's expectation and reality or an error committed by them, PEPs provide detailed information about the internal state of the user [18, 19]. This gain of detailed information distinguishes PEPs from other ERPs. In order to use PEPs for a pBCI, robust detection and classification of this activity pattern must be possible.

2 Theoretical Background

Several EEG studies have revealed that full-body perturbations are reflected by specific cortical activity [20–23]. This activity is called perturbation-evoked potential and usually consists of four different components [19]. PEPs start with a small positive peak P1 that occurs between 30 and 90 ms after perturbation onset. Reported amplitudes of P1 are in the range from 0.2 to 7 μV. This first component is followed by a stronger, negative peak, which is commonly referred to as N1. Most of the research about PEPs focuses on this second component with reported peak latencies between 85 and 163 ms and amplitudes ranging from −0.8 to −80 µV. The last two components, P2 and N2, are together called late perturbation-evoked response. P2 and N2 usually occur 200–400 ms after perturbation onset. P1, P2, and N2 are not always visible in EEG recordings, but the N1 peak can always be found.

The function of the N1 peak is still an object of discussion among balance control researchers. Quant and colleagues suggested that this peak reflects sensory processing of somatosensory inputs [24]. In contrast to that, Adkin and colleagues proposed that N1 is related to error detection. They developed the idea that a mismatch between actual and expected postural state is reflected by this negative peak [22]. Malin and colleagues argued against the role of N1 in error detection. They localized an N1 source generator in the supplementary motor area (SMA). Therefore, their suggestion was that N1 occurs in correspondence with the planning of a motor response that is necessary to compensate the perturbation [25].

Several factors influencing the latency and amplitude of the N1 component have been identified. The amplitude of experienced perturbations has a proportional influence of PEP's amplitude [26]. Other factors are learning effects and predictability of perturbation, among others [27, 28]. However, Mochizuki and colleagues detected that occurrence and shape of PEPs are not affected by the sensory, motor, and postural state of subjects [23]. This robustness to internal parameters of humans makes PEPs a promising activity pattern for HMI research since they can be used as a reliable source of implicit information about the state of a user.

3 Research Model

The use of biosignals in human-machine interaction scenarios is not totally new. Though several examples exist, as described above, the transfer of these concepts into real-world settings is still to be done. In this work we want to present a seldomly used brain signal and give an idea of an experimental setup as it could enrich besides others also the NeuroIS-field.

There are human-machine scenarios where the information about balance control loss can be beneficial to improve interactions. Below, we are briefly describing three fields that could profit from this knowledge.

- **Clinical**: The information provided by PEPs can be used in the context of wearable robotics or exoskeletons, either during the rehabilitation process (e.g. gait training in a reha-robot after stroke) or in the assistive device context (walking exoskeletons for paraplegic people with SCI). Whenever the patient loses balance, the detection of the PEP could give the system supplementary knowledge of the user's state in addition to kinetic/kinematic measures of sensor data and help the system to better interpret the current situation.
- **Autonomous driving/flying**: Whoever has experienced glider flying knows that the body, being tightly fixed into the seat of the cockpit and therefore entirely connected with the plane, feels every single movement in the air. Pilots seeking

for thermals need to react, beside the visual information of the environment, to the movements the plane and therefore the body receives by rising air. This slight imbalance gives the sign to turn into a curve to catch updraft. By detecting the PEP, ideally lateral to the movement, the turn could be initiated faster since it needs reaction time and movement of the pilot in the first place.

- **Virtual reality (VR)**: Due to the mismatch in vestibular and visual sensory information during the usage of a VR-headset, a loss of balance control can occur. Information about the precise time when a user perceives loss of balance control could be used to adapt VR scripts on the fly to compensate this loss by using alternative ways of visualization.

4 Experimental Setup

Experiments to investigate the usability of PEPs for pBCI will be conducted with a car seat that was mounted on a custom-made tilting appliance (see Fig. 1). Participants will be secured with a belt and equipped with a 32-channel EEG cap. The used system is a LiveAmp (BrainProducts, Gilching, Germany) combined with actiCap slim electrodes (BrainProducts, Gilching, Germany). The amplifier will be attached to the chair in order to measure movements of the chair with an in-built accelerom-

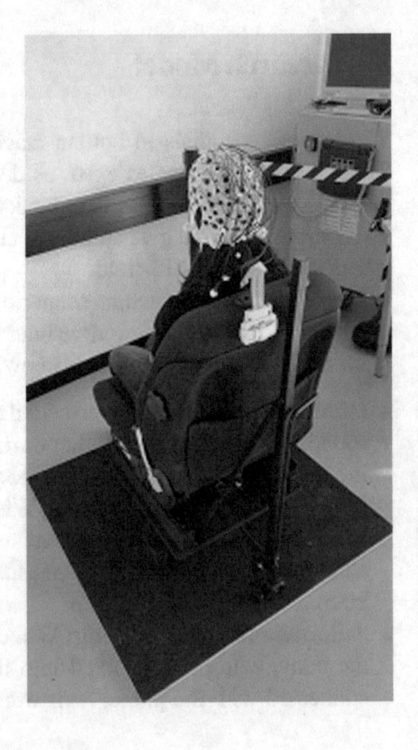

Fig. 1 Picture of the custom-built chair used to induce perturbations. The chair can be tilted using the handle seen behind the backrest. The amplifier is attached at the top of the backrest in order to record movements of the chair with the in-built accelerometer

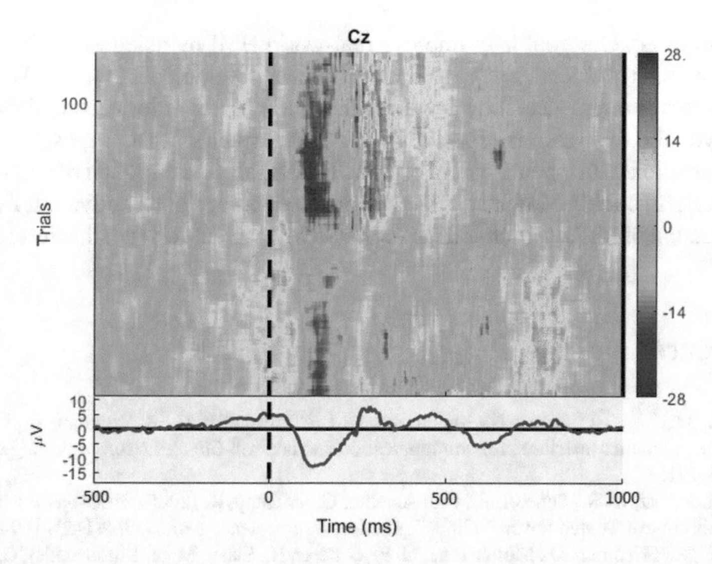

Fig. 2 Response to perturbation of one subject recorded at electrode Cz. Data bandpass filtered between 0.3 and 30 Hz and the curve at the bottom is averaged over 120 trials. The N1 component of PEP can be seen around 100 ms

eter. Thus, the accelerometer data used to determine the perturbation onset will be perfectly synchronous with EEG recordings.

Preliminary recordings with the chair reveal that the tilt angle with a maximum of 5° suffices to elicit a PEP (see Fig. 2). In order to test the ability to evoke PEPs of our experimental setup, four subjects were recorded. All four of them showed negative peaks around 100 ms after the perturbation onset. Data was filtered between 0.3 and 30 Hz using an infinite impulse response (IIR) bandpass filter and epoched around the perturbation onset. Epochs started 500 ms before the perturbation onset and ended 1000 ms after the onset. These epochs were averaged over all trials. For future analysis, independent component analysis (ICA) could be used to on the one hand to remove artifacts introduced by body and eye movement and on the other hand to investigate the sources of the PEP.

5 Conclusion and Outlook

Our first recordings using the tilting chair indicated that it is possible to study PEPs using our setup. Next steps will include the recording of more subjects to create a data pool with workable size, investigating time-series analysis methods and machine learning classifiers for a robust online detection and classification of PEPs, and studying whether sideward perturbations induce a lateralization of PEPs. The first step is necessary to ensure proper testing conditions for machine learning algorithms. Training and test set have to be of a sufficient size in order to prevent stability issues like overfitting. Our goal is to improve real-world HMI by using PEPs. Therefore, the second step is crucial, since the identification or development of methods that can achieve robust online detection and classification is essential for accomplishing this objective. The third step is especially important for our second proposed use case. The interaction with a glider would be more beneficial if the direction of perturbation (left vs. right) could be inferred directly from the recorded brain activity. If there is a lateralization of PEPs, this characteristic could be easily used to get that information.

References

1. Wolpaw, J. R., Birbaumer, N., McFarland, D. J., Pfurtscheller, G., & Vaughan, T. M. (2002). Brain–computer interfaces for communication and control. *Clinical Neurophysiology, 113*(6), 767–791.
2. Müller-Putz, G. R., Scherer, R., Pfurtscheller, G., & Rupp, R. (2005). EEG-based neuroprosthesis control: A step towards clinical practice. *Neuroscience Letters, 382*(1–2), 169–174.
3. Leeb, R., Friedman, D., Müller-Putz, G. R., Scherer, R., Slater, M., & Pfurtscheller, G. (2007). Self-paced (asynchronous) BCI control of a wheelchair in virtual environments: A case study with a tetraplegic. *Computational Intelligence and Neuroscience*.
4. Pfurtscheller, G., Müller-Putz, G. R., Scherer, R., & Neuper, C. (2008). Rehabilitation with brain-computer interface systems. *Computer, 41*(10), 58–65.
5. Müller-Putz, G. R., Riedl, R., & Wriessnegger, S. C. (2015). Electroencephalography (EEG) as a research tool in the information systems discipline: Foundations, measurement, and applications. *CAIS, 37,* 46.
6. Bauernfeind, G., Wriessnegger, S., & Müller-Putz, G. (2014). Using near-infrared spectroscopy (NIRS) for brain-computer interface (BCI) systems. In *Human Cognitive Neurophysiology*.
7. Weiskopf, N., Mathiak, K., Bock, S. W., Scharnowski, F., Veit, R., Grodd, W., et al. (2004). Principles of a brain-computer interface (BCI) based on real-time functional magnetic resonance imaging (fMRI). *IEEE Transactions on Biomedical Engineering, 51*(6), 966–970.
8. Brunner, P., Ritaccio, A. L., Emrich, J. F., Bischof, H., & Schalk, G. (2011). Rapid communication with a "P300" matrix speller using electrocorticographic signals (ECoG). *Frontiers in Neuroscience, 5,* 5.
9. Wodlinger, B., Downey, J. E., Tyler-Kabara, E. C., Schwartz, A. B., Boninger, M. L., & Collinger, J. L. (2014). Ten-dimensional anthropomorphic arm control in a human brain−machine interface: Difficulties, solutions, and limitations. *Journal of Neural Engineering, 12*(1), 016011.
10. Fernández, E., Greger, B., House, P. A., Aranda, I., Botella, C., Albisua, J., et al. (2014). Acute human brain responses to intracortical microelectrode arrays: Challenges and future prospects. *Frontiers in Neuroengineering, 7,* 24.
11. Sellers, E. W., & Donchin, E. (2006). A P300-based brain–computer interface: Initial tests by ALS patients. *Clinical Neurophysiology, 117*(3), 538–548.
12. Pereira, J., Ofner, P., Schwarz, A., Sburlea, A. I., & Müller-Putz, G. R. (2017). EEG neural correlates of goal-directed movement intention. *Neuroimage, 149,* 129–140.
13. Müller-Putz, G. R., Scherer, R., Neuper, C., & Pfurtscheller, G. (2006). Steady-state somatosensory evoked potentials: Suitable brain signals for brain-computer interfaces? *IEEE Transactions on Neural Systems and Rehabilitation Engineering, 14*(1), 30–37.
14. Zander, T. O., Kothe, C. A., Welke, S., & Rötting, M. (2008). Enhancing human–machine systems with secondary input from passive brain–computer interfaces. In *Proceedings of the 4th International Brain–Computer Interface Workshop & Training Course* (pp. 144–149). Graz, Austria: Verlag der Technischen Universität Graz.

15. Zander, T. O., Krol, L. R., Birbaumer, N. P., & Gramann, K. (2016). Neuroadaptive technology enables implicit cursor control based on medial prefrontal cortex activity. *Proceedings of the National Academy of Sciences, 113*(52), 14898–14903.
16. Parra, L. C., Spence, C. D., Gerson, A. D., & Sajda, P. (2003). Response error correction-a demonstration of improved human-machine performance using real-time EEG monitoring. *IEEE Transactions on Neural Systems and Rehabilitation Engineering, 11*(2), 173–177.
17. Bos, D. P. O., Reuderink, B., van de Laar, B., Gürkök, H., Mühl, C., & Poel, M., et al. (2010). Brain-computer interfacing and games. In *Brain-computer interfaces* (pp. 149–178). London: Springer.
18. Holroyd, C. B., & Coles, M. G. (2002). The neural basis of human error processing: Reinforcement learning, dopamine, and the error-related negativity. *Psychological Review, 109*(4), 679.
19. Varghese, J. P., McIlroy, R. E., & Barnett-Cowan, M. (2017). Perturbation-evoked potentials: Significance and application in balance control research. *Neuroscience and Biobehavioral Reviews, 83,* 267–280.
20. Dietz, V., Quintern, J., & Berger, W. (1984). Cerebral evoked potentials associated with the compensatory reactions following stance and gait perturbation. *Neuroscience Letters, 50*(1–3), 181–186.
21. Duckrow, R. B., Abu-Hasaballah, K., Whipple, R., & Wolfson, L. (1999). Stance perturbation-evoked potentials in old people with poor gait and balance. *Clinical Neurophysiology, 110*(12), 2026–2032.
22. Adkin, A. L., Quant, S., Maki, B. E., & McIlroy, W. E. (2006). Cortical responses associated with predictable and unpredictable compensatory balance reactions. *Experimental Brain Research, 172*(1), 85.
23. Mochizuki, G., Sibley, K. M., Cheung, H. J., Camilleri, J. M., & McIlroy, W. E. (2009). Generalizability of perturbation-evoked cortical potentials: Independence from sensory, motor and overall postural state. *Neuroscience Letters, 451*(1), 40–44.
24. Quant, S., Adkin, A. L., Staines, W. R., & McIlroy, W. E. (2004). Cortical activation following a balance disturbance. *Experimental Brain Research, 155*(3), 393–400.
25. Marlin, A., Mochizuki, G., Staines, W. R., & McIlroy, W. E. (2014). Localizing evoked cortical activity associated with balance reactions: Does the anterior cingulate play a role? *American Journal of Physiology-Heart and Circulatory Physiology.*
26. Staines, R. W., McIlroy, W. E., & Brooke, J. D. (2001). Cortical representation of whole-body movement is modulated by proprioceptive discharge in humans. *Experimental Brain Research, 138*(2), 235–242.
27. Dietz, V., Quintern, J., Berger, W., & Schenck, E. (1985). Cerebral potentials and leg muscle emg responses associated with stance perturbation. *Experimental Brain Research, 57*(2), 348–354.
28. Quintern, J., Berger, W., & Dietz, V. (1985). Compensatory reactions to gait perturbations in man: Short-and long-term effects of neuronal adaptation. *Neuroscience Letters, 62*(3), 371–375.

Improved Calibration of Neurophysiological Measures Tools

Florian Coustures, Marc Fredette, Jade Marquis, François Courtemanche and Elise Labonté-Lemoyne

Abstract To carry out more reliable UX tests, it is necessary to analyze how neurophysiological measurements are treated. In particular, the use of baseline, the neutral task that compares a subject to itself, has so far been the subject of little study in the literature. The purpose of this paper is to focus on using the baseline in a UX test context. Our main result is that we can improve the baseline procedure, which currently consists of doing one baseline task at the beginning. Indeed, in our research, the validity of this procedure was studied during the calibration of the EDA (Electrodermal Activity). We then discovered that a majority of our subjects had an increasing or decreasing trend throughout the experiment. To answer this problem we propose to move the location of this one baseline or add a second baseline at the end of the experiment.

Keywords Baseline · EDA · Neurophysiological measures · Noise extraction · Calibration · Individual response

1 Introduction

In UX studies, the validity, reliability, and robustness of measurements are critical to achieving quality results [1]. However, in UX experiments, the high variability of neurophysiological signals between participating subjects can disrupt data analysis.

F. Coustures (✉) · M. Fredette · J. Marquis · F. Courtemanche · E. Labonté-Lemoyne
HEC Montréal, Montréal, Canada
e-mail: florian.coustures@hec.ca

M. Fredette
e-mail: marc.fredette@hec.ca

J. Marquis
e-mail: jade.marquis@gmail.com

F. Courtemanche
e-mail: francois.courtemanche@hec.cac

E. Labonté-Lemoyne
e-mail: elise.labonte-lemoyne@hec.ca

F. D. Davis et al. (eds.), *Information Systems and Neuroscience*,
Lecture Notes in Information Systems and Organisation 32,
https://doi.org/10.1007/978-3-030-28144-1_31

In order to compare these neurophysiological signals between subjects participating in the same research project, but sometimes exposed to different interfaces, it is imperative to calibrate these signals, as they can vary considerably from one subject to another. It is therefore essential to integrate one or more identical tasks for all subjects in the experiment in order to establish a point of reference. This eliminates the specific error associated with the subject signal and, for laboratory experiments, also suppresses the in situ effect. These tasks to calibrate neurophysiological measurement tools are known as a baseline.

The purpose of this article is thus to propose ways to improve the use of baselines in UX experiments.

First of all, the definition of a baseline, as it is applied to laboratory experiments, is a reference state allowing the comparison of a subject with himself, during the experiment and for the different conditions tested during the tasks [2]. For the study of a subject, this allows us to conclude that the observed difference between the value of his signal during the task and the value during the baseline comes only from the task. However, the existence of such a reference state does not make consensus [3, 4]. Jennings speaks, for example, of a state of rest that would be unstable and difficult to control and, despite baselines up to 20 min, Jennings said he was not able to achieve a stability of values [2]. This could indicate that, regardless of baseline length, there is potentially a trend within an experiment. This problem can be exacerbated in UX tests today where time constraints are an important issue. However, the baseline is a critical step for the validity and reliability of experiment results [2, 5].

Also, in studies conducted today in psychophysiology, two quite different types of baselines coexist: resting baseline [6–10] and low-cognitive demand baseline [2, 11]. Here we can see the effects of the confusion pointed out by Jennings between state of rest and state of reference. Resting baseline mainly involves asking the participant to sit, without moving. Jennings prefers to have participants perform a task to maintain a constant level of vigilance. He proposes the Vanilla Baseline method (the type of baseline we used for our study) which has the advantage of avoiding sleepiness or boredom [2]. With this method, subjects are asked to observe a video screen and to count the number of times a designated color occupies a rectangle on the screen. Six clearly distinguishable colors are employed and presented randomly with equal probability of occurrence. At the end of the baseline period, subjects report the number of times the designated color was observed.

Finally, a good baseline must help reduce the inter-subject variation as much as possible [2, 12] and to get as close as possible to the real external conditions [13]. The importance of the baseline is often neglected, we can see it because this task is sometimes not described or even taken into account in the presentation of the results [2]. A good future practice would be to bring more transparency and to make this description systematic [3, 14]. This would improve our knowledge of the various practices in terms of baselining and reinforce the methodological rigor of UX professionals.

The various issues raised in the literature have led to questions about the use of baselines. The relative instability of baselines hardly seems compatible with the imperative of reliability in the UX tests. In particular, using only one baseline at the beginning could be problematic if there is a trend. This is how we came to hypothesize that it would be a better practice to use more than one baseline.

2 Research Method

The data collected comes from a study conducted at Tech3Lab with the goal of improving our knowledge of the optimal location and number of baselines. The task of the participants during the experiment was to navigate through a simulated Facebook page. For about a minute, they had to look at this Facebook page, which contained 3 images of the IAPS bank (Fig. 1) [15] whose physiological response was known in advance. This task was repeated 12 times with 36 images randomly scattered on these 12 pages. For each participant we collected the electrocardiogram, the pupil diameter and the electrodermal activity (EDA), which is the signal that we will use for the continuation of our analyses. In the end we were able to exploit the data of 40 participants.

What will interest us for the future is that before each of the 12 tasks, the participants had to do a baseline. The baseline method was a vanilla baseline method as described earlier. Figure 2 illustrates what a participant can see during a baseline. During the experiment, we alternated a "long baseline" with a duration of 40 s and two "short baseline" each with a duration of 20 s, for a total of 12 baselines for each participant.

We chose to concentrate our first research on the study of the EDA. For each participant we had measured every 10 ms, the value of his EDA expressed in microsiemens (μS). Thus, from this dataset, we were able to calculate the average EDA value during each baseline for each participant. It is the analysis of these results that we will present in the next section.

Fig. 1 Example of the 3 IAPS images

Fig. 2 Screen example during vanilla baseline

3 Preliminary Results

We wanted to understand our results with a descriptive analysis, focusing only on average EDA during our 12 baselines and not during the whole experiment. Thus, we classified the different participants using the k-means clustering method applied to curve clustering. In a simple way we were able to categorize our participants in the 3 groups suggested by the clustering procedure and presented in Fig. 3: those for whom the EDA increased, those for whom the EDA decreased, and the others (stability or strong variability).

To improve the visualization, all the curves have been standardized in order that the average of the 12 baselines for each participant is 1. On these graphs, the pale gray curves represent the change in baseline EDA for each participant. The curve in red represents the average of all curves for a given cluster. The main result of this clustering is that there are 27 out of 40 participants for whom there is a trend (17 increasing, all with statistical significance, and 10 decreasing, out of which 7 are with statistical significance) regardless of the tasks performed. This means that for 2/3 of the subjects, the assumption of a unique reference state that allows to compare the subject to himself from a baseline made at the beginning is questionable. Taking

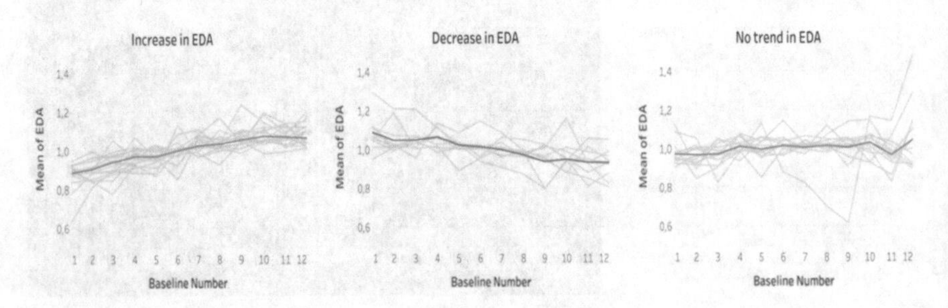

Fig. 3 Result of the k-means clustering

the case of a subject with a rising trend, we will conclude more often that his EDA is above his reference level, which may lead to a bias in the analysis.

As we can see, the main problem is the prediction of the behavior of the EDA from a single baseline done at the beginning for participants with a trend. Solutions would be to modify the location of this baseline or do more than one baseline during the experiment. To improve our knowledge of the natural evolution of a subject's EDA we will consider several scenarios where we could modify the location or the number of baselines. To measure the performance of these different scenarios in the prediction we will use the RMSE (Root Mean Square Error). This indicator calculates the difference between our predicted value and the observed value; a successful model will be closer to zero. The predicted values come from a linear regression using the baseline EDA averages only for selected baselines. The observed values are the values of the EDA for the participant during throughout the whole experiment (during baselines and tasks).

To develop the different scenarios, we used our literature review and our descriptive data analysis. Indeed, the supposed instability of the baseline pushed us to explore the results of a prediction from a baseline in the middle or at the end of the experiment. Above all these elements suggested we not settle for a single baseline and try to measure the additional baseline gain. Our descriptive analysis seemed to show a linear trend, but we also tested a quadratic alternative in case of concave or convex relations using 3 baselines or more. The main results are presented in Table 1.

F means the first baseline of the experiment, M represents the baseline in the middle, and L means the last baseline. Here, for example, we found an RMSE of 1.059 when we made a linear regression with a quadratic trend, using the EDA average for first, middle and last baseline (i.e. 3 points). These results lead us to some initial conclusions.

Table 1 Performance in prediction depending on number and location of baselines

Type of result (baseline location)	Mean of RMSE for 40 participants
Actual procedure: 1 baseline (F)	1.829
Best result with 1 baseline (M)	1.379
Best result with 2 baselines (F-L)	1.144
Best result with 3 baselines–linear trend (F-M-L)	1.071
Best result with 3 baselines–quadratic trend (F-M-L)	1.059
Using all available baselines–linear trend (All 12)	0.996

First, the location of the baseline is important because the more a baseline is placed in the middle of the experiment, the better its prediction performance is (RMSE goes from 1.829 to 1.379). Also, adding a second baseline has a very strong impact on a participant's EDA prediction during the experiment (RMSE drops from 1.829 to 1.144). However, the marginal gain in performance of adding extra baselines beyond this second baseline decreases very quickly. Even if we could do a baseline before each task, our average RMSE would reach the floor of 0.996. This 0.996 is only a reference value that we want to get closer, but reaching it is not a goal. Finally, a simple linear trend seems to be able to adequately explain the evolution of EDA. Tests with other types of trends have little or no improvement in our results (RMSE drops only from 1.071 to 1.059 with a quadratic trend for 3 baselines).

4 Discussion and Conclusion

At this stage we have no physiological explanation for these trends of EDA. However, we observed a significant difference for the EDA average in the first baseline between rising participants and those having a decrease in their EDA during the study. Thus, participants with a higher EDA at the beginning are more likely to see their EDA drop during the experiment. This could mean that other factors (stress, temperature difference) could affect the participants before they find a reference state. It is also possible that the establishment of a reference state in the laboratory is not compatible with what an individual usually lives (in situ effect) [2]. As a limitation, it is possible that the realization of additional baselines compared to the current procedure have affected our results.

To conclude, the current baseline procedure with one baseline at the beginning is not optimal. Thus, we propose the addition of a second baseline at the end, which has the advantage of not being too time consuming and seems a good compromise in terms of the measurement performance. Even if this solution is not perfect, at least it improves our measurement in the case of an increasing or decreasing trend in EDA. This additional baseline can provide additional information, which can lead to more reliable analysis. We thus suggest this practice as an interesting research track for UX professionals. However, if time constraints are too important, moving the baseline to the middle of the experiment could also be a viable option. In both cases, it is necessary to carefully plan the experimental design. Indeed, our study was designed so that the location of the baseline is valid regardless of the task performed, which is not necessarily applicable to all UX tests.

References

1. Riedl, R., Davis, F. D., & Hevner, A. R. (2014). Towards a neurois research methodology: Intensifying the discussion on methods, tools, and measurement. *Journal of the Association for Information Systems*, *15*(10).
2. Jennings, J. R., et al. (1992). Alternate cardiovascular baseline assessment techniques: Vanilla or resting baseline. *Psychophysiology, 29*(6), 742–750.
3. Linden, W., et al. (1997). Physiological stress reactivity and recovery: Conceptual siblings separated at birth? *Journal of Psychosomatic Research, 42*(2), 117–135.
4. Haynes, S. N., et al. (1991). Psychophysiological assessment of poststress recovery. *Psychological Assessment: A Journal of Consulting and Clinical Psychology, 3*(3), 356–365.
5. Cacioppo, J. T., Tassinary, L. G., & Bernston, G. G. (2007). *Handbook of psychophysiology* (3rd ed.). C.U.: Press.
6. Baumgartner, T., Gianotti, L. R. R., & Knoch, D. (2013). Who is honest and why: baseline activation in anterior insula predicts inter-individual differences in deceptive behavior. *Biological Psychology, 94*(1), 192–197.
7. Gamble, K. R., et al. (2018). Different profiles of decision making and physiology under varying levels of stress in trained military personnel. *International Journal of Psychophysiology, 131,* 73–80.
8. Hauschildt, M., et al. (2011). Heart rate variability in response to affective scenes in posttraumatic stress disorder. *Biological Psychology, 88*(2), 215–222.
9. Spangler, D. P., et al. (2018). Resting heart rate variability is associated with ex-gaussian metrics of intra-individual reaction time variability. *International Journal of Psychophysiology, 125,* 10–16.
10. Tracy, L. M., et al. (2018). Intranasal oxytocin reduces heart rate variability during a mental arithmetic task: A randomised, double-blind, placebo-controlled cross-over study. *Progress in Neuro-Psychopharmacology and Biological Psychiatry, 81,* 408–415.
11. Bulut, N. S., et al. (2018). Heart rate variability response to affective pictures processed in and outside of conscious awareness: Three consecutive studies on emotional regulation. *International Journal of Psychophysiology, 129,* 18–30.
12. Fernández, T., et al. (1993). Test-retest reliability of EEG spectral parameters during cognitive tasks: I absolute and relative power. *International Journal of Neuroscience, 68*(3–4), 255–261.
13. Silvia, P. J., Jackson, B. A., & Sopko, R. S. (2014). Does baseline heart rate variability reflect stable positive emotionality? *Personality and Individual Differences, 70,* 183–187.
14. Fairclough, S. H. (2009). Fundamentals of physiological computing. *Interacting with Computers, 21*(1–2), 133–145.
15. Lang, P. J., Bradley, M. M., Cuthbert, B. N. (2008). International affective picture system (IAPS): Affective ratings of pictures and instruction manual. In *Technical Report B-3*.

On Using Python to Run, Analyze, and Decode EEG Experiments

Colin Conrad, Om Agarwal, Carlos Calix Woc, Tazmin Chiles, Daniel Godfrey, Kavita Krueger, Valentina Marini, Alexander Sproul and Aaron Newman

Abstract As the NeuroIS field expands its scope to address more complex research questions with electroencephalography (EEG), there is greater need for EEG analysis capabilities that are relatively easy to implement and adapt to different protocols, while at the same time providing an open and standardized approach. We present a series of open source tools, based on the Python programming language, which are designed to facilitate the development of open and collaborative EEG research. As supplementary material, we demonstrate the implementation of these tools in a NeuroIS case study and provide files that can be adapted by others for NeuroIS EEG research.

Keywords Research methods · Python · Machine learning · Open science · Brain-computer interface

1 Introduction

There has been considerable interest recently in the information systems community concerning tools for conducting interdisciplinary experiments. The motivation for this interest is rooted in the growing acceptance that explicit measures of emotional and cognitive states can capture dimensions of technology use and human computer interaction that users would not be able to self-report [1, 2]. To address this concern, NeuroIS researchers have called for tools that integrate disparate physiological and questionnaire measures [3]. This has led to efforts in the NeuroIS community to develop tools as either proprietary platforms [4] or open source libraries [5]. While these efforts have considerable potential for streamlining NeuroIS research processes in the long run, there has so far been relatively little discussion about existing tools developed in other research communities that might provide validated building blocks that accelerate progress in NeuroIS without reduplication of effort, and at the same

C. Conrad (✉) · O. Agarwal · C. C. Woc · T. Chiles · D. Godfrey · K. Krueger · V. Marini · A. Sproul · A. Newman
Dalhousie University, Halifax, Canada
e-mail: colin.conrad@dal.ca

F. D. Davis et al. (eds.), *Information Systems and Neuroscience*,
Lecture Notes in Information Systems and Organisation 32,
https://doi.org/10.1007/978-3-030-28144-1_32

time allowing NeuroIS to more fully integrate with related research enterprises such as cognitive neuroscience. Beyond simply being able to implement NeuroIS research in as efficient and validated a manner as possible, there is growing recognition across scientific disciplines that research processes and results be open, transparent, and fully reproducible [6–8]. Our goal is to describe a software pipeline that enables this.

In this paper, we introduce and demonstrate the utility of a set of open source tools that are relevant to many NeuroIS research projects—particularly those which use EEG to validate elements of user experience, but extending to other physiological measures including eye tracking, heart rate, skin conductance, MEG, and near-infrared brain imaging (fNIRI). All of the tools discussed are written in the Python programming language, are readily inter-operable, are freely available under the Python Software Foundation's open source license. The result is a set of tools that are both powerful and flexible, that can also be adapted to extend beyond traditional EEG analysis. We illustrate the use these tools, which we will henceforth refer to as the "Python stack", through a brief case study provided online as a supplement to this paper. The case study uses the Python stack to build elements of a P3 speller brain computer interface (BCI) [9, 10] and includes data collected in this paradigm as part of a neurotechnology hackathon at Dalhousie University, Halifax, Canada in 2019. This application was chosen because of its applicability to attention-related IS constructs that have been described in the IS literature [11–14]. The source files for this case and the related tutorial are provided publicly and can be retrieved from GitHub at https://github.com/cdconrad/py-bci.

2 Python Tools for Experiment Design and EEG Processing

Table 1 lists the Python packages that comprise the stack we use in our experimental protocols. The base of this is Anaconda [15], a collection of Python packages designed for scientific computing, which come bunded with a "package manager": a tool for installing and updating packages that ensures that all are compatible and inter-operable with each other. The value of Anaconda is that by downloading and installing this single package, the user is readily equipped with a wide variety of Python packages that will work together, without the overhead of identifying the necessary set of packages for a task and resolving compatibility issues.

The second tool highlighted in Table 1 is Jupyter [16]. This is a scientific "notebook" application which allows the user to write and execute code, view and save the results, and write rich-text documentation using Markdown formatting, all in a single file that is accessed via a Web browser. This has significant advantages over other approaches to using Python or other programming languages, such as interacting with a command line or using an integrated development environment; because all elements of the process are encapsulated in a single file, it is very easy to share and reproduce analysis pipelines across experiments and between labs.

Another advantage of Jupyter notebooks is that, once a pipeline has been implemented in a notebook, e.g., for the processing of EEG data from an individual par-

Table 1 Description of the recommended stack of Python tools for EEG analysis

Tool name	Developers	Description
Anaconda	Anaconda Inc. [15]	A distribution of the Python programming language for scientific computing
Jupyter	Pérez et al. [16]	A notebook format for sharing code and computational narratives
Matplotlib	Hunter et al. [17]	A 2D graphics package for the creation of publication-quality images
NumPy	van der Walt et al. [18]	A library for scientific computing and analysis
Pandas	McWinney et al. [19]	A data library optimized for manipulating large and time series data
PsychoPy	Peirce et al. [20]	An application and library used to run psychology and neuroscience experiments
MNE-Python	Gramfort et al. [21]	A library for preparing, analyzing and visualizing MEG, EEG and other related data
Scikit-learn	Pedregosa et al. [22]	A machine learning library

ticipant, notebooks can simply be copied and re-run for each additional participant, and/or easily adapted to new experiments. This means that while proficiency in the Python language is necessary to build the pipelines in the first place, users little to no programming expertise can readily adapt and run these notebooks for new participants or groups of participants. This makes these an excellent entry point for new researchers who wish to engage in NeuroIS research without first learning Python programming—not only because the learning curve is less steep, but because the notebook format makes it easy for senior lab members to audit others' work to ensure quality control. In this regard it is notable that our lab moved to this pipeline several years ago from the Matlab-based EEGlab platform, which is also widely used in cognitive neuroscience research [23]. While EEGlab offers a menu-driven graphical user interface (GUI), users have to choose the appropriate menu items and manually enter the appropriate parameters each time, and these settings are not all recorded in the output—making the process both more error-prone and more difficult to audit.

The other packages listed in Table 1 include a set of very widely-used tools for scientific computing (Matplotlib, NumPy, and Pandas), PsychoPy for experimental programming and data collection, MNE-Python (hereafter referred to as MNE) for EEG data preprocessing and analysis, and scikit-learn for machine learning. In what follows we describe the steps involved in running and analyzing the results of an EEG experiment using these packages. Note that the Matplotlib, NumPy, and Pandas packages are not explicitly described but are used by the tools that are described.

2.1 Experimental Protocol and Data Collection

An EEG experiment begins with presentation of stimuli to a participant, time-locked with collection of behavioral, EEG, and possibly other physiological measures. Software capable of precise time-locking is essential here, because measures such as EEG have temporal precision on the order of milliseconds. Some mode of inter-device communication is also required, because physiological data such as EEG is typically recorded on a separate device from that controlling stimulus presentation. The PsychoPy library [19] provides an environment for the presentation of a wide range of stimuli such as images, sounds, and movies, as well as collection of behavioral and vocal responses, and the ability to send precisely time-locked "trigger codes" to other hardware such as EEG data collection systems. These trigger codes are essential for later data analysis as they store, in the EEG data file, both the precise timing of events of experimental interest (e.g., stimulus onset, response times), and the identity of these events (e.g., stimulus type, correct vs. incorrect response). PsychoPy offers the ability to build experiments using either a GUI, which translates the user's design into Python code, or writing Python code directly. This again allows users with varying levels of expertise to participate fruitfully in the research enterprise. In addition to output sent to other devices, PsychoPy will save the precise timing of all events in the experiment to a text file for later analysis in any package the user desires.

2.2 Preprocessing EEG Data in Python

Following data collection, EEG data must be preprocessed and analyzed. Preprocessing involves a number of steps designed to improve the signal-to-noise ratio of the data and increase the ability to detect experimental effects, if they are present. In our pipeline, EEG preprocessing and analysis are performed using MNE. MNE provides a collection of data reading and conversion utilities which can be used to import and prepare data from a variety of hardware systems, including most common EEG and MEG systems. MNE converts data into a `mne.raw` object which includes the raw time course data, time-locked trigger codes, and metadata such as participant ID, date and time of data collection, the labels for each data channel, etc.

Common preprocessing steps for EEG data [24, 25] include: band-pass filtering; removal of data channels (electrodes) and trials contaminated with excessive noise; correction of other well-defined artifacts such as eye blinks, eye movements, and muscle noise; and re-referencing EEG data to an electrode(s) appropriate to the experiment. MNE provides functions dedicated to each of these tasks, which have been designed to implement best practices in EEG/MEG research (e.g., the choice of filter type). This relieves the user of the need to extensively research all of their preprocessing parameter choices yet allows—through the use of command-line options—control over common parameter choices (e.g., filter bandwidth). As data are processed, the data are converted from raw format (continuous EEG data) to MNE's epochs format (segments of data time-locked to experimental events of interest) and finally to MNE's evoked format (averages across all epochs of a given category).

While MNE is under active development, at this time it has a wide variety of tools implementing common functions in EEG preprocessing, such as independent components analysis (ICA) for artifact removal. The supplementary material for this paper includes Jupyter notebooks demonstrating our EEG preprocessing pipeline, including the specific MNE functions and associated parameters used, and documentation elaborating on usage and choice of parameters.

2.3 Analysis and Classification

Finally, after preprocessing the data, users can visualize data at the individual or group level, perform analyses to determine if hypothesized effects are present, and/or attempt classification of data based on machine learning. MNE provides tools for the visualization of EEG/MEG data in the time and frequency domains, as both waveform plots at individual or clusters of channels, and as scalp topographic maps. It also includes algorithms for source localization, allowing visualization of data on the cortical surface. MNE also provides some tools for statistical analysis—including parametric (*t*-tests, linear regression) and non-parametric (*t*-test, clustering) approaches and methods for multiple comparison correction—and machine learning decoders. However, perhaps one of the most powerful features of the Python stack is the compatibility between the data formats and machine learning libraries; because MNE is built on the NumPy and Pandas packages, it is easy to convert MNE data to these packages' data objects. This allows the use of a wide variety of other packages in Python, such as scikit-learn, a widely-used package implementing a wide variety of machine learning tools. As well, MNE data objects can be readily exported for use in other statistical packages such as R.

3 Conclusion

As the NeuroIS discipline develops, there will be a greater need for knowledge transfer and open science. Python tools provide an open source alternative for NeuroIS researchers, which has the added benefit of being curated by the Psychology and Neuroscience community. We hope that this technology can be leveraged to benefit the NeuroIS discipline and advance the community's capabilities for EEG research.

References

1. de Guinea, A. O., & Webster, J. (2013). An investigation of information systems use patterns: Technological events as triggers, the effect of time, and consequences for performance. *MIS Quarterly*, *37*(4), 1165–1188.
2. de Guinea, A. O., Titah, R., & Léger, P-M. (2017). Explicit and implicit antecedents of users' behavioral beliefs in information systems: A neuropsychological investigation. *Journal of Management Information Systems*, *30*(4), 179–210.
3. Léger, P-M., Courtemanche, F., Fredette, M., & Sénécal, S. (2019). A cloud-based lab management and analytics software for triangulated human-centered research. In F. D. Davis, R. Riedl, J. vom Brocke, P. M. Léger, & A. B. Randolph (Eds.), *Information systems and neuroscience. Lecture notes in information systems and organisation* (pp. 93–99). Springer International Publishing.
4. Courtemanche, F., et al. (2018). *Method of and system for processing signals sensed from a user*, US Patent US230180035886A1.
5. Michalczyk, S., Jung, D., Nadj, M., Knierim, M. T., & Rissler, R. (2019). BrownieR: the R package for neuro information systems research. In F. D. Davis, R. Riedl, J. vom Brocke, P. M. Léger, & A. B. Randolph (Eds.), *Information systems and neuroscience. lecture notes in information systems and organisation* (pp. 101–109). Springer International Publishing.
6. Baker, M. (2016). Is there a reproducibility crisis? In *nature, Vol. 533*.
7. Toelch, U. & Ostwald, D. (2018). Digital open science—Teaching digital tools for reproducible and transparent research. *PLoS Biology*, *16*(7), e2006022.
8. van der Aalst, W., Bichler, M., & Heinzl, A. (2016). Open research in business and information systems engineering. *Business & Information Systems Engineering*, *58*(6), 375–379.
9. Farwell, L. A., & Donchin, E. (1988). Talkiing off the top of your head: toward a mental prosthesis utilizing event-related brain potentials. *Electroencephalography and Clinical Neurophysiology*, *70*(6), 510–523.
10. Donchin, E., Spencer, K. M., & Wijesinghe, R. (2000). The mental prosthesis: Assessing the speed of a P300-based brain-computer interface. *IEEE Transactions on Rehabilitation Engineering*, 8(2), 174–179.
11. Léger, P. M., Davis, F. D., Cronan, T. P., & Perret, J. (2014). Neurophysiological correlates of cognitive absorption in an enactive training context. *Computers in Human Behavior, 34*, 273–283.
12. Conrad, C. D., & Bliemel, M. (2016). Psychophysiological measures of cognitive absorption and cognitive load in e-learning applications. In *ICIS 2016 Proceedings: 37th International Conference on Information Systems*, December 11–14, 2016, Dublin, Ireland.
13. Conrad, C., & Newman, A. (2019). Measuring the impact of mind wandering in real time using an auditory evoked potential. In F. D. Davis, R. Riedl, J. vom Brocke, P. M. Léger, & A. B. Randolph (Eds.), *Information systems and neuroscience. lecture notes in information systems and organisation* (pp. 37–45). Springer International Publishing.
14. Gwizdka, J. (2019). Exploring eye-tracking data for detection of mind-wandering on web tasks. In F. D. Davis, R. Riedl, J. vom Brocke, P. M. Léger, & A. B. Randolph (Eds.), *Information systems and neuroscience. lecture notes in information systems and organisation* (pp. 47–55). Springer International Publishing.
15. Anaconda, Inc. *Anaconda destruction: The world's most popular python/r data science platform*. Retrieved from https://www.anaconda.com/distribution/.
16. Perez, F., & Granger, B. *Project jupyter: Computational narratives as the engine of collaborative data science*. Retrieved from http://archive.ipython.org.
17. Hunter, J. D. (2007). Matplotlib: A 2D graphics environment. *Computing in Science & Engineering*, *9*(3), 90–95.
18. Van Der Walt, S., Colbert, S. C., & Varoquaux, G. (2011). The NumPy array: A structure for efficient numerical computation. *Computing in Science & Engineering*, *13*(2).
19. McKinney, W. (2011). Pandas: A foundational Python library for data analysis and statistics. *Python for High Performance and Scientific Computing*, *14*.

20. Peirce, J. W. (2007). Psychopy—Psychophysics software in Python. *Journal of Neuroscience Methods*, *162*(1–2), 8–13.
21. Gramfort, A., Luessi, M., Larson, E., Engemann, D. A., Strohmeier, D., Brodbeck, C., et al. (2014). MNE software for processing MEG and EEG data. *Neuroimage, 86,* 446–460.
22. Pedregosa, F., Varoquaux, G., Gramfort, A., Michel, V., Thirion, B., Grisel, O., et al. (2011). Scikit-learn: Machine learning in Python. *Journal of Machien Learning Research, 12,* 2825–2830.
23. Delorme, A., & Makeig, S. (2004). EEGLAB: An open source toolbox for analysis of single-trial EEG dynamics including independent component analysis. *Journal of Neuroscience Methods, 134*(1), 9–21.
24. Luck, S. (2014). *An introduction to the event-related potential technique* (2th ed.). MIT Press.
25. Newman, A. (2019). *Research methods for cognitive neuroscience*. SAGE Publications.

Brand Visual Eclipse (BVE): When the Brand Fixation Spent is Minimal in Relation to the Celebrity

Wajid H. Rizvi

Abstract This study investigates overshadowing effect of a celebrity. A term brand visual eclipse (BVE) is coined, the BVE occurs when brand fixation spent is minimal of the celebrity fixation spent. High BVE occurs when brand receive less than or equal to twenty percent fixation spent of the celebrity; moderate BVE occurs when the brand receives more than twenty percent and less than or equal to eighty percent fixation spent of the celebrity fixation spent and low BVE occurs when brand receive more than eighty percent fixation spent of the celebrity fixation spent. Two product categories were analyzed (Detergent Bar Brand and Mobile Brand), based on eye tracking data, the results suggest that high eclipse was observed in both categories. The BVE can be used as an advertising effectiveness indicator based on biometric data. The data was collected (n = 30) using Tobii-X30 Hz and was analyzed using iMotions biometric platform.

Keywords Celebrity-brand attention · Brand visual eclipse · Eye tracking · Celebrity endorsement

1 Introduction

Celebrity endorsement is considered as one of the most essential factors within marketing communication [1, 2]. The celebrity endorsement has a positive impact on advertising attitude and purchase intentions and it is also associated with higher recall [3, 4]. Generally, it is presumed that likeability of a celebrity would transfer towards a brand [5]. However, that is not always the case, the positive relationship between celebrity endorsement and brand is questioned and it is suggested that selection of celebrity should account for other factors such as brand and celebrity attachment [6, 7].

It is argued that sometimes the celebrity endorsement can be problematic for a brand and the celebrity can eclipse the brand [8]. Thus it can adversely influence

W. H. Rizvi (✉)
Institute of Business Administration, Karachi, Pakistan
e-mail: wrizvi@iba.edu.pk

F. D. Davis et al. (eds.), *Information Systems and Neuroscience*,
Lecture Notes in Information Systems and Organisation 32,
https://doi.org/10.1007/978-3-030-28144-1_33

effectiveness of an advertisement. To justify advertising outflows, it is of a prime importance to specify advertising effectiveness and substantiate it with data [9]. According to Ilicic and Webster [8] celebrity can overshadow brand, they termed it as celebrity eclipsing, higher celebrity attention would result in higher eclipsing and simultaneous attention on both celebrity and brand would result in lower visual eclipsing. To measure effectiveness of advertising and to delineate the celebrity effect is very challenging specially when it is measured through conventional self-reported methods. The celebrity endorsement attracts lots of attention so it is essential to measure respondents' attention through non-conventional methods like eye-tracking to specify level of attention either celebrity or brand [6, 10].

To assess advertising effectiveness, it is essential to measure relative attention towards brand and celebrity using eye-tracking. To isolate celebrity effect using eye-tracking method is scant and it is gaining attention of both academics and practitioners [11]. In this study eye-tracking technology (Tobii X-30 eye tracker and iMotions biometric platform) is used to assess respondents' attention (i.e. time spent fixation) towards celebrity and brand.

In this study a new term "brand visual eclipse" is coined: The Brand Visual Eclipse (BVE). The BVE occurs when a brand receives minimal attention (i.e. fixation time spent) in relation to a celebrity. When a brand receives 20% or less attention of the celebrity attention, it would be called a high visual eclipse. A moderate visual eclipse occurs when the brand receives greater than 20% of the celebrity attention but less than or equal to 80%. No visual eclipse occurs when brand receives greater than 80% of the celebrity attention.

Focus of this paper is to detect the BVE in advertising of two different product categories (Detergent Bar and Mobile). The BVE can be considered as an advertising effectiveness indicator. Since both brands are well-known in the market so to hide their identity no heat maps will be published in this study and the brands are identified as detergent bar brand (DBB) and Mobile brand (MB).

2 Research Questions and Hypotheses

Millions of dollars are spent by companies on celebrity endorsement to gain quick attention and the positive association (of celebrity likeability) towards the brand [5]. There is a need to take a guarded decision regarding selection of celebrity endorsement as it can be problematic specifically when the celebrity overshadows the brand [8]. To detect the BVE two moving area of interest (AOI) were specified from TV commercials (TVCs) for both product categories in which both the brand and celebrity are framed.

This study deals with two major questions:

Do Celebrities gain greater attention than brands (i.e. BVE)?
Does the BVE exist across FMCG (fast moving consumer goods) and non-FMCG products?

So within the defined AOIs we hypothesize that

H1a Celebrity will get higher Fixation Spent than the brand (Detergent Bar—Brand)
H1b Celebrity will get higher Fixation Spent than the brand (Mobile—Brand)

3 Method

In the eye tracking study 30 respondents (n = 30) took part 15 males and 15 females average age of the participants was 26. The study was implemented on iMotions platform and Tobii X-30 Hz was used. The respondents were briefed about the study and their right of data protection was also briefed afterwards they signed a consent form to participate in the study.

In the study participants watched seven video advertisement, out of seven five were filler and two were target advertisement. The target advertisement include a detergent bar brand and a mobile brand, because of the anonymity the brand names are not disclosed. The sequence of the advertisements was randomized for each respondent. Within each targeted advertisement two moving AOIs were created that showing both the celebrity and the brand. The moving AOIs were created using iMotions biometric platform. Further data was analyzed using SPSS.

Duration of both moving AOIs celebrity and brand in the DBB was same (2302 ms). Similarly, duration of both AOIs the celebrity and the brand in the MB was same (2381 ms). Since size of the AOIs of the celebrity and the brand can have an implication so only most visited part of the celebrity (i.e. face) was selected to maintain the equivalence. In the DBB size of both AOIs were roughly same. Whereas in the MB category the celebrity AOIs were relatively larger in size as two celebrities were involved. For better results the duration and size equivalence of the moving AOIs are warranted for within and between category analysis. To measure the BVE Time Spent Fixation (TSF) both moving AOIs (Brand and Celebrity) was used. The TSF data is in milliseconds (duration). In addition to the moving AOIs two static heat maps were generated to observe the relative fixation of brand to celebrity.

The celebrity endorsement is very essential to seek consumer attention and it is presumed that likeability and attachment of a celebrity would translate into a brand. If a celebrity receives attention then it is not necessarily a bad thing, however, it can be problematic when most of the attention is on the celebrity. Moreover, both top of mind recall and brand association are essential factors that are reinforced by advertising and are considered as key advertising effectiveness indicators and both require sufficient attention to form the association. So, it is proposed that brand attention time spent should not be less than 20% of the celebrity time spent. For example, if total spent on celebrity is 2000 ms then less than 400 ms would be very less, thus it will be called a high BVE condition and it may not create a positive association between the brand and the celebrity. The desirable outcome would be when the brand receives at least 80% of the celebrity time spent (1600 ms) and it will be called low BVE. In this case still the celebrity receives higher time spent but the brand receives sufficient time to generate the desired association between the two.

Since heat-maps of the advertisements are not provided because of the anonymity. So, it is important to provide a brief description of the advertisements. It will help to interpret the results and to figure out extent of the eclipsing effect. In the DBB subpart of the advertisement was analyzed in which exposure time of the brand and celebrity were equal, the celebrity was female and she was actively interacting with the brand (holding it and pointing towards it). Similarly, in the MB subpart of the advertisement was analyzed in which exposure time of the brand and the celebrities were equal, the both male and female celebrities were shown. There was active interaction with the brand as they were tacking a selfie.

3.1 Measures

The measures within the defined AOIs were calculated through iMotions software. The software provides wide variety of measures like, time spent fixation (TSF), FC (Fixation count), TTFF (time taken first fixation) and etc.

Since the BVE reflects time spent fixation towards brand in relation to celebrity. So it is measured in relation to celebrity fixation time rather than duration of the moving AOI. To calculate the BVE we need to observe how many milliseconds (ms) time spent the brand has received within the defined moving AOI then we need to calculate what percent that number is of the celebrity fixation time spent. For example, to calculate the BVE in terms of time spent fixation, assuming have aggregated fixation for brand is 190 ms and celebrity aggregated fixation is 1100 ms then by calculating 190/1100 * 100 we get 17.27% of the celebrity fixation.

It is proposed of a brand receive less than or equal to 20% (<=20%) fixation spent of the celebrity fixation then it will be called a high BVE of a brand receive greater than 20% and less than or equal to 80% ($>$ 20% or $<=$ 80%) then it will be called a moderate BVE. If brand receive greater than 80% (>80%) fixation of the celebrity fixation, then it will be called a low BVE, it may reach to 100% when the fixation towards the celebrity and the brand is exactly the same. In low BVE condition, say brand receives greater than 80% of the celebrity fixation. In this case the celebrity still receives higher fixation but the brand receives fair amount of time which reinforce plausible likeable association of the celebrity towards the brand. The BVE can be calculated other way round when a brand gains greater fixation than the celebrity which is highly unlikely.

4 Results

In this study two TV commercials (TVCs) were analyzed, so major part of analysis is based on moving AOIs of both brands in relation to the celebrities. Two static frames were also analyzed to assess aggregated indices.

4.1 *Static Heat Maps*

The results of static heat maps indicate that the celebrity overshadowed the brand across the category. For instance, in the DBB, the time taken first fixation (TTFF) for the celebrity was 0.6 s whereas for the brand it was 2.0 s. Time spent on the celebrity was 1.0 s whereas for the brand it was 0.1 s. In terms of the respondent ratio, 28 out of 30 respondents paid attention to the celebrity, whereas only 7 out of 30 paid attention to brand. In total there were 100 fixations on the celebrity whereas there were only 12 fixations on the brand. Similarly, in MB TTFF for female celebrity was 0.8 s, for male it was 0.8 s, whereas for brand it was 1.4 s and for logo it was 1.5 s. Time spent on Female was 0.3 s, on male it was 0.2 s, whereas on brand and logo it was 0.0 s. The respondent ratio for female was 18 (out of 30) for male it was 20, whereas for brand it was just 3 and for logo it was 0. The fixations showed the same trend, female fixation was 41, for male it was 34, whereas for brand it was 6 and for logo it was 0. The trend in both brand categories suggests that the celebrity overshadowed the brand.

4.2 *Moving AOIs*

Graph 1 shows mean time spent fixation on celebrity and on brand for both categories. In DBB the Celebrity time spent fixation is 1177 (ms) whereas on brand is only 173.8 (ms). In MB the celebrity time spent fixation is 735.6 (ms) whereas on brand is only 71.5 (ms).

Graph 2 shows mean time spent fixation on celebrity and on brand for both categories in terms of gender. In DBB the Celebrity TSF of male respondents is 1395 (ms) and female fixation is 959.9 (ms), whereas Brand TSF of male is only 43.8 (ms) and female fixation is 303.9 (ms). In MB the Celebrity TSF of male respondents is 640.9 (ms) and female fixation is 830.3 (ms), whereas Brand TSF of male is only 37.73 (ms) and female fixation is 105.3 (ms).

Table 1 highlights BVE across category. Based on the guidelines to measure the BVE suggested in the measure section. In the case of DBB, the fixation spent of the brand of the celebrity is 14.76% (173.8/1177 * 100), hence there is a high eclipse as it is less than 20%. Similarly, in the case of MB the relative fixation is 9.71% (71.5/735.6 * 100), hence there is a high eclipse. The results of pair t test also affirm

Graph 1 Celebrity-brand TSF (ms)

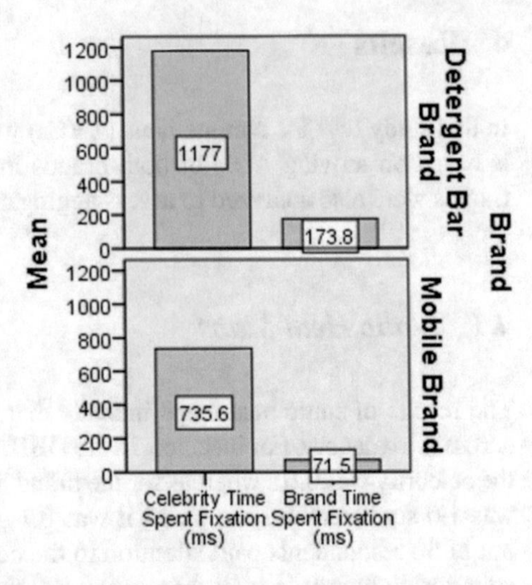

Graph 2 Gender-wise celebrity-brand TSF (ms)

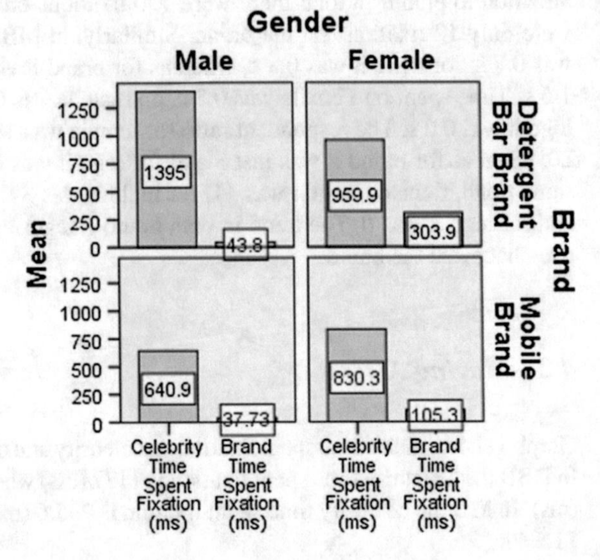

Table 1 Brand visual eclipse across category

Product-brand	Mean TSF (ms)		Brand % of celebrity TSF (%)	(BVE)	Pair t test	Hypothesis affirmation
DBB	Celebrity 1177	Brand 173.8	14.76	High BVE	t (29) = 9.28, p < 0.001	H1a affirmed
MB	Celebrity 735.6	Brand 71.5	9.71	High BVE	t (29) = 6.31, p < 0.001	H1b affirmed

Table 2 Brand visual eclipse across category gender-wise

Product-brand	Gender	Mean TSF (ms)		Brand % of celebrity TSF (%)	Brand visual eclipse (BVE)
DBB	Male	Celebrity 1395	Brand 43.8	3.13	High BVE
DBB	Female	Celebrity 959.9	Brand 303.3	31.59	Moderate BVE
MB	Male	Celebrity 640.9	Brand 37.73	5.88	High BVE
MB	Female	Celebrity 830.3	Brand 105.3	12.68	High BVE

that there is statistically significant difference between celebrity and brand fixation across the categories.

Table 2 highlights BVE across category in terms of gender. The results suggest that there is the brand visual eclipse across category. In the DBB male respondents showed high BVE whereas female respondents showed moderate BVE. Whereas, in the MB both male and female respondents showed high BVE.

5 Conclusion

Main focus of this study was to detect the brand visual eclipse (when celebrity eclipses the brand) as in indicator of advertising effectiveness. Use of the celebrity endorsement is widespread in marketing communication apart from its advantages this study indicates possible disadvantages of the celebrity endorsement. The results sought in this study based on moving AOIs suggest that there is high brand visual eclipse in both brand categories (i.e. DBB and MB) that means the celebrity completely overshadows the brand. However, when the results were further extracted in terms of gender in both brand categories there was a moderate brand visual eclipse in the female sample, whereas there was high brand visual eclipse in the male sample. The moderate visual eclipse also suggests stronger overshadowing of the brand. Low visual eclipse is desirable when brand fixation spent is at least greater than 80% of the celebrity fixation.

Within the advertising literature there is evidence to suggest consumers' involvement may vary due to different types of product. For instance, FMCG is considered as low involvement product whereas non-FMCG products such as mobile is considered high involvement product. So, it was essential to isolate if there is difference between the two in terms of visual attention towards a brand with respect to celebrity? The results of pair t-test also indicate that was statistically significant difference between the celebrity fixation and the brand fixation and high BVE was observed across both categories.

From practitioners' perspective it has always been challenge to delineate the celebrity effect on brand. This study provides insights for practitioners to delineate the celebrity effect by detecting the brand visual eclipse based on eye tracking data. The brand visual eclipsing can be used as advertising effectiveness indicator using the eye tracking data.

5.1 Limitation and Future Directions

The results published in this paper is a small portion of large study, so many factor were not isolated. For instance, to further isolate the celebrity eclipsing factor, non-celebrity AOIs were not compared. The moving AOIs of the both categories were different, so it was not possible to control other factors like background, size, color and overall making of the advertisements. This study only isolates relative visual attention on the brand with respect to the celebrity. However, it does not necessarily equate with the purchase decision or with level of brand loyalty. The visual attention can be a necessary condition but not a sufficient condition for the purchase decision. So it is essential to further look into other factors such as brand attachment and celebrity attachment. Further post study interview can be employed to substantiate the results in terms of top of mind recall and purchase decision. The strength of the association between the celebrity and the brand in low, moderate and high BVE cannot be affirmed because post study survey was not employed to assess the relationship. In future it will be interesting to find optimal threshold of the brand attention with respect to the celebrity to have strong or positive association between the two.

References

1. Zafer Erdogan, B. (1999). Celebrity endorsement: A literature review. *Journal of Marketing Management, 15*(4), 291–314.
2. Kaikati, J. G. (1987). Celebrity advertising: A review and synthesis. *International Journal of Advertising, 6*(2), 93–105.
3. Petty, R. E., Cacioppo, J., & Schumann, D. (1983). Central and peripheral routes to advertising effectiveness: The moderating role of involvement. *Journal of Consumer Research, 10,* 135–146.
4. O'Mahoney, S., & Meenaghan, T. (1998). The impact of celebrity endorsements on consumers. *Irish Marketing Review, 10*(2), 15–24.
5. Walker, M., Langmeyer, L., & Langmeyer, D. (1992). Celebrity endorsers: Do you get what you pay for? *Journal of Services Marketing, 6,* 35–42.
6. Keel, A., & Nataraajan, R. (2012). Celebrity endorsements and beyond: New avenues for celebrity branding. *Psychology & Marketing, 29,* 690–703.
7. Thomson, M. (2006). Human brands: Investigating antecedents to consumers' strong attachments to celebrities. *Journal of Marketing, 70,* 104–119.
8. Ilicic, J., & Webster, C. M. (2014). Eclipsing: When celebrities overshadow the brand. *Psychology & Marketing, 31*(11), 1040–1050.

9. Mehta, A. (1994). How advertising response celebritying (ARM) can increase ad effectiveness. *Journal of Advertising Research, 34,* 62–74.
10. Pieters, R., Wedel, M., & Batra, R. (2010). The stopping power of advertising: Measures and effects of visual complexity. *Journal of Marketing, 74*(5), 48–60.
11. Pileliene, L., & Grigaliunaite, V. (2017). The effect of female celebrity spokesperson in FMCG advertising: Neuro-marketing approach. *Journal of Consumer Marketing, 34*(3), 202–213.

The Impact of Associative Coloring and Representational Formats on Decision-Making: An Eye-Tracking Study

Djordje Djurica, Jan Mendling and Kathrin Figl

Abstract This research-in-progress paper presents our ongoing work on the effects of representational options of decision models on decision performance. With the help of eye-tracking, we want to investigate the influence of the color design of textual, graphical and tabular decision models on decision accuracy, efficiency, and cognitive load. For this purpose, we design a controlled experiment to test whether associative color-highlighting (red for negative decision outcome, green for a positive result) makes decision models easier to understand than monochromatic or non-associative color-highlighting.

Keywords Eye-tracking · DMN · Decision trees · Decision modeling

1 Introduction

Every day, managers make thousands of decisions. Some of those are important for achieving relevant business goals. While making decisions, they also make cognitive errors due to a number of reasons, including a high mental effort that some decision-making tasks require. A strategy to reduce cognitive effort is to use a suitable visual representation of the decision, which also engages our associative system in a meaningful way [1]. In this paper, we argue that color can be used to help people reduce the number of errors while making decisions since the color is one of the most cognitively effective visual variables [2]. The human visual system is highly sensitive to variations in color and can quickly and accurately distinguish between them [3, 4].

D. Djurica (✉) · J. Mendling
Vienna University of Economics and Business (WU Wien), Vienna, Austria
e-mail: djordje.djurica@wu.ac.at

J. Mendling
e-mail: jan.mendling@wu.ac.at

K. Figl
University of Innsbruck, Innsbruck, Austria
e-mail: kathrin.figl@uibk.ac.at

F. D. Davis et al. (eds.), *Information Systems and Neuroscience*,
Lecture Notes in Information Systems and Organisation 32,
https://doi.org/10.1007/978-3-030-28144-1_34

Previous research has compared different types of representational formats with the ambition to identify those that are better suited for a certain type of task. On the one hand, Huysmans et al. [5] present evidence emphasizing the benefits of decision tables over binary decision trees and textual formats in terms of comprehension accuracy and efficiency. On the other hand, Boritz et al. [6] conducted an experiment in which textual representations resulted in higher comprehension efficiency. A recent overview of findings in favor and against certain representations points to the potential of rather investigating in which way representations can be designed to facilitate accurate and efficient cognitive processing, e.g. by the help of secondary notation manipulations [7].

In this paper, we investigate whether decision trees, decision tables or textual representations are better suited for certain decision-making tasks. Using a theoretical foundation grounded in the feature integration theory, dual processing, cognitive fit, cognitive load theory and works by Tversky and Kahneman [8] and Kahneman [9], we analyze conditions upon which color highlighting of the representational formats can be beneficial. We utilize eye-tracking to closely monitor the visual perception of the mentioned representations. The prospective contribution of this experiment will be empirical evidence for or against the benefits of color highlighting and of different representational formats for decision-making tasks.

This paper is a work-in-progress paper and is organized as follows. The next section introduces representational formats and an example of a color-highlighted decision model. After that, we summarize prior research upon which we derive our hypotheses. Finally, we present an experimental design of our research.

2 Theoretical Background

2.1 A Decision Model Example

For modeling decision rules there are three typical representational formats—decision tables, decision trees, and text, which are presented in the below. The Object Management Group [10] recently introduced the standard Decision Model and Notation (DMN). It has been designed to represent operational decisions that are frequently taken and repetitive in nature [11]. DMN provides three parts for the specification of business rules with one of them being decision tables—our focus in the following.

Table 1 is a decision table, adapted from Huysmans et al. [5], extended with associative color highlighting. It shows the rules for granting a scholarship. Input columns are *income, age*, and *GPA*, while the output column is *outcome [of the application]*.

Decision trees are another representational format often used to present decision rules. In our experiment, we will use binary decision trees. An example of such a tree derived from our decision table example is depicted in Fig. 1.

Table 1 Example of the DMN decision table with the associative color highlighting

U, C	Netto Income	Age	GPA	Outcome
A	<700	<25	<1.5	Approved
B	<700	<25	>=1.5	Rejected
C	<700	>=25	–	Rejected
D	>=700	–	–	Rejected

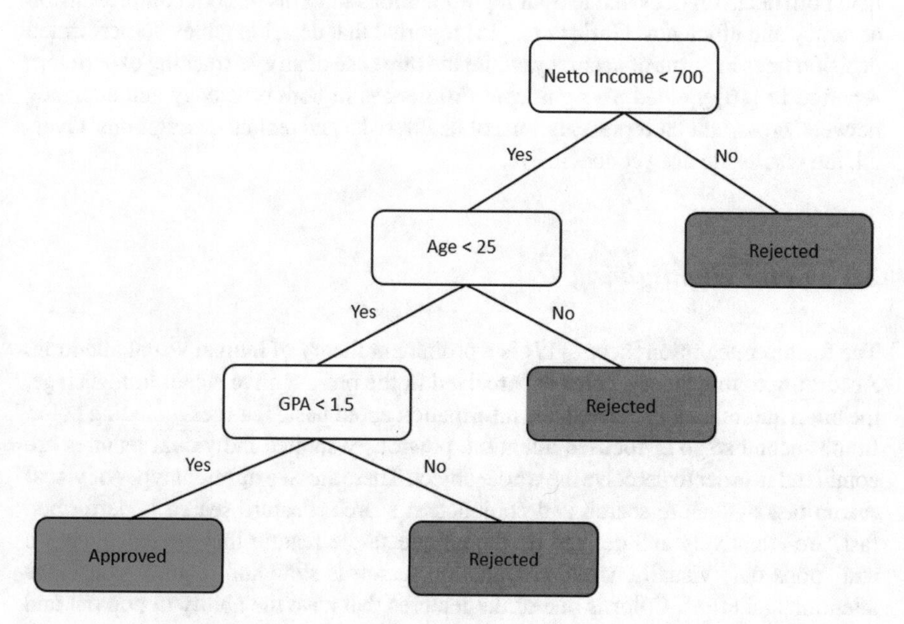

Fig. 1 Example of the decision tree with the associative color highlighting

The most common type of textual representation formats is propositional if-then rules. Figure 2 gives an example of mutually exclusive rules and is based on previously discussed examples from Table 1 and Fig. 1.

IF (NETTO INCOME < 700 and AGE < 25 and GPA < 1.5) THEN APPROVED
IF (NETTO INCOME < 700 and AGE < 25 and GPA >= 1.5) THEN REJECTED
IF (NETTO INCOME < 700 and AGE >= 25) THEN REJECTED
IF (NETTO INCOME >= 700) THEN REJECTED

Fig. 2 Example of the if-then rules with the associative color highlighting

2.2 Representational Formats

From the late 20th century on, researchers started to compare different representational formats for decision rules in terms of comprehension accuracy and efficiency. The experiments conducted during the late 80s and early 90s of the last century show evidence that decision tables are inferior in terms of accuracy when compared to decision trees and structured text [12, 13]. When it comes to comprehension efficiency, results were similar and pointed towards decision tables being less efficient than decision trees [14]. A more recent research stream contradicts the previous one, as e.g. Huysmans et al. [5] provide evidence that decision tables appear to be better than both decision trees and textual representations in terms of both comprehension accuracy and efficiency. Gorla et al. [15] reported that decision tables outperformed decision trees in terms of accuracy, while the rare case of an eye-tracking experiment reported in [16] yielded no significant differences in both efficiency and accuracy between cross-tabular representations of health risks and textual descriptions. Overall, the results are not yet conclusive.

2.3 Color Highlighting

The feature integration theory [17] is a prominent theory of human visual attention. According to this theory, color is perceived in the pre-attentive stage. In this stage, the brain automatically collects the information about basic features including color. In the second stage of focused attention, previously spotted individual features are combined in order to perceive the whole object. There are two different types of visual search tasks—feature search and conjunction search. Feature search is performed fast, pre-attentively and defined by the unique single feature that attracts attention and "pops out" visually, while conjunction search is slow and requires conscious attention and effort. Color is one of the features that have the ability to pop out and therefore can be identified during feature search.

Color conventions are also important in this context, i.e. certain colors are associated with certain meanings, as e.g. colors in traffic-lights with "stop" and "go". According to Mackinlay [3], color is a prominent cognitively effective variable and humans are able to quickly distinguish it. Colors are also remembered easier than symbols and they are detected three times faster than shapes [18, 19].

Despite the fact that colors can have meanings, and our ability to recognize them quickly, they are rarely used in software engineering notations and are prohibited in UML [2]. Therefore, previous research in color highlighting as a secondary notation manipulation yielded inconclusive results. Research by Reijers et al. [20] and Te'eni [21] suggests that color highlighting can reduce visual search in a diagram as it leads to a reduction in cognitive load, which translates into better comprehension accuracy. However, Kummer et al. [22] and Petrusel et al. [23] do not observe a direct influence of colors on comprehension accuracy.

2.4 Research Framework and Hypotheses Development

Based on the theoretical assumptions and the previous research described above, we will now discuss the anticipated effects of the decision model representations and the use of color highlighting on the decision accuracy, efficiency, and cognitive load. We summarize our expectations in the research model shown in Fig. 3.

First, we build our hypothesis upon cognitive fit theory [24]. *The cognitive fit theory* postulates that a fit between the task type and the information emphasized in the visual representation leads to more effective and efficient problem solving. Such a fit leads to the formulation of a consistent mental representation, without the need to transform or align the mental representation to that of the problem. Vessey [24] postulates that for tasks that are symbolic in nature, symbolic (tabular) representation provide a better cognitive fit. Overall, prior empirical results on which representational format is better suited for decision-making tasks are contradictory and inconclusive. While first studies pointed towards decision trees being better in terms of accuracy and decision making [12–14], more recent research works argue that decision tables are better [5, 15]. In line with recent research findings and the cognitive fit theory, we hypothesize that:

H1: Decision-making tasks will be solved with higher accuracy, efficiency, and lower cognitive load when using decision tables, than when using decision trees and textual representations.

Next, we turn to the effects of associative color highlighting (for decision outcomes). We opted for coloring output columns since we believe that the coloring of the input columns would decrease salience and possibly neutralize the ability of the color to "pop-out". We concentrate on the color red, which is rather associated with a negative decision result, and green, which is associated with a positive result. Earlier research in the field of financial losses and profits underpins this color association [25]. In addition, the color red can indicate danger and, for example, cause people to increase their vaccination intentions if they are primed with it [26]. Therefore, we hypothesize:

H2: Decision-making tasks will be solved with higher accuracy, efficiency, and lower cognitive load when using associative color-highlighting (red and green for decision outcomes) in decision representations.

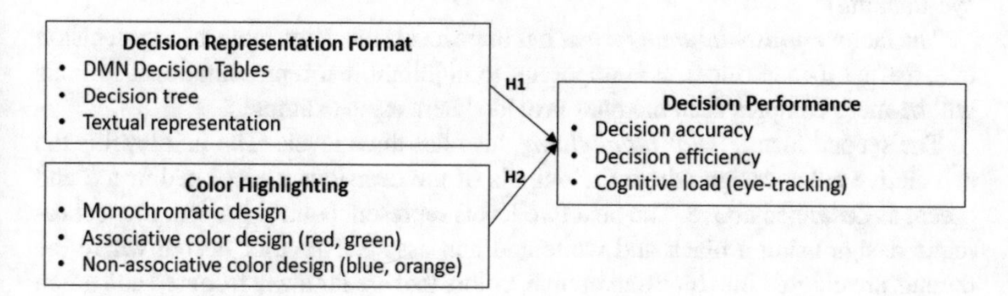

Fig. 3 Research framework

Finally, we focus on the possible combination of color and representational formats and their impact on the decision-making process. We build our hypothesis upon the knowledge about different types of attention and implicit processing of the colors that carry associative meaning. First, we consider visual spatial attention that allows people to process visual inputs selectively and prioritize specific area within the visual field [27]. In the context of our research, this would mean that participants will focus their attention on different types of business decision representations. Next, thanks to the introduction of color, their feature-based attention will take over, allowing them to focus on the particular feature property in the visual field, independent to the spatial attention [28]. In the end, the introduction of colors that carry associative meaning to people such as red and green, will trigger System 1 and allow participants to process the information faster. However, we do not expect all representational formats to benefit equally from the introduction of associative colors. According to MacEachren [29], space is perceptually dominant and allows better discrimination of values and picking out patterns. Therefore, we can say that decision trees already have an advantage over other representational formats due to their large space between decision outputs. For that reason, we think that decision trees might not benefit as much from color coding as text and decision tables could, where different decision outputs are visually close to each other. Hence, we hypothesize that:

H3: When decision tables are used, the positive effect of color highlighting on the accuracy, efficiency, and the cognitive load is stronger than in any other combination of representational format and color highlighting.

3 Method

In order to empirically test our hypotheses, we plan to conduct an experimental study in which we will utilize eye-tracking. While conducting the experiment, it is important to maintain control over potentially confounding external factors. To accomplish this, we will use a within-subjects study design. Our laboratory design features 2 within-subject factors (representational format and color highlighting) and 3 dependent variables (accuracy, efficiency, and cognitive load measured using eye-tracking).

The factor *representational format* has three levels (decision table, binary decision tree, textual if-then rules). It is important to highlight that representational formats will be more complex than the ones provided here as an example.

The second factor, *color highlighting*, also has three levels. The first level is the associative color design where the outputs of the decisions are colored in red and green, as described above. The next two levels represent control groups: monochromatic design using a black and white and non-associative color design where the outputs are colored into blue and orange, colors that are far away from red and green on the color wheel. We believe that the meaning of colors across cultures will not affect our results, as the colors we choose are perceived uniformly across cultures [30].

The first dependent variable is *decision accuracy*. We intend to ask participants to complete a number of decision tasks. Here we provide an example of a decision task based on the decision representations in Table 1, Figs. 1 or 2: "*Andrea is born in 1996. Her GPA is less than 1.5. She is also employed and earns €800, brutto, with a tax of 20%. Will she get the scholarship?*". We refrain from using too simple tasks of low complexity because they might lead to a ceiling effect with high decision accuracy and efficiency and low variance regardless of which representational format was used.

The second dependent measure is the *decision efficiency* which is measured by the time taken for deciding from the moment the task is shown to the participant, to the moment where the participant submitted the answer.

Finally, the third dependent variable is the *cognitive load*. The cognitive load will be measured using eye-tracking metrics following the guidelines proposed in the paper by Zagermann et al. [31]. In our experiment, the values of the cognitive load will be obtained by measuring pupil dilation. We further intend to examine overall patterns of eye movement during decision making and the number and duration of fixations to shed light on how the different visual representations work.

The experiment will be structured as follows. At the beginning of the experiment, participants will be asked to solve two numerical problems. One task will be easy to solve and require little mental effort ("*Please calculate how much is 2 * 2*"), while another task will be more complicated, harder to solve, and will require larger mental effort ("*Please calculate how much is 16 * 131*"). The aim of this is to determine a baseline of for cognitive load of our participants, to which we can compare pupil dilation measures from main experimental tasks to draw conclusions on cognitive load. In the main part of the experiment, we ask participants to provide answers to a number of decision tasks. Here, participants will be shown all combinations of color highlighting and business decision representations coupled with scenario-like questions similar to the one presented above. Finally, the last part of the experiment will include the survey that will capture demographic and other important information about participants, such as their previous experience in modeling as control variables.

It is also important to emphasize that before the beginning of the experiment, participants will be tested for the red-green, blue-yellow, and total color blindness using the Ishihara test [32]. In case the test determines that a participant has red-green color blindness, the participant has to be excluded from the experiment.

4 Conclusion

As far as we know, no empirical evaluation of the impact of associative coloring on decision-making has so far been undertaken. Aiming to provide a contribution to the closure of this research gap, we have developed a research model that describes the effects of business decision representations and color highlighting. The expected results of our planned experiment constitute implications for both academia and practice. For academia, we intend to extend the current body of knowledge, providing further insight into the research of secondary notation manipulation and decision representations. For practice, our results might contribute to the improvement of decision representation design which can have a lasting impact on decision accuracy, efficiency, and cognitive load. In this regard, we expect the results of our experiment to provide important guidance on how to effectively use and design such business decision representations.

References

1. Sloman, S.A.: The empirical case for two systems of reasoning. *Psychological Bulletin* (1996).
2. Moody, D. (2009) The physics of notations: Toward a scientific basis for constructing visual notations in software engineering. *IEEE Transactions on Software Engineering*.
3. Mackinlay, J. (1986). Jock: Automating the design of graphical presentations of relational information. *ACM Transactions on Graphics, 5,* 110–141.
4. Winn, W. (1993). An account of how readers search for information in diagrams. *Contemporary Educational Psychology, 18,* 162–185.
5. Huysmans, J., Dejaeger, K., Mues, C., Vanthienen, J., & Baesens, B. (2011). An empirical evaluation of the comprehensibility of decision table, tree and rule based predictive models. *Decision Support Systems*.
6. Boritz, J. E., Borthick, A. F., & Presslee, A. (2012). The effect of business process representation type on assessment of business and control risks: Diagrams versus narratives. *Issues in Accounting Education, 27,* 895–915.
7. Ritchi, H., Jans, M. J., Mendling, J., & Reijers, H. (2019). The influence of business process representation on performance of different task types. *Journal of Information Systems*. https://doi.org/10.2308/isys-52385.
8. Tversky, A., & Kahneman, D.: Judgment under uncertainty: Heuristics and biases. *Science* (1974).
9. Kahneman, D. (2012). *Thinking, fast and slow*. London: Penguin Books.
10. Object Management Group. (2015). Decision Model and Notation v.1.0. 172.
11. Figl, K., Mendling, J., Tokdemir, G., & Vanthienen, J. (2018). What we know and what we do not know about DMN. *Enterprise Modelling and Information Systems Architectures (EMISAJ), 13,* 2:1–16-2:1–16.
12. Vessey, I., & Weber, R. (1986). Structured tools and conditional logic: An empirical investigation. *Communications of the ACM*.
13. Subramanian, G. H., Nosek, J., Rahunathan, S. P., & Kanitkar, S. S. (1992). A comparison of the decision table and tree. *Communications of the ACM*.
14. Gorla, N., Pu, H. C., & Rom, W. O. (1995). Evaluation of process tools in systems analysis. *Information and Software Technology*.
15. Gorla, N., Chiravuri, A., & Meso, P. (2013). Effect of personality type on structured tool comprehension performance. *Requirements Engineering*.
16. Smerecnik, C. M. R., Mesters, I., Kessels, L. T. E., Ruiter, R. A. C., De Vries, N. K., & De Vries, H. (2010). Understanding the positive effects of graphical risk information on comprehension: Measuring attention directed to written, tabular, and graphical risk information. *Risk Analysis: An International Journal* (2010).
17. Treisman, A. M., & Gelade, G. (1980). A feature-integration theory of attention. *Cognitive Psycholog, 12,* 97–136.
18. Lohse, G. L. (1993). A Cognitive model for understanding graphical perception. *Human-Computer Interaction, 8,* 353–388.

19. Treisman, A. (1982). Perceptual grouping and attention in visual search for features and for objects. *Journal of Experimental Psychology: Human Perception and Performance, 8,* 194–214.
20. Reijers, H. A., Freytag, T., Mendling, J., & Eckleder, A. (2011). Syntax highlighting in business process models. *Decision Support Systems, 51,* 339–349.
21. Te'eni, D. (2001). Dov: Review: A cognitive-affective model of organizational communication for designing IT. *MIS Quarterly, 25*, 251.
22. Kummer, T. F., Recker, J., & Mendling, J. (2016). Enhancing understandability of process models through cultural-dependent color adjustments. *Decision Support Systems.*
23. Petrusel, R., Mendling, J., & Reijers, H. A. (2016). Task-specific visual cues for improving process model understanding. *Information and Software Technology, 79,* 63–78.
24. Vessey, I. (1991). Cognitive fit: A theory-based analysis of the graphs versus tables literature. *Decision Sciences, 22,* 219–240.
25. Kliger, D., & Gilad, D. (2012). Red light, green light: Color priming in financial decisions. *The Journal of Socio-Economics, 41,* 738–745.
26. Gerend, M. A., & Sias, T. (2009). Message framing and color priming: How subtle threat cues affect persuasion. *Journal of Experimental Social Psychology, 45,* 999–1002.
27. Heuer, A., & Schubö, A. (2016). Feature-based and spatial attentional selection in visual working memory. *Memory & Cognition, 44,* 621–632.
28. Theeuwes, J. (2013). Feature-based attention: It is all bottom-up priming. *Philosophical Transactions of the Royal Society of London. Series B, Biological sciences, 368,* 20130055.
29. MacEachren, A. M. (1995). *How maps work: Representation, visualization, and design.* Guilford Press.
30. Berlin, B., & Kay, P. (1969). *Basic color terms.* Berkeley and Los Angeles: University of California Press.
31. Zagermann, J., Pfeil, U., & Reiterer, H. (2016). Measuring cognitive load using eye tracking technology in visual computing. In *Proceedings of the Sixth Workshop on Beyond Time and Errors on Novel Evaluation Methods for Visualization.* ACM.
32. Clark, J. H. (1924). The Ishihara test for color blindness. *American Journal of Physiological Optics.*

Impact of Physical Health and Exercise Activity on Online User Experience: Elderly People and High Risk for Diabetes

Harri Oinas-Kukkonen, Li Zhao, Heidi Enwald, Maija-Leena Huotari, Riikka Ahola, Timo Jämsä, Sirkka Keinänen-Kiukaanniemi, Juhani Leppäluoto and Karl-Heinz Herzig

Abstract This article studies how an individual's physical wellbeing contributes to one's online user experience. The study subjects were elderly people at high risk for type 2 diabetes. The results suggest that the web usage experience of these prediabetic individuals is related to their physical health status and level of physical activity. Those with a better physical health status were more likely to feel ease of orientation in their web usage, and those with more frequent regular physical activity were more likely to perceive pleasure in navigating the web. In practice, variation in physical health and activity levels between individuals could, and should. be addressed in designing systems and services. In more general, studying user experience on par with biochemical measurements provides an exciting combination of research methods and paves the way for new design practices.

Keywords User experience · Flow · Webflow · Physical health · Physical exercise · Type 2 diabetes

H. Oinas-Kukkonen (✉) · L. Zhao
Faculty of Information Technology and Electrical Engineering, Oulu Advanced Research on Service and Information Systems, University of Oulu, Oulu, Finland
e-mail: Harri.Oinas-Kukkonen@oulu.fi

H. Enwald · M.-L. Huotari
Faculty of Humanities, Information Studies, University of Oulu, Oulu, Finland

R. Ahola · T. Jämsä
Medical Research Center, Research Unit of Medical Imaging, Physics and Technology, Oulu University Hospital and University of Oulu, Oulu, Finland

S. Keinänen-Kiukaanniemi
Institute of Health Sciences, Oulu University Hospital and University of Oulu, Oulu, Finland

J. Leppäluoto · K.-H. Herzig
Medical Research Center, Institute of Biomedicine and Biocenter Oulu, Physiology, Oulu University Hospital and University of Oulu, Oulu, Finland

F. D. Davis et al. (eds.), *Information Systems and Neuroscience*, Lecture Notes in Information Systems and Organisation 32,
https://doi.org/10.1007/978-3-030-28144-1_35

1 Introduction

Globally, hundreds of millions of people suffer from diabetes [1–7]. In health promotion and lifestyle counseling, perceived personal relevance of health information may help engage individuals and create opportune conditions for influencing a user [8–12]. In other words, the aim is to produce individualized communication so that a user could think 'This applies to me' [9, p. 55]. Tailoring enhances cognitive conditions for human information processing and acceptance, and a typical aim of it is simply to increase attention and comprehension [10]. A previous study among individuals at high risk for diabetes showed that those whose physical health status was poor desired receiving tailored information on nutrition and physical activity more frequently than those with a better physical health status [13, 14]. This article investigates the relationship between physical health status, physical exercise, and online user engagement.

2 Background

In order for interventions to be genuinely influential, carefully designed and tailored behavior change support systems with the ultimate aim of engaging users in their daily lives and persuading them to adopt and maintain healthier behaviors are called for [15–21]. There is growing interest towards using the flow concept for understanding user experience in information systems' (IS) (see, e.g., [22–26]). Originally, flow was described as a holistic sensation that people feel when they act with total involvement [23]. More recently, flow has been defined as a state, which occurs when navigating in an information space and which is intrinsically enjoyable, self-reinforcing and accompanied by a loss of self-consciousness; it can exist in both experimental and goal-oriented types of behavior [17, 27]. The flow user experience can be made better through interactive relationships between a user's individual characteristics, the characteristics of the artifact, and the characteristics of the primary task [18]. Support for orientation and navigation capabilities have been suggested as key for positive user experience [28]. In addition, an immersive use of such services requires the user's focused attention [29].

Specifically, in the field of healthcare, continuous measuring of physical health status and physical activity adds to the rapid growth of health-related big data. Large data archives can provide information about an individual's health status, which may be objectively measured physiologically and biochemically. In many cases, these objective measurements provide more reliable information about health status than self-reported subjective data [30, 31]. Health behavior change support systems [15] can now be built on top of objective measurements; for example, objective physical activity measurements with accelerometers and pedometers can be used to determine patterns of physical activity behavior [32, 33]. A systematic review by Broekhuizen et al. [34], compare with Kroeze et al. [35] shows a noticeable increase in the amount of objectively measured physical activity in these types of interventions.

3 Research Model and Hypotheses

Online users' feeling of being fully focused and perception of ease of orientation and pleasure of navigation are key components of the Webflow model [21, 28]. These user experience constructs have been adopted for the study here in the context of web-based health applications. An individual's physical health status and physical exercise are then incorporated into in the research model, together with these Webflow components. (Table 1; Fig. 1).

User experience. Previous Webflow studies have found that perceived ease of orientation and perceived pleasure of navigation have a direct effect on gaining an optimal user experience, whereas, somewhat surprisingly, perceived ease of use and usefulness [cf. 36] were not found to have a direct influence on user experience [28]. If an individual feels that s/he can easily be oriented in web navigation, it is more likely for him/her to enjoy the web navigation. Otherwise, if s/he feels it is difficult to do so, that will negatively affect the perceived pleasure of navigation. Focused attention [17, 27] influences the ease of orientation and pleasure of navigation. In the contemporary information and socially-laden web, cognitive overload has become a greater challenge than ever before, resulting in attention focus becoming less and less of a commodity for users.

Physical health status. When an individual is in a bad health status, it will be difficult for him/her to focus his/her attention on a particular task, because his/her bad health status (such as the pain) may distract him/her from the task. Known risk factors for type 2 diabetes and cardiovascular diseases are obesity, high level of low-density lipoprotein (LDL) and low level of high-density lipoprotein (HDL)

Table 1 Research hypotheses

H1: An individual who feels ease of orientation in web usage is more likely to perceive pleasure in navigating the web
H2: An individual who is able to focus on his/her primary activity when using the web is more likely to feel ease of orientation in web usage
H3: An individual who is able to focus on his/her primary activity when using the web is more likely to perceive pleasure in navigating the web
H4: An individual with a poor physical health status is more likely to have difficulty focusing on his/her task when using web services
H5: An individual with a poor physical health status is less likely to feel ease of orientation in web services
H6: An individual who does physical exercise regularly is more likely to have a good physical health status
H7: An individual who does physical exercise regularly is more likely to perceive pleasure in navigating the web
H8: The positive relationship between focused attention and the perceived pleasure in web navigation is contingent upon how much physical exercise an individual regularly does, in such a way that this positive relationship will be stronger for the individual who does less physical exercise and weaker for the individual who regularly does more physical exercise

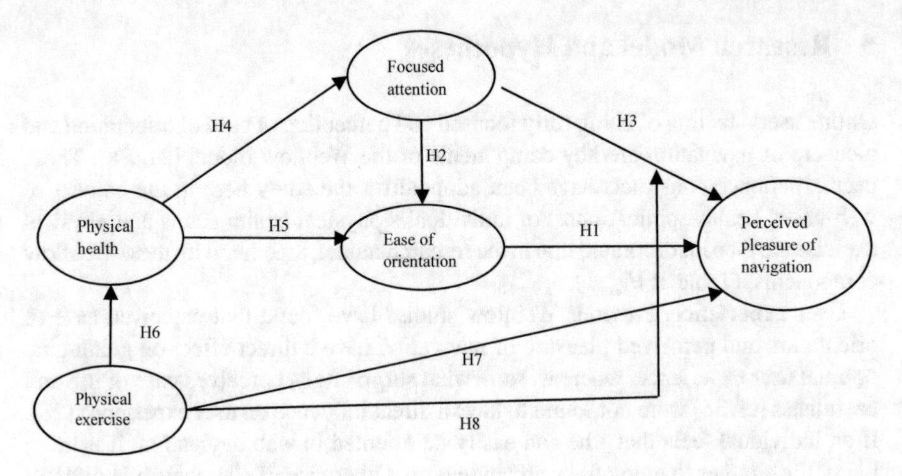

Fig. 1 Research model

cholesterol. Lipid markers are well-established predictors of vascular disease. The most frequently measured lipid variables are total cholesterol, HDL cholesterol, LDL cholesterol, and triglycerides. By following and preventing abnormalities in this lipid homeostasis, diseases such as heart diseases and type 2 diabetes can be prevented. In this study, the individuals' physical health status is indicated by the state of their lipid metabolism, technically by levels of total cholesterol and total triglycerides, as well total fatty acids measured from a blood sample.

Physical exercise activity level. It is commonly believed that regular physical exercise, such as brisk walking, peaceful swimming, performing fitness gymnastics, brisk cycling, ice skating, or skiing, is good for one's health. Prior studies have shown that lifestyle changes such as increased physical exercise and weight loss can reduce the risk of diabetes by approximately 58% [5, 37]. It has been found that an adequate level of regular moderate-intensity physical exercise reduces the risk of numerous chronic diseases, preserves health and functioning (both physical and mental) into old age, and extends longevity [e.g., 38]. Aerobic exercise of 60 min three times a week for 12 months can even improve maximal oxygen uptake and leg muscle strength, and decrease waist circumference, LDL cholesterol, and total cholesterol in premenopausal women [39]. Thus, a person who does physical exercise regularly is more likely to improve his/her physical health status. After doing such exercise people feel refreshed and revitalized. Thus, this individual is more likely to perceive pleasure in navigating the web. If a person exercises regularly, he or she is more likely to feel good, and consequently more likely to perceive pleasure in navigating the web, as noted above. In this case, the effect of focused attention becomes less critical in resulting in enjoying web navigation. In other words, the impact of focused attention on the perceived pleasure in web navigation becomes less salient. Therefore, the physical exercises moderate the impact of focused attention on a user's perception of web navigation (Table 1; Fig. 1).

4 Research Setting

The empirical study was conducted among individuals with a high risk for type 2 diabetes, who in volunteered to participate a physical activity intervention trial known as PreDiabEx carried out in Northern Finland [40]. Study subjects were recruited from outpatient diabetes clinics in the Oulu Deaconess Hospital and the City of Oulu. The survey was conducted at the same time when conducting the biochemical measurements.

The user experience data reported here were collected through a questionnaire survey among individuals with a high risk for type 2 diabetes. The data related to the flow user experience focus on recognizing and explaining the antecedents of a positive user experience. The questionnaire also assessed background information (age, gender, marital status, education, employment) and information about physical activity (specifically, self-reported frequency of moderate-intensity physical exercise). The blood tests for lipid markers were conducted at the same time. The responses for the questionnaires were collected on this occasion by nurse researchers, and when necessary, they also helped the participants fill in the questionnaires. Based on previous studies [21, 28], the first author designed the survey questions related to navigation, orientation, and focused attention.

Of the 72 original participants, 69 responded to the survey. However, 16 survey participants did not answer the questions about Perceived Ease of Orientation, Pleasure of Navigation and Focused Attention. Thus, the final sample size is 53, for a response rate of 73.6%. The average age of the respondents was 60 years. The majority (62%) of the respondents were 60 years or older, and 72% were women. In addition, 79% were married or living with a partner. Nearly half only had primary level education (45%), 30% had completed secondary education, and about 25% higher education. The desired level of total cholesterol for pre-diabetic individuals was considered to be less than 4.5 mmol/l and the desired level for total triglycerides was 2.0 mmol/l, whereas for total fatty acids there is no similar generally agreed threshold. The measured average serum cholesterol levels were 5.08 mmol/l, and total triglycerides and total fatty acids were 1.35 mmol/l and 10.79 mmol/l, respectively. Thus, the average for total cholesterol exceeded the desired level. The self-reported frequency (times/week) of moderate-intensity physical exercise per week was used as a measure of this variable.

5 Data Analysis and Results

Because the sample size for this study is small, partial least squares (PLS) was used to test the research model and hypotheses [41], given that PLS is less affected by small sample sizes [42]. PLS is a structural equation modeling technique that simultaneously evaluates the reliability and validity of the measures of theoretical constructs and tests the relationships among constructs [42]. In the research model, the dependent latent variable with the largest number of independent latent variables impacting it is "Perceived Pleasure of Navigation," which is impacted by three independent variables (Focused Attention, Ease of Orientation, and Physical Exercise). Therefore, the least sample size is 30 (equal to 10 times 3). In what follows, the PLS model is analyzed and interpreted in two stages: the assessment of the measurement model and the assessment of the structural model.

Measurement model PLS was used to calculate the composite scale reliability (CR) [42–44] and average variance extracted (AVE) [42, 43]. CR was used to assess the inter-item reliability by measuring the internal consistency of a given block of indicators [44]. The AVE was used to examine the convergent validity of the constructs, which attempted to measure the amount of variance that a latent variable component captured from its indicators relative to the amount due to measurement error. Cronbach's alphas exceeded 0.70. The lowest CR was 0.87, compellingly exceeding the recommended "0.70" threshold value [43]. The AVE of all measures was much higher than the cut-off value of 0.50 [43] with the lowest AVE of 0.69. These results demonstrate the inter-item reliability and convergent validity of the measures. Moreover, the AVE of each construct exceeds the intercorrelations of the construct with the other constructs in the model, in support of discriminant validity [43, 45]. Additionally, the discriminant validity can also be assessed by inspecting the cross-loadings, which are not substantial in magnitude compared with the loadings [42, 46, 47].

Structural model and hypothesis testing The explanatory power of the research model was evaluated by looking at the R^2 value in the dependent variable–perceived pleasure of navigation. The R^2 value indicates that the research model explained 37.2% of the variance for perceived pleasure of navigation.

Hypothesis 1 is significant, suggesting that an individual who feels ease of orientation in web services is more likely to perceive navigating web services as enjoyable. Hypothesis 2 is also significant, indicating that an individual who is able to focus on his/her primary task at hand when using web services is also more likely to feel ease of orientation in web services. Hypothesis 5 is significant, suggesting that an individual with a poor physical health status is less likely to feel ease of orientation in the web. In addition, Hypothesis 7 is supported, indicating that an individual who does physical exercises regularly is more likely to perceive navigating web services as enjoyable. The significance level for Hypothesis 3 is $p < 0.1$, which is close to the conventional significance level. Given that the sample size is small, this relationship has the potential to be significant if the sample size is larger.

Most hypotheses are supported by the data collected in this study. Although Hypothesis 4 and Hypothesis 8 are not supported, the direction of the effects is consistent as hypothesized. One plausible explanation for the two insignificant results lies in the small sample size used in this study. With respect to Hypothesis 6, a likely explanation for the insignificant result is as follows. It is widely believed that moderate-intensity physical exercise is good for health. In other words, if an individual does moderate-intensity physical exercise regularly, s/he would be healthier than would otherwise be. However, it will take a longer period of time for the influence of physical exercises on an individual's health status to show up.

Control variables Further analysis was carried out to assess the impacts of the control variables, in order to make sure the significant results were not due to covariation with these variables. The demographic information (gender, age, education, marital status, and employment) acts as control variables. After adding the control variables into the research model, the significance level for the relationships in the research model are similar as before (they only change very slightly). The significant paths are still significant and the insignificant paths remain insignificant. None of the control variables have a significant effect on the dependent variable. These results suggest that the research model is stable and independent of control variables.

6 Discussion

This study has many research implications. It is one of the first to explore in an interdisciplinary manner the potential to combine objective biochemical health measurements, self-reported physical activity, and perceptions of user experience. The findings that a user's focused attention facilitates feeling of ease of orientation in web usage and that the perceived ease of orientation influences a user's perception of navigating the web as more pleasurable, were expected and in line with previous findings [cf. 28].

However, the results, which relate to the role of physical health status and physical activity level for user experience perception, are novel. This study shows that the user experience of individuals with a high risk for type 2 diabetes differs according to their physical health status and amount of physical activity. Those with better lipid homeostasis (better physical health status) are more likely to feel ease of orientation in their web usage, and those are more frequent in their physical exercise are more likely to perceive pleasure in navigating the web. This relationship between health and orientation in the web as well as between physical activity and web navigation is an interesting finding, and there may be a natural explanation for it. After all, since a human is a psychosomatic whole, health and psychology are necessarily related, and navigating the web, at least ideally speaking, is similar to performing a physical activity.

The practical implications of these findings are important. Healthcare professionals and service providers in public and private sectors, business people, and software practitioners could, and should, modify their design approaches and business strategies based on them. As a societal implication, better user experiences should be provided for individuals whose health status is poor or who are physically less active. Provision of tailored and/or personalized solutions could improve these experiences.

Admittedly, there are limitations in this explorative study. The participants were mostly over 60 years of age, and the greatest relative increase in type 2 diabetes prevalence is expected to be in the over-65 age group [48]. Cross-validation is needed with a younger population as well as with other groups of people, other than individuals with a high risk for type 2 diabetes. Seventy-two percent of the participants were women, which might reflect the previous findings that women are generally more health-oriented than men [49], or that women are more likely than men to participate in health promotion programs [50, 51]. This being said, from a health promotion perspective, it is important for more men to participate in this type of interventions. In general, gender differences in using such systems earn more attention. Besides the participants' physical health, their knowledge of type 2 diabetes and familiarity with web-based intervention(s) are likely to affect their perceptions of ease of orientation and navigational pleasure as well. Also, because all participants were self-selected volunteers, they may be more receptive to using new technologies and perhaps also more active in their daily lives.

There were also some methodological limitations. The study relied partly on self-reported survey data. Self-reported levels of physical activity are usually over-estimated, and we had no means to adjust for this. The sample size was relatively small, so larger studies would be needed to replicate and confirm the findings. Because of the cross-sectional nature, longitudinal studies would be needed to enable studying sustainable change. Also, the approach adopted here was exploratory as all the topics were assessed with one or only a few questions. For these reasons, repetition of studies is warranted in future work. Adding some basic neuro-IS sensors to the study setting would be an interesting opportunity.

7 Conclusion

This article sought to explore the influence of one's physical health and physical activity on a web experience. User experience perceptions were investigated in relation to the self-reported amount of physical exercise and changes in lipid homeostasis. The results of the study demonstrate that perceived user experience of individuals at high risk for type 2 diabetes differs according to their physical health status and amount of physical activity. Those with a better physical health status are more likely to feel ease of orientation in their web usage, and those who take more frequent regular physical exercise are more likely to perceive pleasure in navigating the web. These results open up a new avenue of research. Taking into consideration the status of a user to render customized web-based services in order to increase the chance of intervention success is of growing importance, and utilizing objective physical measures and self-reports instruments provides a new pragmatic approach to monitor and quantify users on par with sensory technology. A new design paradigm and new business models for developing software, systems, and services based on end-users' physical health and physical activity should now be developed.

References

1. Danaei, G., Finucane, M. M., Lu, Y., Singh, G. M., Cowan, M. J., Paciorek, C. J., et al. (2011). National, regional, and global trends in fasting plasma glucose and diabetes prevalence since

1980: Systematic analysis of health examination surveys and epidemiological studies with 370 country-years and 2.7 million participants. *Lancet, 378*(9785), 31–40.
2. WHO. (1999). *Definition, diagnosis and classification of diabetes mellitus and its complications. Part 1: Diagnosis and classification of diabetes mellitus* (WHO/NCD/NCS/99.2). Geneva, Switzerland: World Health Organization.
3. Uusitupa, M., Tuomilehto, J., & Puska, P. (2011). Are we really active in the prevention of obesity and type 2 diabetes at the community level? *Nutrition, Metabolism and Cardiovascular Diseases, 21*(5), 380–389.
4. Farmer, A. J., Levy, J. C., & Turner, R. C. (1999). Knowledge of risk of developing diabetes mellitus among siblings of type 2 diabetes patients. *Diabetic Medicine, 16*(3), 233–237.
5. Tuomilehto, J., Lindström, J., Eriksson, J. G., Valle, T. T., Hämäläinen, H., Ilanne-Parikka, P., et al. (2001). Prevention of type 2 diabetes mellitus by changes in lifestyle among subjects with impaired glucose tolerance. *The New England Journal of Medicine, 344,* 1343–1350.
6. van Esch, S. C., Cornel, M. C., & Snoek, F. J. (2006). Type 2 diabetes and inheritance: What information do diabetes organizations provide on the Internet? *Diabetic Medicine, 23*(11), 1233–1238.
7. Satterfield, D., Jenkins, C., Bodnar, B., Constance, A., & Sisson, E. (2008). Diabetes education and public health. *Diabetes Educator, 34*(1), 45–48.
8. Lustria, M. L., Cortese, J., Noar, S. M., & Glueckauf, R. L. (2009). Computer-tailored health interventions delivered over the web: Review and analysis of key components. *Patient Education and Counseling, 74*(2), 156–173.
9. Yap, T. L., & Davis, L. S. (2008). Physical activity: The science of health promotion through tailored messages. *Rehabilitation Nursing, 33*(2), 55–62.
10. Hawkins, R. P., Kreuter, M., Resnicow, K., Fishbein, M., & Dijkstra, A. (2008). Understanding tailoring in communicating about health. *Health Education Research, 23*(3), 454–466.
11. Rimer, B. K., & Kreuter, M. W. (2006). Advancing tailored health communication: A persuasion and message effects perspective. *Journal of Communication, 56,* 184–201.
12. Enwald, H. P. K., & Huotari, M. L. A. (2010). Preventing the obesity epidemic by second generation tailored health communication: An interdisciplinary review. *Journal of Medical Internet Research, 12*(2), e24.
13. Enwald, H., Niemelä, R., Keinänen-Kiukaanniemi, S., Leppäluoto, J., Jämsä, T., Herzig, K. H., et al. (2012). Human information behaviour and physiological measurements as a basis to tailor health information. An explorative study in a physical activity intervention among prediabetic individuals in Northern Finland. *Health Information and Libraries Journal, 29*(2), 131–140.
14. Enwald, H., Kortelainen, T., Leppäluoto, J., Keinänen-Kiukaanniemi, S., Jämsä, T., Oinas-Kukkonen, H., et al. (2013). Perceptions of fear appeal and preferences for feedback in tailored health communication. An explorative study among prediabetic individuals. *Information Research: An International Electronic Journal, 18*(3), 584.
15. Oinas-Kukkonen, H. (2013). A foundation for the study of behavior change support systems. *Personal and Ubiquitous Computing, 17*(6), 1223–1235.
16. Kraft, P., Drozd, F., & Olsen, E. (2009). ePsychology: Designing theory-based health promotion interventions. *Communications of the Association for Information Systems, 24*(24).
17. Hoffman, D., & Novak, T. (1997). A new marketing paradigm for electronic commerce. *The Information Society, 13,* 43–54.
18. Finneran, C. M., & Zhang, P. (2003). A person-artifact-task (PAT) model of flow antecedents in computer-mediated environments. *International Journal of Human-Computer Studies, 59,* 475–496.
19. Kamis, A., Koufaris, M., & Stern, T. (2008). Using an attribute-based DSS for user-customized products online: An experimental investigation. *MIS Quarterly, 32*(1), 159–177.
20. Oinas-Kukkonen, H., & Harjumaa, M. (2009). Persuasive systems design: Key issues, process model, and system features. *Communications of The Association For Information Systems, 24*(28), 485–500.
21. Oinas-Kukkonen, H. (2000). Balancing the vendor and consumer requirements for electronic shopping systems. *Information Technology and Management, 1*(1&2), 73–84.

22. Case, D. O. (2012). *Looking for information. A survey of research on information seeking, needs, and behavior* (3rd ed.). Bingley, UK: Emerald.
23. Csikszentmihalyi, M. (1977). *Beyond boredom and anxiety*. San Francisco, CA: Jossey-Bass.
24. Choi, D. H., Kim, J., & Kim, S. H. (2007). ERP training with a web-based electronic learning system: The flow theory perspective. *International Journal of Human-Computer Studies, 65,* 223–243.
25. Lu, Y., Zhou, T., & Wan, B. (2009). Exploring Chinese users' acceptance of instant messaging using the theory of planned behavior, the technology acceptance model, and the flow theory. *Computers in Human Behavior, 25*(1), 29–39.
26. Jung, Y., Perez-Mira, B., & Wiley-Patton, S. (2009). Consumer adoption of mobile TV: Examining psychological flow and media content. *Computers in Human Behavior, 25*(1), 123–129.
27. Hoffman, D., & Novak, T. (1996). Marketing in hypermedia computer-mediated environments: Conceptual foundations. *Journal of Marketing*, (July), 50–68.
28. Oinas-Kukkonen, H., Räisänen, T., Leiviskä, K., Seppänen, M., & Kallio, M. (2009). Physicians' user experiences of mobile pharmacopoeias and evidence-based medical guidelines. *International Journal of Healthcare Information Systems and Informatics, 4*(2), 57–68.
29. Parvinen, P., Oinas-Kukkonen, H., & Kaptein, M. (2015). E-selling: A new avenue of research for service design and online engagement. *Electronic Commerce Research and Applications.*
30. Adamo, K. B., Prince, S. A., Tricco, A. C., Connor-Gorder, S., & Tremblay, M. (2009). A comparison of indirect versus direct measures for assessing physical activity in the pediatric population: A systematic review. *International Journal of Pediatric Obesity, 4*(1), 2–27.
31. De Cocker, K., Spittaels, H., Cardon, G., De Bourdeaudhuij, I., & Vandelanotte, C. (2012). Web-based, computer-tailored, pedometer-based physical activity advice: Development, dissemination through general practice, acceptability, and preliminary efficacy in a randomized controlled trial. *Journal of Medical Internet Research, 14*(2), e53.
32. Plotnikoff, R. C., & Karunamuni, N. (2011). Steps towards permanently increasing physical activity in the population. *Current Opinion in Psychiatry, 24,* 162–167.
33. Short, C. E., James, E. L., Plotnikoff, R. C., & Girgis, A. (2011). Efficacy of tailored-print interventions to promote physical activity: A systematic review of randomized trials. *International Journal of Behavioral Nutrition and Physical Activity, 8,* 113.
34. Broekhuizen, K., Kroeze, W., van Poppel, M. N. M., Oenema, A., & Brug, J. (2012). A systematic review of randomized controlled trials on the effectiveness of computer-tailored physical activity and dietary behavior promotion programs: An update. *Annals of Behavioral Medicine, 44,* 259–286.
35. Kroeze, W., Werkman, A., & Brug, J. (2006). A systematic review of randomized trials on the effectiveness of computer-tailored education on physical activity and dietary behaviors. *Annals of Behavioral Medicine, 31*(3), 205–223.
36. Venkatesh, V., Morris, M. G., Davis, G. B., & Davis, F. D. (2003). User acceptance of information technology: Toward a unified view. *MIS Quarterly, 27*(3), 425–478.
37. Knowler, W. C., Barrett-Connor, E., Fowler, S. E., Hamman, R. F., Lachlin, J. M., Walker, E. A., et al. (Diabetes Prevention Program Research Group). (2002). Reduction in the incidence of type 2 diabetes with lifestyle intervention or metformin. *The New England Journal of Medicine, 346*, 393–403.
38. Blair, S. N., & Morris, J. N. (2009). Healthy hearts—and the universal benefits of being physically active: Physical activity and health. *Annals of Epidemiology, 19,* 253–256.
39. Vainionpää, A., Korpelainen, R., Kaikkonen, H., Knip, M., Leppäluoto, J., & Jämsä, T. (2007). Effect of impact exercise on physical performance and cardiovascular risk factors. *Medicine and Science in Sports and Exercise, 39,* 756–763.
40. Herzig, K. H., Ahola, R., Leppäluoto, J., Jokelainen, J., Jämsä, T., Keinänen-Kiukaanniemi, S. (2013). Light physical activity determined by a motion sensor decreases insulin resistance, improves lipid homeostasis and reduces visceral fat in high risk subjects: PreDiabEx study RCT. *International Journal of Obesity,* November 28. https://doi.org/10.1038/ijo.2013.224.
41. Ringle, C. M., Wende, S., & Will, A. (2005). SmartPLS Version 2.0 M2. http://www.smartpls.de.

42. Chin, W. W. (1998). The partial least squares approach to structural equation modeling. In G.A. Marcoulides (Ed.), *Modern methods for business research* (pp. 295–336). Mahway, NJ: Lawrence Erlbaum Associates Inc.
43. Fornell, C., & Larcker, D. F. (1981). Evaluating structural equation models with unobservable variables and measurement error: Algebra and statistics. *Journal of Marketing Research, 18*(3), 383–388.
44. Werts, C. E., Linn, R. L., & Jöreskög, K. G. (1974). Intraclass reliability estimates: Testing structural assumptions. *Educational and Psychological Measurement, 34*(1), 25–33.
45. Gefen, D., Straub, D. W., & Boudreau, M. C. (2000). Structural equation modeling techniques and regression: Guidelines for research practice. *Communications of the Association for Information Systems, 4*(7), 1–79.
46. Fornell, C., & Bookstein, F. L. (1982). Two structural equation models: LISREL and PLS applied to consumer exit-voice theory. *Journal of Marketing Research, 19*(4), 440–452.
47. Hulland, J. (1999). Use of partial least squares (PLS) in strategic management research: A review of four recent studies. *Strategic Management Journal, 20*(2), 195–204.
48. Wild, S., Roglic, G., Green, A., Sicree, R., & King, H. (2004). Global prevalence of diabetes: Estimates for the year 2000 and projections for 2030. *Diabetes Care, 27*(5), 1047–1053.
49. Bech-Larsen, T., & Scholderer, J. (2010). Functional foods in Europe: Consumer research, market experiences and regulatory aspects. *Trends in Food Science & Technology, 18,* 231–234.
50. Assaf, A. R., Parker, D., Lapane, K. L., Coccio, E., Evangelou, E., & Carleton, R. A. (2003). Does the Y chromosome make a difference? Gender differences in attempts to change cardiovascular disease risk factors. *Journal of Women's Health, 12*(4), 321–330.
51. Goh, J.M,. & Agarwal, R. (2008). Taking charge of your health: The drivers of enrollment and continued participation in online health intervention programs. In *Proceedings of the 41st Hawaii International Conference on System Sciences 2008.*

The Effect of Body Positions on Word-Recognition: A Multi-methods NeuroIS Study

Minah Chang, Samuil Pavlevchev, Alessandra Natascha Flöck and Peter Walla

Abstract Numerous studies have shown that different body postures and positions can differently influence one's physiology, motivation and emotion processes, and behavioral and cognitive performances. The varying level of autonomic activation in various body positions may explain the differing level of arousal in supine (*Lay*), seated (*Sit*), and standing (*Stand*) body positions. The present study compared for the first time the effect of these three body positions on word-recognition performance (i.e., response accuracy and response time) while simultaneously using electroencephalography (EEG) to watch the brain at work. No significant difference was found among the three body positions for response accuracy. However, the mean response time was significantly faster for the *Hit* (correct judgment of the repetition of a word-stimulus) response outcome category in the *Sit* position compared to the *Stand* position. Moreover, the mean response time was slower for the *Miss* (incorrect judgment of the repetition of a word-stimulus) category in the *Lay* position compared to the *Sit* position. EEG data analysis further revealed interesting trends for the brain potential amplitudes at 180 ms post-stimulus. The amplitude corresponding to the *Miss* category was significantly different from the *Correct rejection* category in the *Stand* position at this timepoint. More importantly though, the differences in brain potential amplitudes between the *Miss* and the other three categories (*Hit*, *Correct rejection*, and *False alarm*) in the *Lay* position around this timepoint (180 ms) could be interpreted as a potential neurophysiological correlate underlying the prolonged response time for the *Miss* category in the *Lay* position. Possibly, this phenomenon can be linked with the well-known N200 event-related potential (ERP) component.

Keywords EEG · ERP · N200 · Word-recognition · Word-processing · Memory recall · Response time · Body position · Body posture · Neuroimaging

---

M. Chang · S. Pavlevchev · A. N. Flöck · P. Walla
CanBeLab, Department of Psychology, Webster Vienna Private University, Praterstrasse 23, 1020 Vienna, Austria

P. Walla (✉)
School of Psychology, Newcastle University, Callaghan, Newcastle, NSW, Australia
e-mail: peter.walla@webster.ac.at

F. D. Davis et al. (eds.), *Information Systems and Neuroscience*,
Lecture Notes in Information Systems and Organisation 32,
https://doi.org/10.1007/978-3-030-28144-1_36

1 Introduction

The physiological and cognitive effect of body postures (e.g., leaning forward, leaning back) and positions (e.g., supine, seated, standing) is an important area of scientific investigation whose advancement could potentially impact not only specialized fields such as space flight and neuroimaging, but also more widely applicable fields such as therapy, patient-care, office working conditions and, of course, NeuroIS. In a study utilizing startle reflex and EEG by Price et al. [1], the participants reacted with weaker startle reflex responses to erotic images while leaning forward than while reclining back on a chair. Such a result falls in line with the popular notion that leaning forward heightens the state of approach motivation, which in turn attenuates the reflexive response and increases the processing of emotive stimuli. The EEG data of the same study also showed greater late positive potentials in the 300–1000 ms range at site Pz (i.e., midline parietal electrode) to erotic stimuli in the leaning forward than in the reclining position ($p = .006$). In another EEG study by Harmon-Jones and Peterson [2], anger-induced brain activity in the left front cortical area was found to be attenuated while sitting reclined with the back against the chair compared to sitting erect.

To study how cognitive functions might differ dependent on one's body position, Muehlhan et al. [3] investigated the effects of supine and seated positions on working memory performance of 25 participants. They found no significant difference in performance between the two different body positions for neither the reaction times nor accuracy rates. However, the heart rate variability measurement indicated a higher level of parasympathetic activity in the seated compared to the supine position. Jutras et al. [4] also found that the cognitive performance of 40 computer-interface users was unaffected as a result of direct correlation to the type of workstation setting (i.e., physically active versus inactive). On a similar note, Labonté-LeMoyne et al. [5] found a delayed short-term improvement in memory recall performance of a group using a standing treadmill desk. However, the results are likely attributable mainly to having performed a physical exercise than to the body position. Spironelli and Angrilli [6] considered age as a potential moderator for the effect of different body positions on cognitive performance. However, they found that body positions (i.e., seated and supine) do not affect reaction times and accuracy rates of a word-recognition performance across groups of young and elderly women. In the current study, we recorded behavioral and EEG data of participants as they performed a word-encoding task, followed by a word-recognition task in three different body positions: lying down (*Lay*), sitting down (*Sit*), and standing up (*Stand*).

2 Materials and Methods

2.1 Participants

A total of 25 volunteers (13 females, 12 males) between the ages 18 and 31 (*M* = 21.64, *SD* = 3.01) participated in the present study. All of the participants were college students living in Vienna, Austria, and had normal or corrected-to-normal vision. All but one of the participants reported to having been right-handed since birth. Nine of the participants reported to be native English speakers or considered English to be the most proficient language for them, and the other 16 participants reported to having been fluent in English for an average of 8.97 years (*SD* = 4.68). None reported to having any current physical or psychological health issues.

2.2 Stimuli

The word stimuli were programmed and presented with the Psychology Software Tool (PST) E-Prime 2.0® software on a Dell E2214hb 21.5″ widescreen LED LCD monitor. All word stimuli were displayed in white font against black background. In each of the three body positions, the participants were shown a different list of nouns that were between 4 and 7 letter strings long, considered to be concrete and neutral, and low-frequency according to the SUBLEXus database [7]. One of the three lists of 100 words was shown in random order in each of the three body positions. In order to establish the 100 words into repeated versus new categories, the participants were first exposed to 50 of the 100 words immediately prior to the word-recognition task in a separate word-encoding task. In these first tasks, the participants had to decide whether the first and the last letter of the words were in alphabetical order or not (i.e., low level alphabetical encoding; see Rugg et al. [8]). For both tasks, the screen display order (i.e., one trial) was: blank (1 s), fixation cross (1 s), blank (1 s), word stimulus (1 s), and blank (infinite duration until response). As a control for excessive outliers, we limited the maximum possible response time to 2000 ms.

2.3 Data Collection

All responses were collected with a PST Serial Response Box™, and the behavioral responses and response times were recorded by E-Prime 2.0® software. Electrical brain activity of each participant was acquired with a Geodesic EEG™ System 400 with a HydroGel Geodesic Sensor Net of 64 electrodes embedded with silver chloride sensors. The potential changes were continuously sampled at the rate of 1000 Hz with the EGI (Electrical Geodesics, Inc.) Net Amps 400 amplifier with a built-in

Intel chip under an applied online low-pass filter of 30 Hz. The continuous EEG data was displayed and stored by EGI Net Station 5.4 software.

In order to allow the EEG electrodes to remain unperturbed while lying down, a custom-made neck support cushion was placed under the participant's neck in this position. The neck support cushion was carved to gently support the back of the neck and to hold up the head slightly above the ground, thereby allowing the body to assume a horizontal position without any inclination or reclination of the head in respect to the torso.

2.4 Procedure

Once the participants arrived at the CanBeLab (Cognitive & Affective Neuroscience and Behavior Lab) at the Webster Vienna Private University campus, they were given a brief orientation regarding the experiment and how to minimize EEG artefacts. After filling out the demographics and consent form, the participants were given a visual acuity test, and a short test composed of 20 words in order to ensure their proficiency in recalling the correct sequence of alphabetical order. Subsequently, the EEG net was applied over the whole scalp, and the electrodes were connected to the ground, referenced to the Cz point, and kept below 50 kΩ impedance. The participants were first given a practice task identical to the actual word-encoding task, but composed of only 15 words. Once the participants got used to pressing the buttons corresponding to their respective answers using their index fingers (left button pressed with the left index finger if the first letter of the word comes first in the alphabet, and right button pressed with the right index finger if the last letter of the word comes first in the alphabet), they were given the actual word-encoding task with 50 words, immediately followed by the word-recognition task with 100 words with half of the words being new and half of the words being repeated from the first task. For the second task, the participants were told to press the left button with their left index finger if the word is a repetition, and to press the right button with their right index finger if the word is new. Together, the two tasks lasted an average of 15 min. The tasks were then repeated in the other two body positions, all in a counterbalanced order.

2.5 Data Analysis

Each of the participants' response to the word-recognition task was classified as one of the four following response outcome categories: *Hit* (correct judgment of the repetition of a word-stimulus), *Miss* (incorrect judgment of the repetition of a word-stimulus), *Correct rejection* (correct judgment of the newness of a word-stimulus), and *False alarm* (incorrect judgment of the newness of a word-stimulus). The response times corresponding to each of the four categories were averaged for each data set, and then re-averaged across all participants. Analytical statistics was then applied to resulting data.

The EEG signal processing and extraction were carried out with the EEG DISPLAY 6.4.9 software [9]. For each EEG data set, an offline bandpass filter from 0.1 to 30 Hz was applied before generating epochs from 100 ms before the stimulus onset to 1000 ms post-stimulus onset. The duration of 100 ms before stimulus onset was used as baseline. All epochs contaminated by obvious visible artefacts were manually selected and discarded, and those with electrooculogram (EOG) amplitude variations outside of ±75 mV were automatically discarded. The ensemble average of each data set was re-referenced to the common average across all electrode sites. Analytical statistics was then applied to resulting data.

3 Results

3.1 Behavior

Seven out of the 25 participants demonstrated poor behavioral performance when corrected for guessing (as determined by scoring below the value of 10 when the total number of *False alarms* is subtracted from the total number of *Hits*) in at least one of the body position conditions. For better reliability, we excluded the data of those 7 participants for the following behavioral analysis. The grand averages of the mean number of responses and response times per response outcome category for the 18 remaining participants are reported in Table 1.

Table 1 The mean number of responses and response times ($n = 18$) corresponding to each of the response outcome categories

Response outcome category	Lay	Sit	Stand
Mean number of responses			
Hit	33	34	34
Miss	17	16	16
Correct rejection	35	36	37
False alarm	15	14	13
Memory performance corrected for guessing	18	20	21
Response times in ms			
Hit	716	626	716
Miss	798	693	702
Correct rejection	723	644	694
False alarm	780	690	759
Total	3017	2653	2871

Repeated measures ANOVA revealed a highly significant main effect of response categories on mean number of responses ($F[1.918, 32.598] = 36.106, p < 0.001, \eta^2 = 0.680$), but no significant interaction effect with body positions ($p = 0.804, \eta^2 =$ 019). Repeated measures ANOVA for response times revealed no significant effect of response outcome categories ($p = 0.191, \eta^2 = 0.094$), nor for the body positions ($p = 0.106, \eta^2 = 0.128$). There was also no significant interaction effect between the two factors ($p = 0.590, \eta^2 = 0.038$). However, paired samples *t*-tests revealed a mildly significant trend for different response times in the *Hit* category between the *Sit* position and the *Lay* position ($t[17] = 1.901, p = 0.074$) and a significant difference between the *Sit* and the *Stand* position ($t[17] = -2.351, p = 0.031$). A mildly significant trend also occurred for response times in the *Miss* category between the *Lay* and *Sit* body positions ($t[17] = 1.867, p = 0.079$).

3.2 Electroencephalography (EEG)

Figure 1 shows event-related potentials (ERPs) related to all response outcome categories in the *Lay* and *Stand* body positions. As can be seen at 180 ms post-stimulus, the *Miss* category elicited less negative going brain activity compared to all other categories. This effect occurred most dominantly at left occipito parietal and left anterior frontal cortical areas. An ANOVA of mean amplitude values at 180 ms post-

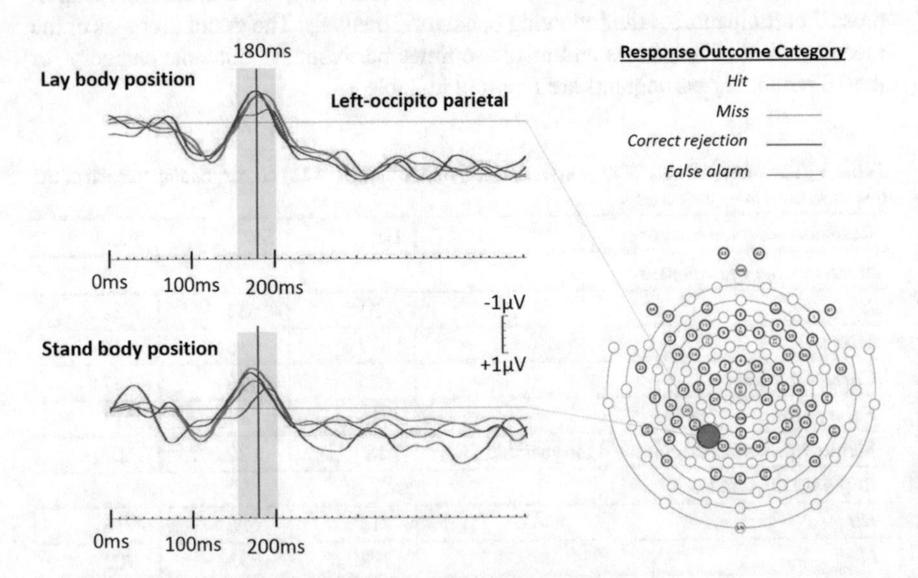

Fig. 1 Event-related potentials (ERPs) at the left occipito-parietal cortical area for the *Lay* and *Stand* body positions. It is clearly visible that the brain activity for the *Miss* response outcome category has a lower amplitude compared to the other three categories. Note also that in both body positions, the peak amplitude elicited by the *Miss* category seems delayed compared to all other categories. However, this effect has not been further analyzed at this stage

stimulus at the left occipito-parietal area did not reveal any main factor effect for body positions ($p = 0.678$, $\eta^2 = 0.160$), nor for response outcome categories ($p = 0.245$, $\eta^2 = 0.056$). There was also no significant interaction effect between the two factors ($p = 0.722$, $\eta^2 = 0.022$). Paired samples *t*-tests among amplitude values of the four categories in the *Sit* position did not reveal any significant trend. However, we found strong trends in the *Lay* position between the *Miss* and *Correct rejection* categories ($t[24] = -1.817, p = 0.082$) and between the *Miss* and *False alarm* categories ($t[24] = -1.760, p = 0.091$). In the *Stand* body position, the *Miss* and *Correct rejection* categories elicited significantly different brain potentials ($t[24] = -2.111$, $p = 0.045$).

4 Discussion

According to Rice et al. [10], cortical activity may be affected in the supine position by the gravity induced changes in the cerebrospinal fluid (CSF) layer thickness (changes by approximately 1 mm) when compared to being in an upright position. Because CSF is up to ten times more conductive than white and gray matter, a 30% shift (3 mm to 2 mm) of CSF can significantly affect EEG signal magnitudes. In magnetic resonance imaging (MRI), a decrease of 1 mm in CSF may correspond to a 30–58.8% increase in scalp power [10]. Using magnetoencephalography (MEG), Thibault et al. [11] found increased left-hemisphere high-frequency oscillatory activity over common speech areas while seated in comparison to lying reclined at 45° or supine. Their findings in the change of baseline brain activity as a result of change in one's body position warrant caution when comparing data across different neuroimaging modalities or acquired while in different body positions (e.g., EEG data reflect implicated gamma band activity while the BOLD signal of fMRI data correlates tightly with synchronized gamma activity) [11].

In contrast to these studies, the present findings provide empirical evidence that there could be more to the effects of different body positions than just general brain activity differences. Despite the lack of significant differences found in corrected for guessing word-recognition memory performance, we found significant response time differences between the three body positions (supine, seated and standing). The results show that for all response outcome categories (*Hit*, *Miss*, *Correct rejection*, and *False alarm*), the response times were longer in both supine and standing positions compared to the seated position. Most notably, the participants took more than 100 ms longer on average to respond to the *Miss* category in the supine position compared to the seated position. In other words, it took the participants more than 100 ms longer to incorrectly report that the word they were seeing was new when it actually was not.

Analysis of brain imaging data revealed an interesting result that could be linked to this behavioral effect. The averaged brain potential amplitudes elicited in all four response outcome categories at 180 ms post-stimulus correspond to the well-known N200 ERP component [12]. At this timepoint, *Hit*, *Correct rejection,* and *False alarm* categories elicited more negative going brain potentials than the *Miss* category, indicating differences in brain activity levels in both the supine and standing body positions. The N200 ERP component has been linked to visual search and controlling of incorrect responses [12], and the present brain imaging findings could reflect a neurophysiological correlate of the causality for found response time differences. More specifically, the current findings indicate a prolonged processing time for the *Miss* category which at 180 ms post-stimulus elicited the lowest brain potential amplitudes compared to all other categories (see Fig. 1). In other words, less brain activity resulted in slower response time. Nonetheless, it has to be emphasized that at this stage our interpretation is rather speculative than based on sound scientific evidence, but our findings might lead to a new hypothesis for future studies.

In conclusion, our findings suggest that body position does not impair actual word-recognition memory performance, response accuracy (as determined by the number of responses for the *Hit* and *False alarm* categories) does not seem to be directly affected by body position. However, body position seems to influence response times in word-recognition, particularly in the *Miss* category that seems to be most affected according to the results of this study. The results also indicate that we might have found a neurophysiological correlate for this behavioral effect. Quite possibly, the N200 ERP component might be linked to neural activity reflecting functions that check below the conscious surface for inconsistency as the conscious mind declares, "No, I have not seen this word before." To the best of our knowledge, the current study is the first to discover a relationship between response time and underlying brain activity in the context of word-recognition performance in different body positions. Further research is needed for more conclusive results.

References

1. Price, T. F., Dieckman, L. W., & Harmon-Jones, E. (2012). Embodying approach motivation: Body posture influences startle eyeblink and event-related potential responses to appetitive stimuli. *Biological Psychology, 90*(3), 211–217. https://doi.org/10.1016/j.biopsycho.2012.04.001.
2. Harmon-Jones, E., & Peterson, C. K. (2009). Supine body position reduces neural response to anger evocation. *Psychological Science, 20*(10), 1209–1210. https://doi.org/10.1111/j.1467-9280.2009.02416.x.
3. Muehlhan, M., Marxen, M., Landsiedel, J., Malberg, H., & Zaunseder, S. (2014). The effect of body posture on cognitive performance: A question of sleep quality. *Frontiers in Human Neuroscience, 8*(171), 1–10. https://doi.org/10.3389/fnhum.2014.00171.
4. Jutras, M.-A., Labonté-LeMoyne, E., Sénécal, S., Léger, P.-M., Begon, M., & Mathieu, M.-È. (2017). When should I use my active workstation? The impact of physical demand and task difficulty on IT users' perception and performance. *Special Interest Group on Human-Computer Interaction 2017 Proceedings, 14*. Retrieved from https://aisel.aisnet.org/sighci2017/14.
5. Labonté-LeMoyne, É., Santhanam, R., Léger, P.-M., Courtemanche, F., Fredette, M., & Sénécal, S. (2015). The delayed effect of treadmill desk usage on recall and attention. *Computers in Human Behavior, 46*, 1–5. https://doi.org/10.1016/j.chb.2014.12.054.

6. Spironelli, C., & Angrilli, A. (2017). Supine posture affects cortical plasticity in elderly but not young women during a word learning-recognition task. *Biological Psychology, 127,* 180–190. https://doi.org/10.1016/j.biopsycho.2017.05.014.
7. Brysbaert, M., & New, B. (2009). Moving beyond Kucera and Francis: A critical evaluation of current word frequency norms and the introduction of a new and improved word frequency measure for American English. *Behavior Research Methods, 41*(4), 977–990. https://doi.org/10.3758/BRM.41.4.977.
8. Rugg, M. D., Mark, R. E., Walla, P., Schloerscheidt, A. M., Birch, C. S., & Allan, K. (1998). Dissociation of the neural correlates of implicit and explicit memory. *Nature, 392*(6676), 595–598. https://doi.org/10.1038/33396.
9. Fulham, W. R. (2015). EEG display (Version 6.4.9) [Computer software]. Callaghan, New South Wales: Functional Neuroimaging Laboratory, University of Newcastle.
10. Rice, J. K., Rorden, C., Little, J. S., & Parra, L. C. (2012). Subject position affects EEG magnitudes. *NeuroImage, 64,* 476–484. https://doi.org/10.1016/j.neuroimage.2012.09.041.
11. Thibault, R. T., Lifshitz, M., & Raz, A. (2015). Body position alters human resting-state: Insights from multi-postural magnetoencephalography. *Brain Imaging and Behavior, 10*(3), 772–780. https://doi.org/10.1007/s11682-015-9447-8.
12. Folstein, J. R., & Van Petten, C. (2008). Influence of cognitive control and mismatch on the N2 component of the ERP: A review. *Psychophysiology, 45*(1), 152–170. https://doi.org/10.1111/j.1469-8986.2007.00602.x.

The Relationships Between Emotional States and Information Processing Strategies in IS Decision Support—A NeuroIS Approach

Bin Mai and Hakjoo Kim

Abstract In this Work-in-Progress report, we describe an innovative experiment design for investigating the potential relationships between IS users' emotional states (positive vs. negative) and their information processing strategies (automatic processing vs. controlled processing) during decision makings in an IS decision support environment. In the extant literature studying this topic, the users' emotional states are usually determined by self-report or mental cue induction, and their information processing strategies by self-report or experiment task performance. In this paper, we describe an experiment design that utilizes neural and psychophysiological signals from the users to infer their emotional states and information processing strategies in real time. Our results will provide additional empirical evidence that are objective and accurate to this significant open research question.

Keywords Emotional state · Information processing · Facial expression analysis · EEG · Decision support

1 Introduction

Human emotional state has long been recognized as a significant factor in human-computer interaction in general [1], and in human information processing in particular [2, 3]. For example, Wyer et al. [2] proposed that emotional states can be associated with information processing in two distinctive ways:

1. "Positive and negative affect can be learned responses to external or internal stimuli. Once acquired, these responses may become preconditions for cognitive operations … and govern [specific] behavior" (p. 2);

B. Mai (✉) · H. Kim
Texas A&M University, College Station, USA
e-mail: binmai@tamu.edu

H. Kim
e-mail: skyler0605@tamu.edu

F. D. Davis et al. (eds.), *Information Systems and Neuroscience*,
Lecture Notes in Information Systems and Organisation 32,
https://doi.org/10.1007/978-3-030-28144-1_37

2. "The affective reactions that one experiences at a given moment can be used as information about one's attitude toward either oneself, other persons, situations with which one is confronted, the outcomes of behavior in these situations, and the appropriateness of certain strategies for attaining specified processing objectives. Consequently, affect can influence both judgments of these entities and behavioral decisions that concern them" (p. 2).

Similarly, Schwarz [3] pointed out that emotional states may "influence which information comes to mind and is considered in forming a judgment, or serve as a source of information in their own right" (p. 546), and "inform us about the nature of the current situation" (p. 547).

Walla [4] made a distinction among the terms of "affect", "feeling" and "emotion", in an attempt to clarify the definitions of these three terms, which are often used inter-changeably in existing literature about human emotions. According to Walla, "affective processing" refers to "neural activity coding for valence", "feeling" refers to "felt bodily response arising from supra-threshold affective processing", and "emotion" is the "behavioral output of affective processing communicating feelings" ([4], p. 147). We adopt this approach in our research context, and follow [2] in defining emotion as "the positively or negatively valenced subjective reactions that a person experiences at a given point in time. These reactions are experienced as either pleasant or unpleasant feelings." ([2], p. 3).

Information processing strategies are generally categorized into controlled processing and automatic processing (e.g., [5]), where controlled processing is defined as "taking detailed variables such as situational factors into consideration to add correction to an automatic response, and adjusting responses in the more appropriate direction", and automatic processing is defined as "'preparing output which derives from instant response to stimuli without any correction" ([5], p. 142).

Over the years, there have been numerous experimental studies addressing this significant research question (e.g., [6–9]):

- What are the relationships between an IS user's emotional states and his/her information processing strategies when interacting with IS?

In these types of studies, the users' emotional states are usually inferred by self-report or mental cue induction (e.g., [3], p. 535; [5], pp. 143–144), and their information processing strategies are usually inferred by self-report or experiment task performance (e.g., [5] p. 144, p. 148).

Interestingly, with the research over the years investigating this question, a consensus has not been reached. Some of the research (e.g., [6, 7]) have produced results that suggest the users' positive emotions would be associated more with automatic information processing, and their negative emotions would be associated more with controlled information processing, while some others showed that the impact of users' emotional states on their information processing strategies remain unclear (e.g., [8]) or the information processing strategies of the users do not differ between different emotional states (e.g., [9]). The research question remain a wide-open one.

Therefore, in this report, we propose an innovative experiment design for investigating this open research question. The unique characteristic of our experiment design is that we will use neural, behavioral, and psychophysiological signals to infer the users' emotional states and their information processing strategies, instead of relying on the user self-report, cue inductions, and task performance, as are the case in extant research. To the best of our knowledge, our experiment design is the first one that attempts to investigate the relationships between emotional states and information processing strategies based completely on the users' neural, behavioral, and psychophysiological data. The results of our research will provide additional objective and accurate empirical evidence to this significant open research question.

In Sect. 2, we present our research model. In Sect. 3, we illustrate our experiment design and data analysis plan, especially the mechanisms through which neural, behavioral, and psychophysiological data can be analyzed to infer the user's emotional states and information processing strategies. In Sect. 4, we summarize and describe our next steps of the research project.

2 Research Framework

The following Fig. 1 illustrates the basic research framework for our study.

In the model, the users' emotional states are determined by capturing and analyzing their behavioral/psychophysiological signals during their engagement of cognitive tasks. Certain patterns of the signals will indicate either positive or negative emotion of the user at the moment. Similarly, the users' neural signals are also captured and analyzed, and are used to determine the user's information processing strategy at the moment. Subsequently, by analyzing the correlations between the behavioral/psychophysiological signals and the neural signals, we will be able to provide empirical evidence for the potential relationship between the user's emotional states and information processing strategies.

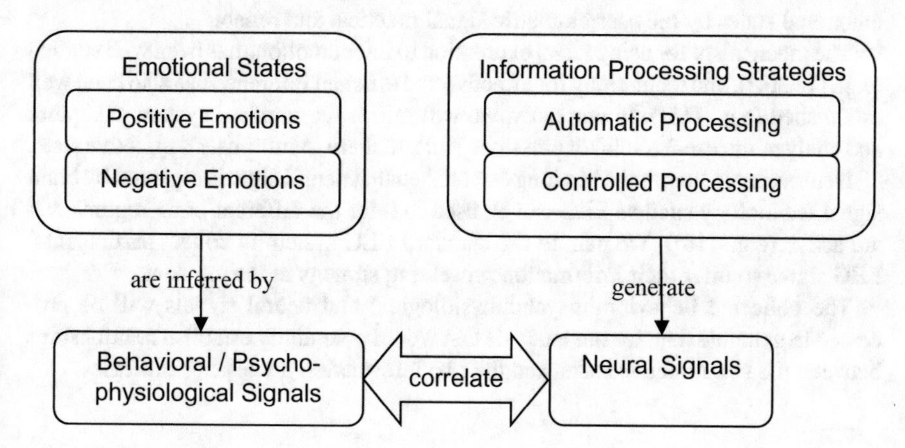

Fig. 1 Research model

3 Experiment Design and Analysis Plan

3.1 Experiment Design

Based on our research model, we will arrange for test subjects to perform a specified cognitive task, during which their behavioral/psychophysiological signals will be detected, recorded and analyzed to infer their emotional states (i.e., positive or negative), and their neural signals generated by their information processing mental activities will also be recorded and analyzed to infer their information processing strategy (i.e., automatic processing or controlled processing). We will then analyze the behavioral/psychophysiological signals and corresponding neural signals to investigate the potential relationships between users' emotional states and their information processing strategies.

For the cognitive task the test subjects will perform, we plan to choose an open-ended mental task that requires the participants to understand a societal issue that is well known, sophisticated, and not yet having a consensus of a solution, and to endeavor to come up with their own conclusions. The participants will be provided a standard computer with internet connection during this mental task for them to utilize for information search and decision support. One potential source for the societal issues we plan to choose for this study could be the Global Issue repository from United Nations [10]. The total time allocated to each participant for the mental task would be 5 min.

To infer the user's emotional states, we plan to observe and capture the user's facial expressions, and through facial expression analysis, infer the current emotion the subject is experiencing.

Using users' facial expression to infer their emotional states is not exactly a new idea (e.g. [11]). The recent advancement of technology has enabled effective and efficient facial recognition and emotion inference (e.g. [12]), making inferring user's emotional states by the user biometric signal practical and reliable.

The mechanism for using facial expression to infer emotions has been well studied (e.g., [13, 14]), and technology for effective and efficient inference has also been well established (e.g., [15]). In our study, we will utilize the standard system to capture and analyze the user's facial expressions and use them to infer user's affective state.

Neuroscience literatures has long established that neural signals captured by brain signal technology such as EEG can be used to infer the different brain regions that are active (e.g., [16]). We plan to use standard EEG system to collect participants' EEG signal to infer their information processing strategy at the moment.

The collected behavioral/psychophysiological and neural signals will be processed to generate data for our analysis that would potentially establish relationships between the two affective states and the two information processing strategies.

3.2 Experiment Plan

The traditional procedure for studying emotions in the laboratory setting is to conduct a three-stage design (e.g., [17]): emotion induction, emotion elicitation, and task completion. In most extant research, the emotion elicitation is usually done by self-reported surveys, primarily PANAS. There are two main problems with this elicitation approach: (1) the users' self-report may not reflect their real emotions; (2) the cognitive task of answering the survey could dilute the emotions induced in the first stage, and interfere with the emotions during the task completion in the third stage. Thus in our experiment, we plan to eliminate the self-report survey stage, and opt to use facial expression recognition and analysis system to infer the users' emotional states in real time.

3.2.1 Emotion Induce Stage

The participants will be randomly assigned to one of two groups: positive and negative. For the participants assigned to the positive group, they will watch a video clip that induces positive emotions; and for the participants assigned to the negative group, they will watch a video clip that induces negative emotions.

3.2.2 Task Completion Stage

After stage 1, the participant will start performing their cognitive task of addressing an assigned significant societal issue.

Emotion Elicitation: During the participant's task performance, a facial expression recognition and analysis system called AFFDEX ([18]) will be used to capture/recognize the participant's facial expression and infer his/her emotions accordingly. This system will be adapted and configured to capture slight changes in facial expressions to infer the current emotion the participant is experiencing. It is completely noninvasive and operates through any standard webcam. It works through keeping a huge database of faces associated with different positive and negative emotions. Once the participant's facial expression is recognized by the software, it automatically compares it with this database to provide an accurate estimate of the participant's current emotional state. This technology operates at 30 Hz, meaning that it generates 30 observations (or likelihood indexes) for each emotion per second.

Information Processing Strategy Elicitation: During the participant's task performance, the participants will also be connected to an EEG system to collect their neural signals during the task.

3.2.3 Data Processing/Analysis Stage

The collected facial expression data and neural signals will be processed to generate data about the participants' emotional states and their respective information processing strategy in real time. We then conduct statistical analysis that would potentially establish relationships between the two emotional states and the two information processing strategies.

4 Summary

While it has been recognized that users' emotional states have significant correlations with their information processing strategies during cognitive tasks, the details of such correlations remain an open research questions. More specifically, there is currently no consensus regarding which emotional state (positive or negative) is correlated with which information processing strategy (controlled process or automatic process). In this work-in-progress report, we propose our innovative experiment design for investigating this significant open question. Compared to extant relevant literature in which the users' emotional states are usually inferred by self-report or mental cue induction, and their information processing strategies are usually inferred by self-report or experiment task performance, our experiment design plans to utilize behavioral, neural and psychophysiological signals from the users to infer their emotional states and information processing strategies in real time. We aim to contribute additional empirical evidences that are objective and accurate to this open research question.

Acknowledgements We thank the two anonymous reviewers who referred us to Walla [4], and provided valuable comments. We also greatly appreciate the comments and suggestions from Marco Palma and Steven Woltering which significantly facilitated the revision of the paper.

References

1. Lopatovska, I., & Arapakis, I. (2011). Theories, methods and current research on emotions in library and information science, information retrieval and human–computer interaction. *Information Processing and Management, 47*(4), 575–592.
2. Wyer, R. S., Clore, G. L., & Isbell, L. M. (1999) Affect and information processing. In M. P. Zanna (Ed.), *Advances in experimental social psychology* (Vol. 31, pp. 1–77). New York: Academic Press.
3. Schwarz, N. (2002). Feelings as information: Moods influence judgments and processing strategies. In T. Gilovich, D. Griffin, & D. Kahneman (Eds.), *Heuristics and biases: The psychology of intuitive judgment* (pp. 534–547). Cambridge: Cambridge University Press.
4. Walla, P. (2017). Affective processing guides behavior and emotions communicate feelings: Towards a guideline for the NeuroIS community. In F. D. Davis, R. Riedl, J. Vom Brocke, P.-M.

Léger, & A. B. Randolph (Eds.), *Information systems and neuroscience: Gmunden retreat on NeuroIS 2017* (pp. 141–150). New York, NY: Springer Berlin Heidelberg.
5. Kitamura, H. (2005). Effects of mood states on information processing strategies: Two studies of automatic and controlled processing using misattribution paradigms. *Asian Journal of Social Psychology, 8*(2), 139–154.
6. Schwarz, N. (1990). Feelings as information: Informational and motivational functions of affective states. In E. T. Higgins & R. M. Sorrentino (Eds.), *Handbook of motivation and cognition* (Vol. 2, pp. 527–561). New York: Guilford.
7. Schwarz, N., Bless, H., & Bohner, G. (1991). Mood and persuasion: Affective states influence the processing of persuasive communications. In M. P. Zanna (Ed.), *Advances in experimental social psychology* (Vol. 24, pp. 161–199). San Diego, CA: Academic Press.
8. Forgas, J. P. (2000). *Feeling and thinking*. Cambridge: Cambridge University Press.
9. Bless, H., Bohner, G., Schwarz, N., & Strack, F. (1990). Mood and persuasion: A cognitive response analysis. *Personality and Social Psychology Bulletin, 16,* 331–345.
10. United Nations Global Issues Overview. http://www.un.org/en/sections/issues-depth/global-issues-overview/.
11. Nguyen, Y., & Noussair, C. N. (2014). Risk aversion and emotions. *Pacific Economic Review, 19*(3), 296–312.
12. Kahyaoglu, M. B., & Ican, O. (2016). Risk aversion and emotions in DoND. *International Journal of Economics and Finance, 9*(1), 32.
13. El Kaliouby, R., & Robinson, P. (2005). Real-time inference of complex mental states from facial expressions and head gestures. In *Real-time vision for human-computer interaction* (pp. 181–200). Springer, Boston, MA.
14. Scherer, K. R., & Grandjean, D. (2008). Facial expressions allow inference of both emotions and their components. *Cognition and Emotion, 22*(5), 789–801.
15. Baltrušaitis, T., McDuff, D., Banda, N., Mahmoud, M., El Kaliouby, R., Robinson, P., et al. (2011, March). Real-time inference of mental states from facial expressions and upper body gestures. In *Face and gesture* 2011 (pp. 909–914). IEEE.
16. Dale, A. M., & Halgren, E. (2001). Spatiotemporal mapping of brain activity by integration of multiple imaging modalities. *Current Opinion in Neurobiology, 11*(2), 202–208.
17. Ifcher, J., & Zarghamee, H. (2011). Happiness and time preference: The effect of positive affect in a random-assignment experiment. *American Economic Review, 101*(7), 3109–3129.
18. Affdex for Market Research. https://www.affectiva.com/product/affdex-for-market-research/.

Improving Knowledge Acquisition from Informational Websites: A NeuroIS Study

Amir Riaz and Shirley Gregor

Abstract This research investigates the relationship between the relevance of website imagery with the given text, emotions and cognition and in return how those constructs influence knowledge acquisition and willingness of website revisits of online users. It particularly examines the moderating role of need-for-cognition on the relationship between website stimuli, emotion and cognition. We conducted lab experiments using electroencephalography (EEG) to investigate the phenomenon. The initial phase of data collection has been completed. We hope to contribute to NeuroIS literature. Specifically, that examines the role of website design for emotional and cognitive responses and their relationship with online users' knowledge acquisition and their willingness to revisit websites.

Keywords Knowledge acquisition · Website revisit · Emotion · Cognition · Electroencephalography

1 Introduction

Online information dissemination using websites is becoming a primary vehicle for governments and organisations for informing the public about specific topics. Although many strategies are being employed to achieve this objective, such as presenting information in more readable and comprehensible ways [1] and also by increasing the information's quality [2], no prior research has investigated if knowledge acquisition from informational websites can be improved through website design features in the context of online information provisions [exceptions are 3, 4]. Further, notwithstanding a large percentage of people turns to informational websites to acquire knowledge, personality attributes of people such as need-for-cognition (NFC) may affect how much knowledge people can acquire and whether

---

A. Riaz (✉) · S. Gregor
The Australian National University, Canberra, Australia
e-mail: amir.riaz@anu.edu.au

S. Gregor
e-mail: shirley.gregor@anu.edu.au

F. D. Davis et al. (eds.), *Information Systems and Neuroscience*,
Lecture Notes in Information Systems and Organisation 32,
https://doi.org/10.1007/978-3-030-28144-1_38

they would like to revisit those websites. For example, according to a survey conducted by Pew [5], 25% of American online users can be considered as individuals with low-NFC because despite going online for information, they are not as engaged with information as the other 22% who can be considered as individual with high-NFC. This finding suggests that information providers should be aware of such differences in personality attributes and develop their informational websites accordingly.

Prior studies have investigated the influence of website design on knowledge acquisition (e.g., [3, 4]). However, these studies did not explore the role of one crucial personality attribute—the need for cognition, that plays a vital role in knowledge acquisition [6]. Moreover, even though prior research has pointed out the critical role that working memory plays in information processing [7] and knowledge acquisition [6], research exploring the influence of website design features on cognitive processes has mainly treated the role of working memory as a black box. Therefore, this study investigates the following research question.

How does the need-for-cognition influence the relationship between the perceived relevance of website imagery with the given text, online users' emotions, and their knowledge acquisition and websites' revisit?

2 Theoretical Background

2.1 Website Imagery: Visual Design Features

Website design features are generally divided into usability and visual features. Usability features help online users to complete tasks (e.g., navigation aids) [8]. While visual features are not directly related to websites' functionality, these are included in the websites for other reasons (e.g., website images to induce emotions in online users) [9]. Prior studies have investigated website images of humans [10], products [11], and animals [3]. Mostly, these images were investigated to see if they induced emotions and, in turn, enhanced online users' e-loyalty [9], online trust [10] or information recall [3]. Other studies examined the role of website images (i.e., images of products) in increasing the number of complementary informational cues and, in turn, in enhancing information recall [11] and the influence of website images that induce emotions and also provide complementary informational cues on information recall [4]. However, there is little research that investigates the role of need-for-cognition for the relationship between website images, their relevance with the text, and emotional and cognitive processes. This study fills this gap.

2.2 Neuro-IS

NeuroIS methods allow "researchers to 'open up' the black box of the brain to further understanding of human cognition and emotional processing" [7, p. 55]. Prior studies encourage researchers to use EEG to measure emotional [9] and cognitive responses [7] as evoked by stimuli.

EEG measures electrical postsynaptic potential from the scalp. These electrical potentials can be analysed from two perspectives: (1) spectral analysis and, (2) event-related potential (ERP) [12]. Spectral analysis is the analysis of brain activity by extracting brain waves from the raw EEG data [12]. These brain waves have two characteristics; namely, power and frequency. Power is measured in decibels (db) and depends on neurons' synchronous activity. The more synchronously the neurons work, the higher the power. Frequency is measured in cycles per seconds (Hz). Brain waves are divided into five categories according to their frequency: delta (<4 Hz), theta (4–8 Hz), alpha (8–13 Hz), beta (13–20 Hz) and gamma (>20 Hz) [13]. ERP is "a depiction of the changes in scalp-recorded voltage over time that reflects the sensory, cognitive-affective and motor processes elicited by a stimulus" [12, p. 4]. We adopt the spectral analysis approach for in this study following [9] to measure emotional responses and [7] to measure cognitive responses.

2.3 Emotional Responses

According to theories of emotion in Evolutionary Psychology [14–16], humans developed emotions to increase their survival chances. They developed positive emotions to help them approach fitness-maximising events such as food sources or potential mates and developed negative emotions to help them avoid potential danger [16]. Another important aspect of early humans' emotional development was that emotions helped them to remember both the fitness-maximising and fitness-impairing events and elements so that they could approach or avoid them in the future [17]. Further, studies in the Neuroscience literature (e.g., [18]) suggest that, when information is presented in the context of emotion-eliciting stimuli, the retention of that information on later stages is enhanced because, at the encoding stage, an emotional tag is added to the information. Therefore, retention of emotion-eliciting images is higher than with neutral images [19]. Even recall of words that were presented in the context of emotion-eliciting sentences has been reported to be higher than recalls of words presented in the context of neutral sentences [20]. Kock and Chetelain-Jardon [15] investigated the impact of negative emotions on the recall of descriptive information. They found that participants who experienced negative emotions could retain more information than those who did not experience them.

We aim to extend this line of work in several ways in this research. First, we investigate both positive and negative emotions elicited by website imagery as well as their relevance with the given text on websites. Second, we investigate neural correlates of emotional as well as cognitive processes and explore if they can be used as predictors of online users' knowledge acquisition and their willingness to revisit the websites. Lastly, we investigate the moderating role of need for cognition for the relationship between website stimuli, emotional and cognitive responses and neural correlates of emotional and cognitive responses. In the NeuroIS domain, other studies have also theorised based on evolutionary psychology theory to study IS constructs (e.g., see [21, 22].

Prior research has shown hemispheric asymmetry in the prefrontal cortex (PFC) regions using EEG. PFC has been argued to be central to emotion detection [23]. EEG has been argued to be sensitive to approach-avoidance states [24]. According to the valence hypothesis, "the left prefrontal cortex is activated during positive emotions whereas the right prefrontal cortex is activated during negative emotions" [25, p. 256]. The valence hypothesis was originated from brain lesions studies (e.g., [26, 27]). According to brain lesions studies, people who have an abnormality in the left hemisphere tend to experience negative emotions whereas people who have an abnormality in the right hemisphere tend to experience positive emotions [28]. Many studies have supported the valence hypothesis, see [29–31]. We adopted the perspective of the valence hypothesis to show brain activity related to emotional responses. Following [25], we measure the emotional response (i.e., valence) with the level of activation in the PFC region (left and right).

2.4 Cognitive Responses

Prior research points out the role of Dorsolateral Prefrontal Cortex (DLPFC) in cognitive processing [32] and suggests that working memory that is highly involved in cognitive processing is located in the DLPFC region of the brain [33, 34]. It also suggests an inverse relationship between cognitive processing and alpha frequency amplitude—in the event of high levels of cognitive processing, there will be low levels of alpha [7]. When an individual comes across stimuli (e.g., text and image in this case), a high level of categorisation of the stimuli is processed as such words and images are differentiated and are processed in language and image regions of the brain accordingly [35]. After the high level of categorisation of the stimuli, a relevance check is performed on the information [36].

Working memory plays a vital role in cognition and is pivotal for information processing [37]. It encodes information coming in from the external environment or stimuli and retrieving relevant information from the long-term memory to make sense of the information being encoded [38]. Therefore, changes in the brain activation of DLPFC have been argued to be representative of changes in the load of working memory [39]. Gevin et al. have argued that an attenuated alpha rhythm over the DLPFC represent high levels of working memory load [40] and increased activity in the DLPFC can be associated with the better cognitive processing [41]. Our objective is to investigate how stimuli that consist of images relevant to the given text and images that are irrelevant to a given text are processed and if this processing differs depending upon the personal attribute of need-for-cognition (e.g., low vs high).

Need-for-Cognition

Need-for-cognition (NFC) represents individuals' motivation to enjoy cognitive activities [42]. Individuals with high-NFC tend to engage in effortful processing [43] and enjoy effortful cognitive tasks [44]. Therefore, individuals with high-NFC are likely to be more interested in performing the task required in this study that may lead to high performance (i.e., high level of knowledge acquisition) [45]. On the contrary, individuals with low-NFC who are also considered as "cognitive misers" [46], may not be that interested in the task and tend to depend on the facilitating attributes of the stimuli (complementary informational cues in this case) [47].

Therefore, we argue that individuals with low-NFC who tend to avoid effortful cognitive processing will find an image-text relevant website more appealing because those websites provide complementary informational cues and therefore will show more activation in the DLPFC as compared to their activation in the DLPFC for the image-text irrelevant website. However, we argue that individuals with high-NFC will not show any significant difference in their brain activation in the DLPFC region for the image-text relevant and irrelevant websites.

2.5 *Hypotheses*

We aim to test several hypotheses. Two of the hypothesis are given below:

Hypothesis 1: Need-for-cognition will moderate the relationship between website stimuli and level of emotions such that;
Hypothesis 1a: Individuals with low-NFC will show higher levels of positive emotions than individuals with high-NFC for the image-text relevant websites
Hypothesis 1b: Individuals with low-NFC will show higher levels of negative emotions than individuals with high-NFC for the image-text irrelevant websites
Hypothesis 2: Need-for-cognition will moderate the relationship between website stimuli and level of PFC activation such that;
Hypothesis 2a: Individuals with low-NFC will show higher levels of activation in the left PFC (higher valence) for emotionally positive image-text relevant websites as compared to the activation of individuals with high-NFC
Hypothesis 2b: Individuals with low-NFC will show higher levels of activation in the right PFC (lower valence) for emotionally harmful image-text irrelevant websites as compared to the activation of individuals with high-NFC.

3 Research Method

We conducted laboratory experiments to investigate the moderating role of need-for-cognition on the relationship between the perceived relevance of website imagery with the given text, online users' emotions, and their knowledge acquisition and websites' revisit. We used EEG to measure neural correlates of emotional and cognitive responses.

Thirty right-handed participants took part in the study. All participants were healthy and had normal or corrected-to-normal vision. We discarded the data from 3e participants because of excessive EEG artefacts. Demographic information showed that all the participants were Internet users and went online daily to look for some information.

Participants performed the experiment individually. When the participants arrived in the lab, they were greeted and given a consent form. Next, they were provided with an instructions sheet that explained the experimental scenario, task, and feedback mechanisms. Participants were told that there would be recall tests and examples of the test questions were given in the instructions sheet. The experiment consisted of three sessions. Participants took part in a practice session first. They performed all the tasks for the study phase, distraction phase and test phase during the practice session so that they could familiarise themselves with the experimental procedures. After the practice session, the EEG section of the experiment started. First, the participant's vertex was measured to place the Cz electrode of the EEG cap accurately. The EEG cap was fitted to participants' heads while ensuring that the Cz electrodes sat on the right place on the scalp. Participants were instructed not to move their heads or other body parts unless required. The chair was fixed on the floor, and the distance from the chair and the 17" monitor was three feet. Participants used their right hand to press keys from a special keypad (such keypads are designed explicitly for EEG experiments: they have less number of keys). Once participants were comfortable, the EEG recording began. Two minutes of EEG was recorded (i.e., one-minute eyes open and one-minute eyes closed) at the start and two minutes at the end of the experiment as a baseline.

After the two minutes of EEG recording, participants entered their responses for "need for cognition" scale. Then they performed five sets of study and test phases. At the end of each session, which included five blocks of study and test phases, each participant rated their mood on a 1–7 scale at the end of each session. Also, after every session, participants were given one minute break to move freely and stretch. At the end of the experiment, participants answered post-experiment questions in a qualitative response booklet. The entire experiment took less than 90 min, of which EEG recording occurred for about one hour. Each participant received a gift voucher of $30 as a token of appreciation for taking part in the experiment.

We used BioSemi (16 electrodes) and positioned the electrodes according to the 10–20 international electrode placement system. We used two electrooculograms (EOG) (one vertical and one horizontal) to record participants' eye movements and blink artefacts so that they could be identified and eliminated. We filtered EEG signals with a 40 Hz high-pass and 0.01 low-pass and digitised them at 500 Hz.

4 Conclusion

We hope to contribute to NeuroIS literature. Specifically, that examines the role of website design for emotional and cognitive responses and their relationship with online users' knowledge acquisition and their willingness to revisit websites.

References

1. Berland, G. K., Elliott, M. N., Morales, L. S., & Algazy. J. I. (2001). Health information on the Internet: accessibility, quality, and readability in English and Spanish. *The Journal of the American Medical Association, 285*, 2612–21.
2. Morahan-Martin, J. M. (2004). How Internet users find, evaluate, and use online health information: A cross-cultural review. *Cyberpsychology Behavior, 7*, 497–510.
3. Riaz, A., Gregor, S., & Lin, A. (2018). Biophilia and biophobia in website design: Improving internet information dissemination. *Information & Management, 55*, 199–214.
4. Riaz, A., Gregor, S., Dewan, S., & Xu, Q. (2018). The interplay between emotion, cognition and information recall from websites with relevant and irrelevant images: A Neuro-IS study. *Decision Support Systems, 111*, 113–123.
5. Pew Research Center, http://www.pewinternet.org/2017/09/11/how-people-approach-facts-and-information/.
6. Day, E. A., Espejo, J., Kowollik, V., Boatman, P. R. B., & McEntire, L. E. (2007). Modeling the links between need for cognition and the acquisition of a complex skill. *Personality and Individual Differences, 42*, 201–212.
7. Minas, R. K., Potter, R. F., Dennis, A. R., Bartelt, V., & Bae, S. (2014). Putting on the thinking cap: Using NeuroIS to understand information processing biases in virtual teams. *JMIS, 30*, 49–82.
8. Eroglu, S. A., Machleit, K. A., & Davis, L. M. (2001). Atmospheric qualities of online retailing: A conceptual model and implications. *Journal of Business Research, 54*, 177–184.
9. Gregor, S., Lin, A., Gedeon, T., Riaz, A., & Zhu, D. (2014). Neuroscience and a nomological network for the understanding and assessment of emotions in information systems research. *Journal of Management Information Systems, 30*, 13–47.
10. Cyr, D., Head, M., Larios, H., & Pan, B. (2009). Exploring human image in website design: A multi-method approach. *MIS Quarterly, 33*, 539–566.
11. Jiang, Z., & Benbasat, I. (2007). The effects of presentation formats and task complexity on online customers' product understanding. *MIS Quarterly, 31*, 475–500.
12. Luck, S. J., & Kappenman, E. S. (Eds.). (2012). *The oxford handbook of event-related potential components*. Oxford University Press.
13. Harmon-Jones, E., & Peterson, C. K. (2009). Electroencephalographic methods in social and personality psychology. In E. Harmon-Jones & J. S. Beer (Eds.), *Methods and social neuroscience* (pp. 170–197). New York: Guilford Press.
14. Kock, N. (2009). Information systems theorizing based on evolutionary psychology: An interdisciplinary review and theory integration framework. *MIS Quarterly, 33*(2), 395–418.
15. Kock, N., & Chatelain-Jardon, R. (2011). Four guiding principles for research on evolved information processing traits and technology task performance. *Journal of the Association for Information Systems, 12*(10), 684–713.

16. Ulrich, R. S. (1993). Biophilia, biophobia, and natural landscapes. In S. R. Kellert & E. O. Wilson (Eds.), *The biophilia hypothesis* (pp. 73–137). Washington, DC: Island Press.
17. Nairne, J. S. (2010). Adaptive memory: Evolutionary constraints on remembering. In B. H. Ross (Ed.), *The psychology of learning and motivation* (Vol. 53, pp. 1–32). Burlington: Academic Press.
18. Hamann, S. B., Ely, T. D., & Kilts, C. D. (1999). Amygdala activity related to enhanced memory for pleasant and aversive stimuli. *Nature Neuroscience, 2*(3), 289–294.
19. Jaeger, A., & Rugg, M. D. (2012). Implicit effects of emotional context: An ERP study. *Cognitive Affective and Behavioral Neuroscience, 12*(4), 748–760.
20. Maratos, E. J., & Rugg, M. D. (2001). Electrophysiological correlates of the retrieval of emotional and non-emotional context. *Journal of Cognitive Neuroscience, 13*(7), 877–891.
21. Riedl, R., Mohr, P. N. C., Kenning, P. H., David, F. D., & Heekeren, H. R. (2014). Trusting humans and avatars: A brain imaging study based on evolution theory. *Journal of Management Information Systems, 30*(4), 83–114.
22. Walden, E., Cogo, G. S., Lucus, D. J., & Moradiabadi, E. (2018). Neural correlates of multidimensional visualizations: An fMRI comparison of bubble and three-dimensional surface graphs using evolutionary theory. *MIS Quarterly, 42*(4), 1097–1116.
23. Davidson, R. J., Ekman, P., Saron, C., Senulis, J., & Friesen, W. (1990). Approach-withdrawal and cerebral asymmetry: Emotion expression and brain asymmetry. *Journal of Personality and Social Psychology, 58,* 330–341.
24. Mauss, I., & Robinson, M. (2009). Measures of emotion: A review. *Cognition and Emotion, 23,* 209–237.
25. Herrmann, M. J., Ehlis, A. C., & Fallgatter, A. J. (2003). Prefrontal activation through task requirements of emotional induction measured with NIRS. *Biological Psychology, 64,* 255–263.
26. Morris, J. S., Frith, C. D., Perrett, D. I., Rowland, D., Young, A. W., & Calder, A. J. (1996). A differential neural response in the human amygdala to fearful and happy facial expressions. *Nature, 383,* 812–815.
27. Paradiso, S., Chemerinski, E., Yazici, K. M., Tartaro, A., & Robinson, R. G. (1999). Frontal lobe syndrome reassessed: Comparison of patients with lateral or medial frontal lobe damage. *Journal of Neurology, Neurosurgery and Psychiatry, 67,* 664–667.
28. Starkstein, S., Robinson, R., Honig, M., Parikh, R., Joselyn, J., & Price, T. (1989). Mood changes after right-hemisphere lesions. *British Journal of Psychiatry, 155,* 79–85.
29. Onal-Hartmann, C., Pauli, P., Ocklenburg, S., & Güntürkün, O. (2012). The motor side of emotions: Investigating the relationship between hemispheres, motor reactions and emotional stimuli. *Psychological Research, 76,* 311–316.
30. Smith, S. D., & Bulman-Fleming, M. B. (2004). A hemispheric asymmetry for the unconscious perception of emotion. *Brain and Cognition, 55,* 452–457.
31. Roesmann, K., Dellert, T., Junghoefer, M., et al. (2019). The causal role of prefrontal hemispheric asymmetry in valence processing of words—Insights from a combined cTBS-MEG study. *NeuroImage, 191,* 367–379.
32. Dimoka, A., Davis, F. D., Gupta, A., Pavlou, P. A., Banker, R. D., et al. (2012). On the use of neurophysiological tools in IS research: Developing a research agenda for Neuro-IS. *MIS Quarterly, 36*(3), 679–702.
33. Riedl, R., & Leger, P. M. (2016). A primer on neurobiology and the brain for information systems scholars. In M. Reuter & C. Montag (Eds.), *Fundamentals of NeuroIS: Information systems and the brain* (pp. 25–45). Heidelberg: Springer.
34. Rypma, B., & D'Esposito, M. (1999). The roles of prefrontal brain regions in components of working memory: Effects of memory load and individual differences. *Proceedings of the National Academy of Sciences, 96,* 6558–6563.
35. Rohm, D., Klimesch, W., Haider, H., & Doppelmayr, M. (2001). The role of theta and alpha oscillations for language comprehension in the human electroencephalogram. *Neuroscience Letters, 310,* 137–140.
36. Watkins, K., & Paus, T. (2004). Modulation of motor excitability during speech perception: The role of Broca's area. *Journal of Cognitive Neuroscience, 16,* 978–987.

37. Baddeley, A. (1992). Working memory. *Science, 255,* 556–559.
38. Welsh, M. C., Satterlee-Cartmell, T., & Stine, M. (1999). Towers of Hanoi and London: Contribution of working memory and inhibition to performance. *Brain and Cognition, 41,* 231–242.
39. Dimoka, A., Hong, Y., & Pavlou, P. A. (2012). On product uncertainty in online markets: Theory and evidence. *MIS Quarterly, 36,* 395–426.
40. Gevins, A., Smith, A. M. E., McEvoy, L., & Yu, D. (1997). High-resolution EEG mapping of cortical activation related to working memory: Effects of task difficulty, type of processing, and practice. *Cerebral Cortex, 7,* 374–385.
41. Hoptman, M. J., & Davidson, R. J. (1998). Baseline EEG asymmetries and performance on neuropsychological tasks. *Neuropsychologia, 36,* 1343–1353.
42. Cacioppo, J. T., & Petty, R. E. (1982). The need for cognition. *Journal of Personality and Social Psychology, 42,* 116–130.
43. Haugtvedt, C. R., Petty, R., & Cacioppo, R. (1992). Need for cognition and advertising: Understanding the role of personality variables in consumer behaviour. *Journal of Consumer Psychology, 1,* 239–260.
44. Larsen, V., Wright, N. D., & Hergert, T. R. (2004). Advertising montage: Two theoretical perspectives. *Psychology & Marketing, 21,* 1–15.
45. Davis, F. D., Bagozzi, R. P., & Warshaw. P. R. (1992). Extrinsic and intrinsic motivation to use computers in the workplace. *Journal of Applied Social Psychology, 22*, 1111–1132.
46. Chatterjee, S., Heath, T., & Mishra, D. (2002). Communicating quality through signals and substantive messages: The effect of supporting information and need for cognition. *Advances in Consumer Research, 29,* 228–229.
47. Baker, J., Grewal, D., & Parasuraman, A. (1994). The influence of store environment on quality inferences and store image. *Journal of the Academy of Marketing Science, 22,* 328–340.

Adaptation of Visual Attention: Effects of Information Presentation in Idea Selection Processes

Arnold Wibmer, Frederik Wiedmann, Isabella Seeber and Ronald Maier

Abstract Idea submissions in innovation contests contain a variety of information as e.g., the idea description, information on its contributor or feedback by the crowd. Raters might perceive and attend to these sources of information differently, which potentially influences the selection of the best ideas. Up to now, however, we know little about the extent to which the visual attention to such information changes during the idea selection process and what impact such changes might have on the outcome of idea selection. The goal of our experiment is to investigate the effect of two idea presentation modes on changes in visual attention to idea attributes, measured with fixations using eye-tracking over time. Preliminary results on a sample of 30 participants show that visual attention to idea attributes decreases rapidly after participants saw the first 8 ideas.

Keywords Attributes · Decision making · Decision quality · Eye-tracking · Idea selection · Innovation contest · Presentation mode · Visual attention

1 Introduction

Ideas in open innovation contests can comprise a variety of information, such as content-, contributor- and crowd-based information [1]. The content category encompasses textual descriptions of an idea by the contributor as well as pictures or media, but typically consists mainly of unstructured, written text [1]. The contributor cat-

A. Wibmer (✉) · F. Wiedmann · I. Seeber · R. Maier
University of Innsbruck, Innsbruck, Austria
e-mail: arnold.wibmer@uibk.ac.at

F. Wiedmann
e-mail: frederik.wiedmann@uibk.ac.at

I. Seeber
e-mail: isabella.seeber@uibk.ac.at

R. Maier
e-mail: ronald.maier@uibk.ac.at

F. D. Davis et al. (eds.), *Information Systems and Neuroscience*,
Lecture Notes in Information Systems and Organisation 32,
https://doi.org/10.1007/978-3-030-28144-1_39

egory relates to information on the person or group who developed the idea and presents for example their success of submitted ideas in the past. A rich history of successful ideas can indicate idea quality as some people were found to generate better ideas than others [1, 2]. The crowd category, often labeled as the "wisdom of the crowd" [1, 3], expresses the opinions of a community about the idea (i.e. likes).

The selection of the best ideas from open innovation contests is typically performed by a jury of expert raters [4] or crowd raters [5, 6] who use information from these three categories to base their selection decisions upon. Similar to the product characteristics in consumer choices [7, 8], content-based, crowd-based or contributor-based information sources [1, 9, 10] can also be considered as attributes of an alternative.

Up to now, however, we have little understanding of how ideas with several idea attributes should be presented to raters in order to support effective decision making. On the one side, presenting more ideas at a time could be beneficial to raters as it allows them to compare more ideas at a time. On the other side, increasing the number of alternatives by presenting many ideas at a time increases the complexity of the decision task [11], which was found to reduce the amount of information that is searched [12–15]. Moreover, evidence from multi-attribute choice experiments suggests that decision makers sometimes skip or even systematically ignore attributes [16, 17]. When decision makers do not attend to attributes this could imply three things: a decision maker could (1) visually ignore attributes, (2) ignore the information of an attribute by not processing it or (3) ignore the attribute in determining a choice, i.e. the attribute's information played no role when a decision was made [16]. Our study focuses on the visual attention on attributes as this kind of attention is closely coupled to eye movements [18] and can tell about a decision maker's goal-driven attention. In decision making, goal-driven attention is paid to stimuli that are perceived as relevant for the task, while stimuli that appear less relevant to make a decision are ignored [19]. Hence, understanding visual attention to idea attributes gives insights into which idea attributes decision makers find more or less relevant during idea selection. However, visual attention to idea attributes is likely to change over time as raters become familiar with the task. According to the information-reduction hypothesis, people become better at distinguishing between task-relevant and task-redundant information and thus become more selective in their visual attendance, indicated by a greater number of fixations on relevant information [20, 21]. Hence, over time raters will be able to encapsulate the information they need and ignore other irrelevant attributes [22].

We have little knowledge to what extend raters pay attention to the presented idea attributes over time given different idea presentation modes. Hence, the goal of this study is to investigate the effects of two different modes of presentation that vary the number of ideas presented at a time on the attendance to idea attributes over time. Consequently, this study addresses the research question: "*How do variations of idea presentation and the progress of idea selection over time change individuals' attendance to idea attributes?*" We designed a repeated-measures laboratory experiment using eye-tracking to capture the visual attendance of participants to idea attributes. Preliminary findings indicate support for both, a reduced visual attendance to idea attributes when presenting more ideas at a time and a decreasing visual attendance to idea attributes as raters progress in the idea selection task.

2 Theoretical Background and Hypotheses Development

Human decision makers often do not include all available information into their decisions [23]. A decisive factor in this context is the complexity of a task, which is determined by the number of available alternatives and the number of available attributes [11]. Increasing the task complexity by presenting more alternatives and/or more attributes was found to reduce the amount of information that was searched [8, 11, 12, 14]. Accordingly, Svenson [24] inferred based on an analysis of previous studies that the proportion of information that is searched by the decision maker could be understood as a function of the number of attributes and the number of alternatives, where the amount of information searched decreases as one or both rise. Lohse and Johnson [13] report of a similar effect using eye-tracking that both, more attributes and more alternatives induce participants to visually examine less presented information. The relationship between a higher task complexity in choice experiments and the amount of information searched can be explained by the assumption of an adaptive decision maker who constructs decision strategies, which are sensitive to the local problem structure [25, 26]. A higher task complexity can be related to a higher cognitive effort for decision makers [17]. Some decision strategies require less effort under high task complexity by not considering all information, while still achieving a comparable decision accuracy [27]. Hence, an adaptive decision maker would change to a decision strategy that is accompanied with a selective search for information. Similarly, we suggest in the context of idea selection that presenting more ideas at a time creates more cognitive effort and induces raters to adapt to the problem structure. As a result, we predict a tendency to visually examine idea descriptions less profoundly and to attend to fewer idea attributes, resulting in a lower attendance of both, idea descriptions and idea feedback.

H1: Participants presented with more ideas at a time have lower visual attendance of (a) idea descriptions and (b) idea feedback than participants presented with fewer ideas at a time.

In contrast to stimulus-driven attention, where participants increasingly attend to stimuli that have a higher visual saliency, participants are more likely to attend to stimuli with higher task relevance in the case of goal-driven attention [19]. Therefore, a choice can be affected by previous attributes or the perception of the quality of previous choices, which has been referred to as precedent-dependent order effects [22]. Participants recall preceding tasks and incorporate their effects into their current choice, which highlights this dependency. In addition, decision makers contrast a current decision with a reference developed from the features of previous tasks or experiences [22]. Researchers showed that subjects formed a reference value from earlier decisions, considered as baseline for present decisions so that subjects make increasingly reliable decisions because they become increasingly familiar with the task [28–30].

Contrasting, researchers found growing fatigue to negatively affect respondents' motivation due to high cognitive demands in the course of an exhausting task [22]. Subjects apply different strategies to cope with fatigue, such as greater randomness in choices [31–33], basing decisions only on subsets of attributes [34] or choosing the option with the lowest cost [35, 36].

We argue that familiarity and the identification of relevant information in the choice task as well as fatigue explain why decision makers switch decision strategies and hence adapt the amount of information they search over the progress of selection tasks, represented by a change of attention. Decision makers should apply a decision strategy in the beginning in which they attend to many idea attributes in order to get familiar with the decision environment. Over the course of the idea selection process, decision makers might be able to encapsulate attributes that are relevant for their choice by focusing on them and ignoring others, resulting in a decreasing attendance over time. Moreover, they might get tired and hence switch to less effortful decision strategies that might reinforce the effect of diminishing attention on certain attributes. These less effortful decision strategies should be characterized by attending to fewer idea attributes in the course of the selection task.

H2: Attendance of (a) idea descriptions and (b) idea feedback decreases with participants' progress through the idea selection process.

3 Method

We tested our hypotheses with a between-subject repeated measures experimental design in a laboratory setting using eye-tracking. The task was that raters selected the best ideas given a choice set of 32 ideas gathered from a real idea competition ("OpenIDEO") regarding gratitude at the workplace. The idea descriptions were adapted to fit into 16 screens of two ideas each for the treatment group '2 ideas per screen' and eight screens of four ideas each for the treatment group '4 ideas per screen'. Each idea had the following idea attributes for which we defined non-overlapping Areas of Interest (AOIs): "*idea description*" (incl. title of idea; AOI_descr), "*contributor feedback*" (historical idea score; AOI_his), "*crowd feedback*" (number of likes; AOI_like), as well as "*content feedback*" (creativity score [37] and tags; AOI_cs and AOI_tag) as illustrated in Fig. 1. Data was collected using a Tobii Pro X3-120 eye-tracker with a sample rate of 120 Hz.

Students of an Information Systems Master program were invited to take part in the experiment. Participation was voluntary and students received a class bonus for their contribution. We randomly assigned participants to one of the two treatment groups (two vs. four ideas per screen). After instructions on the general procedure, we asked participants to perform a regulatory focus priming according to Chernev [38] in order to manipulate participant's strategy in decision making. The priming

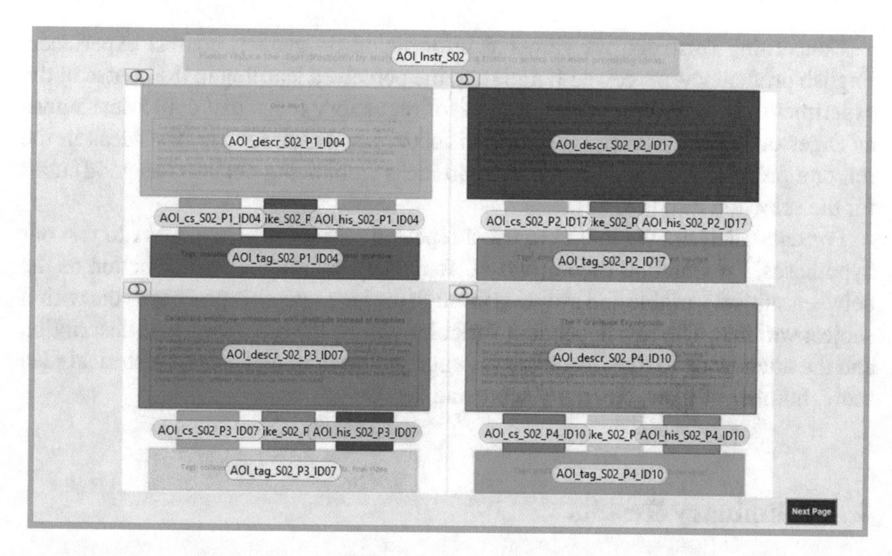

Fig. 1 AOIs for attributes in the treatment condition of four ideas per screen

was performed with the two tasks (1) writing about hopes and aspirations (promotion) versus duties and obligations as graduate students (prevention), similar to [39, 40] and (2) finding a way through a paper maze with either promotion or prevention cues adapted from [41]. Subsequently, we calibrated the eye-tracker to the participant, showed the participant an example idea to introduce the decision environment and tasked them to drastically reduce the ideas by analyzing and comparing them to select the most promising ones. The participants could not go back once they switched screens by clicking on "Next Page". Once the eye-tracking part was finished, participants filled out a survey for collecting data on controls, followed by a debriefing by the experimenter.

We operationalized **processed idea description** with the number of fixations on the AOI_descr. We used a threshold of 251 ms for fixations on idea descriptions since reading texts requires deep information processing compared to a glimpse on for example the number of likes. Our threshold for reading is in line with prior research examining eye movements during reading, which showed an average reading speed of 250 ms for adults [42]. **Attendance to idea feedback** was measured as sum of the feedback attributes (AOI_cs, AOI_his, AOI_like, AOI_tag) fixated at least once divided by the total number of feedback attributes. For example, a value of 0.6 implies that a participant visual attended 60% of the idea feedback and ignored the remaining 40%. We used the Tobii standard value of 60 ms as threshold for fixations which corresponds to Glöckner [43] who associate a lower fixation threshold of less than 250 ms with information scanning. We measured **progress through the idea selection process** for eight equal sections of ideas presented labelled as E1 to E8. Depending on the treatment condition a progress section, E_n, included either one screen (treatment: 4 ideas per screen) or two screens (treatment: 2 ideas per screen).

Concerning **controls**, we gathered information on gender, contest experience, English proficiency, perceived fatigue and the perceived learning in the course of the experiment. Additionally, we controlled for regulatory focus that could determines an eager or vigilant strategy (promotion focus, i.e. search for the best ideas in the set, or a prevention focus, i.e. prevent bad ideas from being declared as good) used for the selection task [38, 44].

For data analysis, we intend to use a repeated measures MANCOVA to test our hypotheses. Presentation mode (two vs. four ideas per screen) will function as the between subject variable and progress through the idea selection process as the within subject variable. The two dependent variables will be the processed idea description and the attendance to idea feedback (as aggregation of attendance to historical idea score, number of likes, creativity score and tags).

4 Preliminary Results

We have already collected data on 30 out of 62 cases planned. A first descriptive inspection of the experiment data showed that visual attendance appears to be higher for idea presentations with fewer ideas (see hypothesis 1). In both treatments, raters that were presented with four ideas per screen processed idea descriptions not as deep and attended to fewer idea feedback attributes than when presented with 2 ideas per screen. Moreover, this preliminary analysis shows that attendance decreased in the course of the experiment (see hypothesis 2). We recognized that the attention for both, processed description and idea feedback attributes decreased (Figs. 2 and 3) in the progress of the selection task, indicating an adjustment of the selection strategy. This could be justified by the adaptation of the decision strategy in dynamic tasks described by [38]. Figures 2 and 3 show the attendance in the progress of the selection task for the treatment presentation mode 2 ideas (16 cases) and 4 ideas (14 cases) per screen.

Fig. 2 Mean fixation counts to idea description

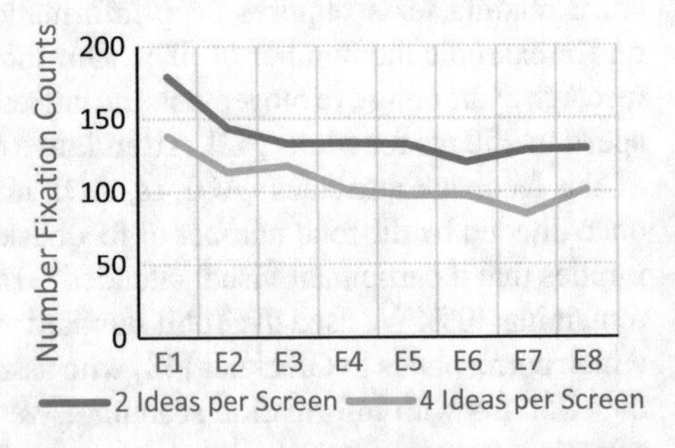

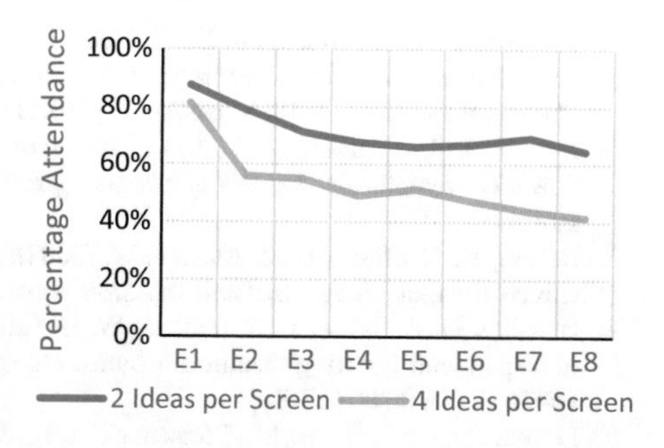

Fig. 3 Mean attendance to idea feedback attributes in percent

5 Conclusion

The first descriptive inspection showed that visual attention is higher when presenting fewer ideas at a time and that visual attention to the description and the idea feedback attributes decreases over the course of the selection task. The decrease of visual attention could be attributed to the increasing fatigue of participants that might have inhibited them from keeping up a cognitively effortful decision strategy. Another explanation could be that participants found it gradually easier to choose ideas, as they determined their current choice based on references to ideas seen before. As the selection task progresses, participants might identify promising attributes as indicators of good ideas, focusing their attention on these particular attributes and ignoring others (selective search). That behavior leads to a reduction of visual attention over time as shown in Fig. 1 as well as Fig. 2, supporting our hypotheses. In other words, the more ideas participants have already seen, the easier it will be for them to carry out idea selection by building up a reference knowledge base.

Our future research will explore whether familiarization (reference effects) or fatigue explains the decrease in visual attention over time and to what extent the idea presentation mode relates to accurate choices. Moreover, future research could also investigate cognitive load to deduce whether information was actually processed and not only visually attended.

References

1. Hoornaert, S., Ballings, M., Malthouse, E. C., & Van Den Poel, D. (2017). Identifying new product ideas: Waiting for the wisdom of the crowd or screening ideas in real time. *Journal of Product Innovation Management, 34*(5), 580–597. https://doi.org/10.1111/jpim.12396.
2. Girotra, K., Terwiesch, C., & Ulrich, K. T. (2010). Idea generation and the quality of the best idea. *Management Science, 56*(4), 591–605. https://doi.org/10.2139/ssrn.1082392.
3. Surowiecki, J. (2004). *The wisdom of crowds: Why the many are smarter than the few and how collective wisdom shapes business, economies, societies and nations little*. London: Little Brown.
4. Bullinger, A. C., & Moeslein, K. (2010). Innovation contests—Where are we ? In *AMCIS 2010 Proceedings*, 1–9.

5. Magnusson, P. R., Wästlund, E., & Netz, J. (2016). Exploring users' appropriateness as a proxy for experts when screening new product/service ideas. *Journal of Product Innovation Management, 33*(1), 4–18. https://doi.org/10.1111/jpim.12251.
6. Kornish, L. J., & Ulrich, K. T. (2014). The importance of the raw idea in innovation: Testing the Sow's ear hypothesis. *Journal of Marketing Research, 51*(4), 14–26. https://doi.org/10.2139/ssrn.2035643.
7. Reisen, N., Hoffrage, U., & Mast, F. W. (2008). Identifying decision strategies in a consumer choice situation. *Judgement and Decision Making, 3*(8), 641–658.
8. Hensher, D. A., Rose, J., & Greene, W. H. (2005). The implications on willingness to pay of respondents ignoring specific attributes. *Transportation, 32*(3), 203–222. https://doi.org/10.1007/s11116-004-7613-8.
9. Leimeister, J. M., Huber, M., Bretschneider, U., & Krcmar, H. (2009). Leveraging crowdsourcing: Activation-supporting components for IT-based ideas competition. *Journal of Management Information Systems, 26*(1), 197–224. https://doi.org/10.2753/mis0742-1222260108.
10. Chan, K. W., Li, S. Y., & Zhu, J. J. (2018). Good to be novel? Understanding how idea feasibility affects idea adoption decision making in crowdsourcing. *Journal of Interactive Marketing, 43*, 52–68. https://doi.org/10.1016/j.intmar.2018.01.001.
11. Kerstholt, J. H. (1992). Information search and choice accuracy as a function of task complexity and task structure. *Acta Psychologica, 80*(1–3), 185–197.
12. Payne, J. W., & Braunstein, M. L. (1978). Risky choice: An examination of information acquisition behavior. *Memory & Cognition, 6*(5), 554–561. https://doi.org/10.3758/BF03198244.
13. Lohse, G. L., & Johnson, E. J. (1996). A comparison of two process tracing methods for choice tasks. *Organizational Behavior and Human Decision Processes, 68*(1), 28–43. https://doi.org/10.1006/obhd.1996.0087.
14. Payne, J. W. (1976). Task complexity and contingent processing in decision making: An information search and protocol analysis. *Organizational Behavior and Human Performance, 16*(2), 366–387. https://doi.org/10.1016/0030-5073(76)90022-2.
15. Swain, M. R., & Haka, S. F. (2000). Effects of information load on capital budgeting decisions. *Behavioral Research in Accounting, 12*, 171–198.
16. Balcombe, K., Fraser, I., & McSorley, E. (2015). Visual attention and attribute attendance in multi-attribute choice experiments. *Journal of Applied Econometrics, 30*(3), 447–467. https://doi.org/10.1002/jae.2383.
17. Hensher, D. A. (2006). How do respondents process stated choice experiments? Attribute consideration under varying information load. *Journal of Applied Econometrics, 21*(6), 861–878. https://doi.org/10.1002/jae.877.
18. Deubel, H., & Schneidert, W. X. (1996). Saccade target selection and object recognition: Evidence for a common attentional mechanism. *Vision Research, 36*(12), 1827–1837.
19. Orquin, J. L., & Mueller Loose, S. (2013). Attention and choice: A review on eye movements in decision making. *Acta Psychologica, 144*(1), 190–206. https://doi.org/10.1016/j.actpsy.2013.06.003.
20. Haider, H., & Frensch, P. A. (1999). Eye movement during skill acquisition: More evidence for the information-reduction hypothesis. *Journal of Experimental Psychology. Learning, Memory, and Cognition, 25*(1), 172–190. https://doi.org/10.1037/0278-7393.25.1.172.
21. Meißner, M., & Decker, R. (2010). Eye-tracking information processing in choice-based conjoint analysis. *International Journal of Market Research, 52*. https://doi.org/10.2501/s147078531020151x.
22. Day, B., Bateman, I. J., Carson, R. T., Dupont, D., Louviere, J. J., Morimoto, S., … Wang, P. (2012). Ordering effects and choice set awareness in repeat-response stated preference studies. *Journal of Environmental Economics and Management*, *63*(1), 73–91. https://doi.org/10.1016/j.jeem.2011.09.001.
23. Simon, H. A. (1955). A behavioral model of rational choice. *The Quarterly Journal of Economics, 69*(1), 99–118.
24. Svenson, O. (1979). Process descriptions of decision making. *Organizational Behavior and Human Performance, 23*(1), 86–112.

25. Payne, J. W., Bettman, J. R., Coupey, E., & Johnson, E. J. (1992). A constructive process view of decision making: Multiple strategies in judgment and choice. *Acta Psychologica, 80*(1–3), 107–141. https://doi.org/10.1016/0001-6918(92)90043-D.
26. Bettman, J. R., Luce, M. F., & Payne, J. W. (1998). Constructive consumer choice processes. *Journal of Consumer Research, 25*(3), 187–217. https://doi.org/10.1086/209535.
27. Johnson, E. J., & Payne, J. W. (1985). Effort and accuracy in choice. *Management Science, 31*(4), 395–414.
28. Isoni, A. (2011). The willingness-to-accept/willingness-to-pay disparity in repeated markets: Loss aversion or "bad-deal" aversion? *Theory and Decision, 71*(3), 409–430. https://doi.org/10.1007/s11238-010-9207-6.
29. Plott, C. R., & Zeiler, K. (2005). The willingness to pay-willingness to accept gap, the "endowment effect", subject misconceptions, and experimental procedures for eliciting valuations. *American Economic Review, 95*(3), 530–545. https://doi.org/10.1257/0002828054201387.
30. Mazumdar, T., Raj, S. P., & Sinha, I. (2005). Reference price research: Review and propositions. *Journal of Marketing, 69*(4), 84–102. https://doi.org/10.1509/jmkg.2005.69.4.84.
31. Bradley, M., & Daly, A. (1994). Use of the logit scaling approach to test for rank-order and fatigue effects in stated preference data. *Transportation, 21*(2), 167–184. https://doi.org/10.1007/BF01098791.
32. Savage, S. J., & Waldman, D. M. (2008). Learning and fatigue during choice experiments: A comparison of online and mail survey modes. *Journal of Applied Econometrics, 23*(3), 351–371. https://doi.org/10.1002/jae.984.
33. Scarpa, R., Notaro, S., Louviere, J., & Raffaelli, R. (2011). Exploring scale effects of best/worst rank ordered choice data to estimate benefits of tourism in alpine grazing commons. *American Journal of Agricultural Economics, 93*(3), 809–824.
34. Tversky, A. (1972). Elimination by aspects: A theory of choice. *Psychological Review, 79*(4), 281–299. https://doi.org/10.1037/h0032955.
35. Hensher, D. A., & Greene, W. H. (2010). Non-attendance and dual processing of common-metric attributes in choice analysis: A latent class specification. *Empirical Economics, 39*(2), 413–426. https://doi.org/10.1007/s00181-009-0310-x.
36. Foster, V., & Mourato, S. (2002). Testing for consistency in contingent ranking experiments. *Journal of Environmental Economics and Management, 44*(2), 309–328. https://doi.org/10.1006/jeem.2001.1203.
37. Toubia, O., & Netzer, O. (2017). Idea generation, creativity, and prototypicality. *Marketing Science, 36*(1), 1–20. https://doi.org/10.1287/mksc.2016.0994.
38. Chernev, A., & Chernev, A. (2004). Goal-attribute compatibility in consumer choice goal—Attribute compatibility in consumer choice. *Journal of Consumer Psychology, 14*(1–2), 141–150. https://doi.org/10.1207/s15327663jcp1401.
39. Freitas, A. L., Higgins, E. T., Freitas, A. L., & Higgins, E. T. (2002). Enjoying goal-directed action: The role of regulatory fit. *Psychological Science, 13*(1), 1–6. https://doi.org/10.1111/1467-9280.00401.
40. Higgins, E. T., Roney, C. J. R., Crowe, E., & Hymes, C. (1994). Ideal versus ought predilections for approach and avoidance: Distinct self-regulatory systems. *Journal of Personality and Social Psychology, 66*(2), 276–286. https://doi.org/10.1037/0022-3514.66.2.276.
41. Friedman, R. S., & Forster, J. (2001). The effects of promotion and prevention cues on creativity. *Journal of Personality and Social Psychology, 81*(6), 1001–1013. https://doi.org/10.1037//0022-3514.81.6.1001.
42. Liversedge, S. P., & Findlay, J. M. (2000). Saccadic eye movements and cognition. *Trends in Cognitive Sciences, 4*(1), 6–14.
43. Glöckner, A., & Herbold, A. K. (2011). An eye-tracking study on information processing in risky decisions: Evidence for compensatory strategies based on automatic processes. *Journal of Behavioral Decision Making, 24*(1), 71–98. https://doi.org/10.1002/bdm.684.
44. Higgins, E. T. (1997). Beyond pleasure and pain. *American Psychologist, 52*(12), 1280–1300.

FITradeoff Decision Support System: An Exploratory Study with Neuroscience Tools

Anderson Lucas Carneiro de Lima da Silva and Ana Paula Cabral Seixas Costa

Abstract Multicriteria problems are present in several contexts and are often complex. To support the decision maker, decision support systems (DSS) have been developed. FITradeoff is a DSS focused on multicriteria additive problems and has a flexible and interactive approach. This study aimed to investigate how engagement and cognitive load is evoked during the use of DSS for different types of problems. It is a work in progress that seeks to raise hypotheses to be tested to promote changes in the DSS and generate recommendations for the decision analyst to make the experience with DSS and the whole decision-making process more efficient and effective.

Keywords Decision support system · FITradeoff · EEG · Decision process

1 Introduction

Decision problems are very common, being part of the everyday lives of all people. These problems involve more than one alternative, often with multiple objectives to be achieved that may be conflicting. They are called multicriteria decision problems, as defined by De Almeida et al. [6].

To support the decision-making process through a structured approach, several methods have been developed. One method that is implemented in a decision support system (DSS) is FITradeoff [5] that aims for the development of an interactive and flexible process with the decision maker, reducing the elicitation time and the cognitive effort demanded. The consequence of this is the reduction of inconsistencies in the results and an improved experience.

The present work in progress aims to identify behavioural aspects, such as engagement and cognitive load, that are present in the preference elicitation process with

---

A. L. Carneiro de Lima da Silva (✉) · A. P. Cabral Seixas Costa
Universidade Federal de Pernambuco-UFPE, Recife, Brazil
e-mail: andersonlucas12@hotmail.com

A. P. Cabral Seixas Costa
e-mail: apcabral@cdsid.org.br

F. D. Davis et al. (eds.), *Information Systems and Neuroscience*,
Lecture Notes in Information Systems and Organisation 32,
https://doi.org/10.1007/978-3-030-28144-1_40

the use of FITradeoff DSS in different multicriteria problems. With the identified results, recommendations can be made to the decision analyst who is responsible for assisting the decision maker during the decision-making process, and changes can be made to the DSS itself to make it more effective. For this purpose, a pilot experiment with 17 subjects was performed using an electroencephalography (EEG) and an eye-tracker. During the experiment each subject applied a self-made decision problem in the DSS. The frontal alpha asymmetry (FAA), the power of the theta band at the parietal electrodes and alpha band at frontal electrodes and the pupil size were analysed. An exploratory analysis was conducted to obtain insight for the development of future research with FITradeoff DSS.

2 FITradeoff Decision Support System

During the construction of multicriteria decision models, especially additives, an important step is the definition of the scale constants of the criteria of the problem. Several methods have been proposed for eliciting preferences and defining these constants. However, the process of preference elicitation and the definition of constants is generally complex and may require time and information that is not available to the decision maker [2, 15]. A poor definition of the constants may result in inconsistency in the results, rendering the developed models unsuitable.

The identification of the aspects that influence the judgement of constants and the generation of inconsistencies has been the focus of behavioural studies, as is the case of the work by Weber and Borcherding [17]. Knowing the levels of cognitive effort and stress becomes very useful in the evaluation of the methods of elicitation since such factors can harm the obtained results, making them unreliable [17].

To make the elicitation process easier and more valid, FITradeoff uses an interactive approach, using partial information of the decision maker's preference. It is based on Tradeoff that has a strong axiomatic structure [5]. With FITradeoff, the decision maker is not asked to answer several questions or provide information about points of indifference for the definition of scale constants. In this sense, FITradeoff presents advantages over other methods, making the experience with the decision maker successful and allowing the reduction of the inconsistency rate. The inconsistency is associated with the complexity demanded by the decision maker in the process, reaching 67% in the Tradeoff according to a study by Weber and Borcherding [17].

The experience with FITradeoff begins with inserting the problem data into the DSS, such as problem criteria, alternatives, and the performance of these alternatives in the criteria, through a spreadsheet. Figure 1 shows the FITradeoff DSS elicitation process that can be divided into four steps: inserting the problem data, ordering the scale constants of the criteria, flexible elicitation, and evaluating the recommendation.

During the ordering of the scale constant, a hypothetical alternative is presented to the decision maker. The performance of the alternative in the criteria established by the decision maker is the lowest according to the set of original alternatives of the problem. Thus, the decision maker is asked which criterion should be raised

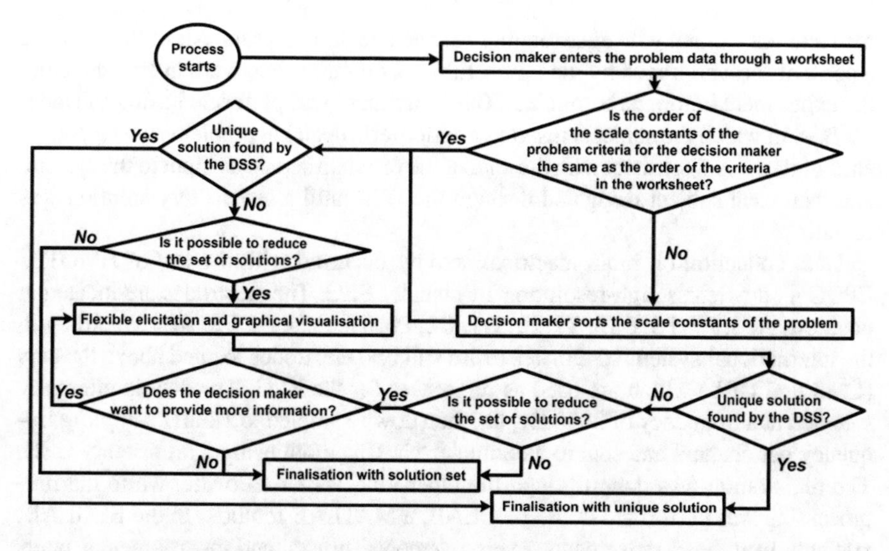

Fig. 1 FITradeoff DSS elicitation process (*Source* Authors)

to the maximum performance, according to the set of original alternatives. Thus, a new alternative is formed, with the maximum performance in the criterion chosen previously and the minimum performance chosen for the other criteria. Again, it is asked which criterion should be improved within the remaining criteria. A new alternative with the highest performance in the criteria chosen beforehand and the minimum performance in the other criteria is formed. Thus, it continues until all criteria are chosen. The selection sequence of the criteria defines the order of the scale constants of the criteria. Already in the step of flexible elicitation, the decision maker is presented with a pair of hypothetical alternatives. Each one has minimal performance in all but one of the criteria. The criterion with performance between the minimum and maximum is different for each hypothetical alternative. The decision maker is asked about his or her preference among the pair. Still, in this step, the decision maker can see the performance in graphs (bar, bubble, and radar) of the set of alternatives that are potential solutions. In the last step, it is up to the decision maker to see the DSS recommendation that he or she can or cannot abide by.

3 Methodology and Preliminary Results

3.1 Experiment Design

The experiments were conducted in the NeuroScience for Information and Decision (NSID) laboratory of the Universidade Federal de Pernambuco (UFPE). The sample of the experiment consisted of 17 PhD and master's students in management engineering at the UFPE. In this study, only the data for 15 participants could be used: five men and ten women with ages ranging between 23 and 35 years. All of them signed a consent form approved by the UFPE Ethics Committee and were instructed about the experiment before its beginning. The experiment was performed using FITrade-off DSS in which the subjects inserted multicriteria decision problems developed by each of them. A worksheet with the data of the problem served as input to the system, and then each subject navigated through the DSS until a satisfactory solution was found.

Data collection on brain electrical activity occurred with the use of EMOTIV EPOC +, a wireless high-resolution 14-channel EEG. The electrodes are located at positions AF3, F7, F3, FC5, T7, P7, O1, O2, P8, T8, FC6, F4, F8, AF4 according to the international system 10–20. There are still two electrodes located above the ears (CMS and DRL) which are used as references for the EEG. The data is internally sampled at a frequency of 2048 Hz, but then down-sampled to 128 Hz sampling frequency per channel and sent to a computer via Bluetooth using a proprietary USB. The registration was done through the OpenVibe Writer software, while the pre-processing was performed in the EEGLAB, a MATLAB toolbox. In the EEGLAB, artefacts from head movements, eye movements, blinks, and environmental interference were corrected through the following procedure: (1) re-referencing based on the average activity calculated along the electrodes (2) filtering the data (high pass: 40 Hz, and lowpass: 0.1 Hz) (3) visual inspection and exclusion of bad portions of data, and (4) correcting artefacts using the Independent Component Analysis (ICA) method. After that, the data were exported to a spreadsheet and analysed using Microsoft Excel software.

Data related to pupil size, fixations, and saccades were also obtained through the Tobii X120 eye-tracker. Pupil size data correction and analysis were performed in Microsoft Excel software. The presentation of the experiment took place through an LCD monitor 48 cm wide by 26.9 cm high with 1280 $\times$ 1024 pixel resolution connected to a 64-bit notebook running Windows 10 with 4 GB of RAM. The notebook was also connected to the eye-tracker and USB of the EEG and had the software installed.

The synchronization between EEG and eye-tracker systems was possible through the application of a trigger responsible for inserting markers related to the stages of the DSS in the two datasets.

3.2 *Analysis and Preliminary Results*

For this study, only the steps of ordering the scale constants of the problem criteria, flexible elicitation, and presentation and evaluation of the recommendation of the DSS were considered. These steps were chosen because they are directly related to the preference elicitation process. With the data captured by the EEG, the frequency bands were analysed. Power spectral density values were obtained for each channel of the EEG during the three main steps of the elicitation process. Those values were averaged for the theta EEG frequency band (4–8 Hz) and alpha EEG frequency band

(8–13 Hz) [13]. Eye-tracker data related to pupil size were also used and a baseline defined at the start of the experiment was applied for analysis.

The decision problems were classified and separated according to the type of criteria: quantitative (7), qualitative (3), and combined (5). The questions asked during flexible elicitation were also classified according to the pair of criteria highlighted: quantitative vs quantitative (42), qualitative versus qualitative (16), and quantitative vs qualitative (31). More details about the decision problems, as the number of questions made in each step and the time in seconds the subjects spent are presented in Table 1.

Regarding the questions asked during flexible elicitation, an analysis of the pupil size was performed (see Fig. 2). The results indicate greater pupil size for questions in which only quantitative criteria are compared, indicating greater cognitive effort [14].

In turn, the results related to pupil size during the elicitation process indicate higher values for problems classified as quantitative and smaller values for problems classified as qualitative (see Fig. 3). In addition, only for the problems classified as quantitative, there is a growth trend of pupil size through the steps.

The power value of the theta band was also analysed along the three main steps at the channel pairs P7 and P8 since there is evidence of a negative correlation between the power of the band in these channels with the cognitive effort [3, 8, 10]. The analysis of the theta band behaviour between the steps for each type of problem showed an increase only in Step 2 in channel P7 for qualitative problems.

The alpha band is also negatively correlated with cognitive effort [11] and the power of this band through the frontal channels was analysed because there is evidence that this region of the brain is related to decision-making process [18]. The results showed an increase in power in Step 2 for qualitative problems in the channels AF4 and F3, and an increase in Step 2 for problems classified as Combined only in the channel FC6.

An analysis of the FAA was still performed for the different types of problems; however, the results did not indicate any suggestive differences. This measure is related to engagement [1, 4, 9] and can be useful to evaluate different types of problems and the interaction of the decision maker with the DSS.

4 Discussion and Next Steps

The identified results are still preliminary but show promise. The pupil size analysis indicates that the questions in which only quantitative criteria are compared during flexible elicitation demand more from decision makers. The same is observed for problems classified as quantitative. In turn, the theta and alpha band values of this study suggest an increase during the flexible elicitation stage for qualitative problems, indicating reduced cognitive effort. Thus, it is hypothesised that quantitative criteria require more of the decision maker. Further experiments are intended to be performed to obtain a larger sample that allows the application of statistical tests.

Table 1 Number of questions and time in seconds of the decision problems steps for each subject

Subject 1			Subject 2			Subject 3			Subject 4		
Step	No.	Time	Step	No.	Time	Step	No.	Time	Step	No.	Time
1	3	38.87	1	5	26.77	1	5	22.33	1	6	69.95
2	3	58.88	2	8	68.72	2	15	148.29	2	4	86.46
3	1	39.80	3	1	5.64	3	1	63.23	3	1	10.99
Subject 5			Subject 6			Subject 7			Subject 8		
Step	No.	Time	Step	No.	Time	Step	No.	Time	Step	No.	Time
1	3	46.15	1	4	26.20	1	7	48.75	1	3	33.00
2	4	92.60	2	6	79.99	2	9	156.58	2	2	33.20
3	1	35.71	3	1	14.74	3	1	92.68	3	1	20.90
Subject 9			Subject 10			Subject 11			Subject 12		
Step	No.	Time	Step	No.	Time	Step	No.	Time	Step	No.	Time
1	3	19.88	1	6	90.17	1	4	34.48	1	3	15.10
2	2	51.94	2	9	141.73	2	1	12.36	2	5	119.45
3	1	2.87	3	1	12.92	3	1	29.17	3	1	14.86
Subject 13			Subject 14			Subject 15					
Step	No.	Time	Step	No.	Time	Step	No.	Time			
1	3	19.03	1	4	24.59	1	6	44.62			
2	3	109.12	2	15	375.53	2	2	35.47			
3	1	21.21	3	1	89.03	3	1	9.78			

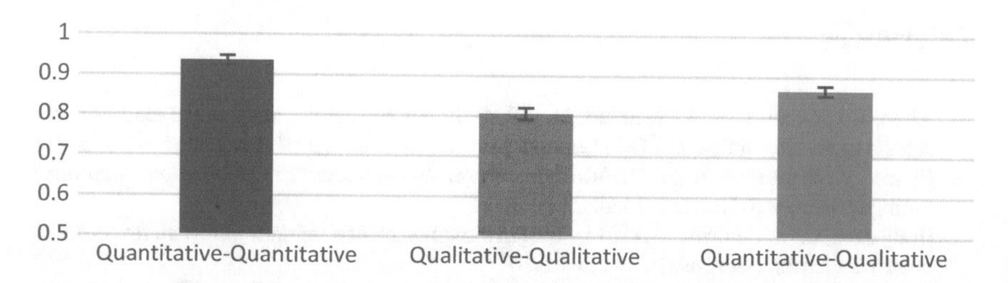

Fig. 2 Mean pupil size and standard error for flexible elicitation questions (*Source* Authors)

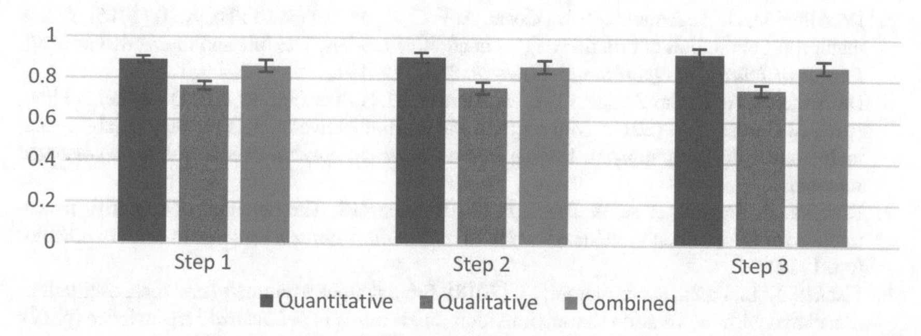

Fig. 3 Mean pupil size and standard error during the steps 1, 2 and 3 for different problems (*Source* Authors)

Investigations in this regard are important because they can generate insight for the decision analyst and the DSS developer to make the decision making and use of DSS more effective and efficient [7]. Identifying whether quantitative criteria actually require more of the decision maker can lead to changes within the DSS in representing the criteria. On the other hand, the analysis of engagement may complement the results because it is expected that there will be high engagement during the elicitation process. Thus, with more data, it is possible to evaluate whether there are differences in engagement for different types of problems and criteria that require changes in the DSS. With the sample obtained here, however, no result can be identified that indicates greater or lesser engagement between stages for different types of problems.

Finally, future studies should analyse other measures, such as the power of beta and gamma bands, which are related to more complex cognitive processes [11], on the frontal brain areas [12, 16, 18] and can generate valuable insight. Data regarding saccades, and fixations will also be analysed for patterns of information search.

Acknowledgements This project was supported by the National Council for Scientific and Technological Development (CNPq) and Coordination for the Improvement of Higher Education Personnel (CAPES).

References

1. Allen, J. J., Coan, J. A., & Nazarian, M. (2004). Issues and assumptions on the road from raw signals to metrics of frontal EEG asymmetry in emotion. *Biological Psychology, 67,* 183–218.
2. Belton, V., & Stewart, T. (2002). *Multiple criteria decision analysis: an integrated approach.* Springer Science & Business Media.
3. Braboszcz, C., & Delorme, A. (2011). Lost in thoughts: Neural markers of low alertness during mind wandering. *Neuroimage, 54,* 3040–3047.
4. Davidson, R. J., Ekman, P., Saron, C. D., Senulis, J. A., & Friesen, W. V. (1990). Approach-withdrawal and cerebral asymmetry: Emotional expression and brain physiology: I. *Journal of Personality and Social Psychology, 58,* 330–341.
5. De Almeida, A. T., Almeida, J. A., Costa, A. P. C. S., & Almeida-Filho, A. T. (2016). A new method for elicitation of criteria weights in additive models: Flexible and interactive tradeoff. *European Journal of Operational Research, 250,* 179–191.
6. De Almeida, A. T., Cavalcante, C. A. V., Alencar, M. H., Ferreira, R. J. P., De Almeida Filho, A. T., & Garcez, T.V. (2015). Multicriteria and multiobjective models for risk, reliability and maintenance decision analysis. In *International series in operations research & management science.*
7. Dimoka, A., Pavlou, P. A., & Davis, F. (2010). NeuroIS: The potential of cognitive neuroscience for information systems research. In *information systems research articles in advance* (pp. 1–18).
8. Fischer, N. L., Peres, R., & Fiorani, M. (2018). Frontal alpha asymmetry and theta oscillations associated with information sharing intention. In *Frontiers in behavioral neuroscience* (p. 12).
9. Gollan, J. K., Hoxha, D., Chihade, D., Pflieger, M. E., Rosebrock, L., & Cacioppo, J. (2014). Frontal alpha EEG asymmetry before and after behavioral activation treatment for depression. *Biological Psychology, 99,* 198–208.
10. Klimesch, W. (1999). EEG alpha and theta oscillations reflect cognitive and memory performance: A review and analysis. *Brain Research Reviews, 29,* 169–195.
11. Müller-Putz, G. R., Riedl, R., & Wriessnegger, S. C. (2015). Electroencephalography (EEG) as a research tool in the information systems discipline: Foundations, measurement, and applications. *CAIS, 37,* 46.
12. Owen, A. M., McMillan, K. M., Laird, A. R., & Bullmore, E. (2005). N-back working memory paradigm: A meta-analysis of normative functional neuroimaging studies. *Human Brain Mapping, 25,* 46–59.
13. Pizzagalli, D. A. (2007). Electroencephalography and high-density electrophysiological source localization. In J. T. Cacioppo, L. G. Tassinary, & G. G. Berntson (Eds.), *Handbook of psychophysiology* (pp. 56–84). Cambridge University Press.
14. Rosch, J. L., & Vogel-Walcutt, J. J. (2013). A review of eye-tracking applications as tools for training. *Cognition, Technology & Work, 15,* 313–327.
15. Salo, A., & Punkka, A. (2005). Rank inclusion in criteria hierarchies. *European Journal of Opera-tional Research, 163,* 338–356.
16. Van der Linden, M., Juillerat, A. C., & Adam, S. (2003). In R. Mulligan, M. Van der Linden, & A. C. Juillerat (Eds.), *Cognitive intervention* (pp. 169–233). Erlbaum.
17. Weber, M., & Borcherding, K. (1993). Behavioral influences on weight judgments in multiattribute decision making. *European Journal of Operational Research, 67,* 1–12.
18. Zhao, Y., & Siau, K. (2018). Cognitive neuroscience in information systems research. In *Applications of neuroscience: Breakthroughs in research and practice* (pp. 158–175). IGI Global.

Printed in the United States
By Bookmasters